T0342465

Statistics in Archaeology

Michael Baxter

Reader in Archaeological Statistics
at Nottingham Trent University, UK

John Wiley & Sons, Ltd

First published in Great Britain in 2003 by
Hodder Arnold, a member of the Hodder Headline Group,
338 Euston Road, London NW1 3BH

John Wiley & Sons Ltd, The Atrium, Southern Gate, Chichester, West Sussex, PO19 8SQ, United Kingdom

For details of our global editorial offices, for customer services and for information about how to apply for permission to reuse the copyright material in this book please see our website at www.wiley.com.

British Library Cataloguing in Publication Data
A catalogue record for this book is available from the British Library

Library of Congress Cataloging-in-Publication Data
A catalog record for this book is available from the Library of Congress

ISBN 978-0-470-71113-2

1 2 3 4 5 6 7 8 9 10

Typeset in 10.5/11 Times by Charon Tec Pvt. Ltd, Chennai, India

Contents

Preface   ix

1 Introduction   1
  1.1 Statistics and archaeology   1
    1.1.1 Introduction   1
    1.1.2 Statistical use in archaeology   2
  1.2 Statistics in archaeology   3
    1.2.1 The background   3
    1.2.2 Landmark and other papers   8
    1.2.3 Other literature   13
  1.3 Simple statistics?   16

2 Data sets and problems   19
  2.1 Introduction   19
  2.2 Data   19
    2.2.1 Lead isotope ratio data   19
    2.2.2 Artifact compositional data 20
    2.2.3 Artifact typology   22
    2.2.4 Assemblage comparison I   23
    2.2.5 Assemblage comparison II   24
    2.2.6 Spatial data   27

3 Kernel density estimates   29
  3.1 Introduction   29
  3.2 Kernel density estimates   30
    3.2.1 Univariate KDEs   30
    3.2.2 Multivariate KDEs   32
  3.3 Applications   33
    3.3.1 Italian Bronze Age cups   33
    3.3.2 The Mask Site bone splinter data   35
    3.3.3 Other applications   36

4 Sampling   **38**
  4.1 Introduction   38
  4.2 Sampling methods   39
    4.2.1 Notation   39
    4.2.2 Simple random sampling   40
    4.2.3 Stratified random sampling   41
    4.2.4 Cluster sampling   41
    4.2.5 Systematic sampling   43
  4.3 Archaeological considerations   43
    4.3.1 Random versus purposive sampling   43
    4.3.2 What is being sampled?   45
  4.4 Adaptive sampling   46
  4.5 Sampling for discovery/rare features   47
    4.5.1 Sampling for discovery   47
    4.5.2 Sampling for 'nothing'   49

5 Regression and related models   **50**
  5.1 Introduction   50
  5.2 Simple linear regression   52
    5.2.1 An example   52
    5.2.2 Models for data   54
    5.2.3 Flake size and flake-size distribution   54
  5.3 Multiple linear regression   55
    5.3.1 The basic model   55
    5.3.2 Polynomial regression and related models   56
    5.3.3 Trend-surface analysis   56
    5.3.4 Regression with 0–1 dependent variables   57
    5.3.5 Regression with 0–1 independent variables 58
  5.4 Generalized linear models   59
    5.4.1 Basic ideas   59
    5.4.2 Logistic regression   60
  5.5 Non-linear regression   62
  5.6 Non-parametric regression   63

6 Multivariate methods – an introduction   **66**
  6.1 Introduction   66
  6.2 Notation and terminology   66
  6.3 The singular value decomposition   68
  6.4 Measures of distance   68
    6.4.1 Manhattan distance   68
    6.4.2 Euclidean distance   69
    6.4.3 Mahalanobis distance   69

7 Principal component analysis and related methods   **73**
  7.1 Introduction   73
  7.2 Aspects of principal component analysis   74
    7.2.1 The mathematics   74
    7.2.2 Transformation   74
    7.2.3 Biplots   78

	7.2.4	Choosing the number of components	79
	7.2.5	Rotation	80
7.3	Factor analysis		83
7.4	Multidimensional scaling		85
	7.4.1	The main ideas	85
	7.4.2	Example: Analysis of ceramic thin-section data	86
7.5	Projection pursuit		88

8 Cluster analysis — **90**

8.1	Introduction		90
8.2	Hierarchical clustering methods		92
	8.2.1	Hierarchical agglomerative clustering	92
	8.2.2	Hierarchical divisive clustering	95
	8.2.3	Hierarchical clustering methods in burial studies	96
8.3	Relocation/partitioning methods		97
	8.3.1	Main ideas	97
	8.3.2	A case study	98
8.4	Model-based methods		99
	8.4.1	Classification maximum likelihood	99
	8.4.2	Mixture maximum likelihood	100
	8.4.3	Archaeological applications	101

9 Discrimination and classification — **105**

9.1	Introduction		105
9.2	Linear and quadratic discriminant analysis		106
	9.2.1	Normal-theory LDA and QDA	106
	9.2.2	Fisher's LDA	107
	9.2.3	Assessing the success of a classification	108
	9.2.4	Two-group discriminant analysis and regression	110
	9.2.5	Variable selection	111
	9.2.6	Quadratic discriminant analysis	111
	9.2.7	Examples	112
9.3	Logistic discrimination		114
9.4	Classification trees		116

10 Missing data and outliers — **119**

10.1	Missing and censored data		119
	10.1.1	Introduction	119
	10.1.2	Missing data	119
	10.1.3	Censored data	121
	10.1.4	An EM algorithm	122
10.2	Outliers		123

11 Analysis of tabular data — **128**

11.1	Introduction		128
11.2	Chi-squared analysis of contingency tables		129
	11.2.1	Two-way tables	129
	11.2.2	Three-way tables	129

11.3 Log-linear models   131
    11.3.1 Three-way tables   131
    11.3.2 Four-way tables – an example   135
11.4 Correspondence analysis   136
    11.4.1 Introduction   136
    11.4.2 Detrended correspondence analysis   139
    11.4.3 Multiple correspondence analysis – example 140
    11.4.4 Some applications   140
11.5 The mathematics of correspondence analysis   143
    11.5.1 Chi-square distance   143
    11.5.2 Correspondence analysis as PCA   144
    11.5.3 Matrix formulations   144
    11.5.4 Multiple correspondence analysis – theory   145

12 Computer-intensive methods   **147**
12.1 Introduction   147
12.2 The bootstrap   148
    12.2.1 Basic ideas   148
    12.2.2 Examples of bootstrapping   148
12.3 The jackknife   153
12.4 Other applications of randomization   154
    12.4.1 Buildings at Danebury   154
    12.4.2 Bronze Age cairns on Mull   155
    12.4.3 Artifacts in graves   155
12.5 Markov chain Monte Carlo   157

13 Spatial analysis   **159**
13.1 Introduction   159
13.2 Spatial clustering   160
13.3 Predictive modeling   162
13.4 Point pattern analysis   163
    13.4.1 Nearest neighbor analysis   163
    13.4.2 Applications of nearest neighbor analysis   165
    13.4.3 Second-order methods   166
    13.4.4 Applications of second-order methods   168
13.5 Spatial autocorrelation   169
    13.5.1 Introduction   169
    13.5.2 The Classic Maya collapse   169
    13.5.3 Spatial patterns among blood types   172
13.6 Shape analysis   173
    13.6.1 Introduction   173
    13.6.2 Post-hole patterns   173
    13.6.3 Comparing shapes   175

14 Bayesian methods   **176**
14.1 Bayes' theorem   176
14.2 Bayesian inference in archaeology – an overview   178
14.3 Bayesian inference in archaeology – examples   179
    14.3.1 Estimating proportions   179

14.3.2 Bayesian clustering 182
14.3.3 A miscellany 185
14.4 Discussion 186

15 Absolute dating – radiocarbon calibration **187**
15.1 Introduction 187
15.2 Combining dates 189
15.2.1 Theory 189
15.2.2 Application 190
15.3 Some Bayesian solutions 191
15.3.1 Calibration of a single date 191
15.3.2 Multiple dates – single phase 192
15.3.3 Multiple dates – ordered 193
15.3.4 Multiple dates – multiple phases 194
15.3.5 Prior assumptions 196
15.4 Other applications 198
15.4.1 Outlier detection 198
15.4.2 Sample selection 198

16 Relative dating – seriation **200**
16.1 Introduction 200
16.2 A brief review 202
16.3 Practical seriation 203
16.3.1 Similarity matrices 203
16.3.2 Correspondence analysis and MDS 204
16.3.3 Bayesian methods 207
16.4 Archaeological considerations 207

17 Quantification **210**
17.1 Introduction 210
17.2 Quantification of vertebrate faunal remains 211
17.2.1 Simple approaches to quantification 211
17.2.2 Models – the Lincoln/Peterson index 212
17.2.3 Maximum likelihood estimation of bone counts 213
17.3 Pottery quantification 215
17.3.1 Introduction 215
17.3.2 Simple measures and estimated vessel equivalents 216
17.3.3 Pottery information equivalents 217
17.4 Other find types and the pie-slice approach 219
17.4.1 Bone 219
17.4.2 Glass 220
17.4.3 Stone tools 221

18 Lead isotope analysis **222**
18.1 Introduction 222
18.2 Statistical issues 223
18.2.1 Data transformation 223
18.2.2 Outliers 224

	18.2.3 Normality	224
	18.2.4 Sample size	226
	18.3 Conclusion	227

19 The megalithic yard — **228**
19.1 Introduction — 228
19.2 Models for the megalithic yard — 230
   19.2.1 The basic model — 230
   19.2.2 Broadbent's method — 230
   19.2.3 Kendall's method — 231
   19.2.4 Freeman's method — 233
19.3 Discussion — 234

20 Comparing assemblage diversity — **236**
20.1 Introduction — 236
20.2 Diversity — 236
20.3 Regression and the sample-size effect — 238
20.4 Simulation and the sample-size effect — 240
20.5 Resampling approaches — 243

21 Shorter studies — **244**
21.1 Introduction — 244
21.2 Artifact classification — 244
21.3 Age estimation — 245
21.4 Particle-size analysis — 246

Appendix – Web resources — **248**
A.1 S-Plus and R — 248
A.2 Other resources — 249

References — **251**

Index — **288**

Preface

The intended readership of this book is archaeologists with an interest in what statistics has to offer their discipline, and statisticians with an interest in how their subject is applied in other disciplines. It is assumed that readers have been exposed to an introductory statistics course or text, so that the focus is on applications at an intermediate or advanced level. Many of the topics covered are not treated in other statistics texts written for archaeologists, and I believe that at least some of these have considerable potential for archaeological data analysis.

Statistics in archaeology is concerned with the collection and interpretation of numerical data in archaeology and those disciplines that feed into it, where the methods of these disciplines have been used to address archaeological problems, and where statistical data analysis has been involved. The scope of archaeology, as stated in the entry for it in *The Oxford Companion to British History*, is that as a discipline archaeology 'attempts to reconstruct the origin, prehistory, and history of the human race using material remains such as artefacts, settlements, earthworks, burials and skeletal remains', using 'evidence for human impact on the natural environment such as pollen, soil erosion, and animal and plant remains', and 'drawing on anthropology and geography for models and analogues, on biology for environmental reconstruction, geology for excavation techniques, and the natural sciences for analytical and dating methods'.

Terms such as 'statistical archaeology', 'archaeostatistics', and 'archaeometrics', have been used to describe the use of statistical methods in archaeology. Archaeometry can be defined, more specifically, as relating to applications of the physical sciences to archaeological problems and issues, and is used in that sense in the book.

The scope of the disciplines of archaeology and statistics is vast, and so, perhaps surprisingly to some, is their intersection. I have had to curb my initial ambitions and be selective in what is presented here. Some of the material excluded from the book is available on the website associated with the book, as described in the Appendix. Most statisticians will not find the level of mathematics used in the text taxing; many archaeologists will. This is, perhaps, inevitable, given the level of statistical training most archaeologists receive, and the fact that books in this series are not intended to be introductory. I think, nevertheless, that many 'complex' statistical methods are increasingly accessible to archaeologists, and I argue this case in Section 1.3.

It often takes a long while for (potentially useful) statistical techniques to permeate the archaeological literature. Thus, some of the topics covered here will be regarded as 'old hat' by statisticians but will be largely unknown to archaeologists. I believe that statisticians can often most usefully aid archaeologists and archaeology by identifying, and helping to implement, potentially useful existing methods for archaeological data analysis that archaeologists may be either unaware of or have difficulty using. Archaeological problems have stimulated the development of novel statistical methodology (witness the chapters on seriation and the megalithic yard), but such instances are few and have not always been well regarded within archaeology. Unfortunately, in Britain at least, there are few statisticians who work with archaeologists on a regular basis, and that number has been declining in recent years.

This is a great shame. I believe that statistics has much to offer archaeology but, in order to winnow the methodological wheat from the chaff, collaboration with archaeologists is essential. I came to the writing of this book with a good general knowledge of how statistics has been used in archaeology, based on over 30 years of interest involving considerable practical archaeological experience and extensive collaboration with archaeologists and archaeological scientists. Even so, I have been surprised by the range of applications that I have found and, as already noted, this could have been a much longer book. If this book encourages archaeologists to seek professional statistical help when required, statisticians to understand the problems that archaeologists face, and future collaboration between the two, it will have achieved something.

It is the diversity of application that makes statistics a fascinating subject, and what attracted me to it as a student. Statistical methods are easily abused and, particularly when a discipline is newly engaged in coming to terms with the possibilities – as happened in archaeology in the 1960s and 1970s – abuses occur. Some have seen this as a reason for turning against the use of statistics in archaeology. This is nonsense. In learning to use 'new' tools mistakes occur, but this does not mean that the tools are useless. I elaborate on this in Section 1.2.1. Readers will find that sometimes I conclude that a particular method has not been 'much used', or has not been used recently. This may be because it is really not of much use to archaeology, or because it has not been adequately appreciated or explored. My attitude is to try things out. Methods may prove to be useful or they may not, but only by accumulating enough evidence can judgments be made about the value of particular methods, and the circumstances in which they can deliver useful insights.

My intellectual debts should be obvious from the text and references. I owe more specific thanks to the following, who at various stages have allowed me to pick their brains (sometimes unwittingly), drawn my attention to unfamiliar material, provided papers or data sets, assisted with analyses, and generally provided encouragement etc. In alphabetical order they are David Adan-Bayewitz, John Aitchison, Robert Aykroyd, Alex Bayliss, David Bellhouse, Christian Beardah, Caitlin Buck, Christopher Bronk Ramsey, M.A. (Cuco) Cau, Hilary Cool, Nick Fieller, Michael Greenacre, Mark Hall, Caroline Jackson, Ken Kvamme, Morven Leese, David Lucy, Andrew Millard, Clive Orton, Jari Pakkanen, Ioulia

Papageorgiou and Mike Shott. I apologize for the fact that, in some cases, considerations of space have not allowed me to use material or treat it at the length it deserved. I'd also like to thank my editors, successively Liz Gooster, Lesley Riddle and Christina Wipf Perry, for guiding me through the vicissitudes of production, from the conception to the birth of this book, and to Brian Everitt for inviting me to attempt it.

Of the above, especial thanks are owed to Caitlin Buck, Morven Leese, Jari Pakkanen and Clive Orton for commenting on draft chapters and, particularly, to Hilary Cool for reading the entire manuscript at a late stage of the proceedings. Needless to say, I accept full responsibility for any misunderstandings and errors in the text, and would be grateful (up to a point, at least) to be notified of these.

Michael Baxter
2003

Arnold Applications of Statistics Series

Series Editor: **BRIAN EVERITT**
Department of Biostatistics and Computing, Institute of Psychiatry, London, UK

This series offers titles which cover the statistical methodology most relevant to particular subject matters. Readers will be assumed to have a basic grasp of the topics covered in most general introductory statistics courses and texts, thus enabling the authors of the books in the series to concentrate on those techniques of most importance in the discipline under discussion. Although not introductory, most publications in the series are applied rather than highly technical, and all contain many detailed examples.

Other titles in the series:

Statistics in Education Ian Plewis

Statistics in Civil Engineering Andrew Metcalfe

Statistics in Human Genetics Pak Sham

Statistics in Finance Edited by David Hand & Saul Jacka

Statistics in Sport Edited by Jay Bennett

Statistics in Society Edited by Daniel Dorling & Stephen Simpson

Statistics in Psychiatry Graham Dunn

1

Introduction

It comes as a surprise to many people that there are any applications at all of statistics and mathematics in archaeology. (Fieller, 1993: 279)

1.1 Statistics and archaeology

1.1.1 Introduction

The aims of this book are discussed at greater length in the Preface. In summary, and notwithstanding the quotation that heads this chapter, the book attempts to review the very extensive application that statistical methods have had in archaeology. It is assumed that readers have a basic knowledge of statistics which, for archaeologists, is covered in texts such as Shennan (1997), so that basic techniques such as the use of descriptive statistics and the simpler forms of hypothesis testing are not covered. Though not introductory, the emphasis in the book is an applied one.

Much of the material here will not be found in other statistical texts written for archaeologists. There is some overlap with specialist texts, on multivariate methods (Baxter, 1994a) and Bayesian methods (Buck *et al.*, 1996), but I have made an effort to reference applications that post-date these books. Orton's (2000a) comprehensive text on sampling in archaeology appeared while this book was being written, and I have truncated my treatment of this topic accordingly.

The book is loosely structured into three parts. This and the following chapter, on data and problems, are introductory, and place what follows in context. Chapters 3 to 14 focus on specific statistical topics, and Chapters 15 to 21 focus on specific archaeological problems. I use the term 'loosely structured' because the 'statistical' chapters contain many examples of archaeological applications, and the 'archaeological' chapters include statistical material not used elsewhere in the book.

Large numbers of books have been written on each of the statistical methods covered, including some specifically for archaeologists, and what is presented here is necessarily selective. I have tried to highlight areas where applications in the literature are widely scattered, and possibly not well known to archaeologists. I was, for example, surprised at the range of applications of regression analysis (Chapter 5) which are not hinted at in the standard archaeological texts that deal with the subject.

Although archaeological problems have given rise to novel statistical methodology (e.g., Chapter 19), my view is that it is often more useful to try and identify existing statistical methodology with the potential for archaeological application. To name just a few, methods such as adaptive sampling (Section 4.4), non-parametric regression (Section 5.6), model-based clustering (Section 8.4), classification tree analysis (Section 9.4), shape analysis (Section 13.6), and Bayesian applications to problems other than those of chronology (Chapter 14) are all of potential value and in need of evaluation. I am currently less convinced of the general utility of other methods discussed, such as projection pursuit (Section 7.5), but would be delighted to be proved wrong. Many of these can be viewed as computer-intensive methods (Chapter 12), and it is in the application of such methods that I expect statistics to have its biggest impact on archaeological data analysis in the foreseeable future.

Some of these methods are readily available in software such as S-Plus and R (see the Appendix), though such software is currently not used much by archaeologists. More complex methods currently require statistical expertise to implement. In the context of Bayesian methods, but with general application, Buck *et al.* (1996: 304) concluded that the 'culture of interdisciplinary collaboration [between archaeologists and statisticians] needs to be fostered and encouraged'. I can only echo this. I hope this book serves to alert archaeologists and statisticians to what is on offer, and encourages such collaboration.

1.1.2   Statistical use in archaeology

Statistical use in archaeology might be categorized in various ways. For example:

1. Standard methods are applied in a 'routine' fashion, such as the use of multivariate statistics to investigate the chemical similarity between artifacts characterized by measurements on their chemical composition.
2. Methods are 'borrowed' from other quantitative disciplines, such as geography and ecology, that address problems perceived to be similar to those arising in archaeology. Examples are early work on spatial analysis in archaeology that was influenced by the geographical literature (Hodder and Orton, 1976), and the study of diversity in archaeological assemblages influenced by the ecological literature (Leonard and Jones, 1989).
3. Methods are developed by archaeologists, designed to be 'concordant' with archaeological data and problems, often as a result of dissatisfaction with the appropriateness of borrowed methodology (Carr, 1985a; Aldenderfer, 1987b).
4. Methods are developed by statisticians in response to specific archaeological problems, such as that of seriating archaeological assemblages on the basis of their similarity (Kendall, 1971a,b) or determining the reality or otherwise of the megalithic yard (Kendall, 1974; Freeman, 1976).
5. Methods are developed in extended collaboration between archaeologists and statisticians, applying 'new' statistical theory to archaeological problems. The development of Bayesian methods for radiocarbon calibration is a particularly good example of this (Buck *et al.*, 1996).

Examples of each kind of use will be found in the chapters to follow. The rest of this chapter provides historical background on the way statistical use has developed in archaeology, and reviews of sources of statistical applications in archaeology.

1.2 Statistics in archaeology

1.2.1 The background

Philosophies of statistics
It can come as a surprise to archaeologists that there are different theories of statistical inference, and that debate about their relative merits has generated considerable heat. Two prominent approaches are *classical* or *frequentist* inference and *Bayesian* inference. Lindley (2000) provides a useful overview of aspects of these approaches which differ, among other things, in the concept of probability used and the way prior information or beliefs are incorporated into data analysis.

Historically the classical approach has been dominant, and many practicing statisticians could be categorized as 'classical' without consciously thinking of themselves as such. Bayesian inference has, by contrast, been a minority pursuit with Bayesian statisticians conscious of the fact that they are 'Bayesian'. Bayesians, at least in the past, might be caricatured as members of an evangelical and fundamentalist sect, in a minority and persecuted, but absolutely convinced of the rightness of their position and the folly of other philosophies of statistical analysis.

Debates between protagonists of the two positions arguably generated more philosophical heat than analytical light. It was possible to get the impression (as I did when a student) that the Bayesian approach was justified on the grounds that it gave rise to the same results as the classical approach, but did so by what was claimed to be a 'purer' route. Practically, there were no obviously compelling reasons to be Bayesian.

The full strength of the approach could not be exploited because of the inadequacy of computational theory and power. This has changed since the 1980s, with the development of Markov chain Monte Carlo (MCMC) methods that have released Bayesians from the constraints imposed by these inadequacies and enabled them to investigate problems not readily handled within a classical framework. This has permeated the archaeological literature with particularly important applications to the calibration of radiocarbon dates (Chapter 15).

Archaeological theory
Although an archaeologist might be surprised to learn that there are competing schools of statistical inference, that they have not always co-existed easily will come as less of a surprise given the existence of, and relationships between, different schools of thought in archaeology. Reading some texts can give the impression that archaeology is both riven and driven by theoretical debate (though most practicing archaeologists probably get by without feeling the need to categorize themselves as adhering to any particular position). For present purposes it suffices to distinguish between the *culture history* approach to archaeology that dominated the 20th century up to about 1960, the *processual* approach (or *New Archaeology*) developed in reaction to culture history from the 1950s, and *post-processual*

archaeologies from the 1980s that reacted against processual archaeology. Briefly, the story is that developments within the New Archaeology fostered the use of statistics in archaeology, and that the reaction of the post-processualists militated against such use, though this reaction occurred by the time statistics was firmly established as a useful discipline for archaeologists to be aware of.

Culture history has been defined by Trigger (1989: 162) as the 'labelling of geographical and temporally restricted assemblages of prehistoric archaeological material as cultures or civilizations and their identification as the remains of ethnic groups'. Johnson (1999: 16–17) characterizes the approach as *normative*, tending to stress the differences between archaeological cultures rather than their similarities, and with an emphasis on cultures as unchanging entities. To explain change within the archaeological record, given this latter view, recourse is had to the idea of external influence in the form of migration or diffusion. Trigger (1989: 195) observes that 'to a large degree American archaeology came to be preoccupied with typologies of artifacts and cultures and working out cultural chronologies'.

The New Archaeology developed as a reaction against, among other things, the narrow focus on typology and chronology, perceived inadequacies in the way change within the archaeological record was explained, and difficulties in linking archaeological cultures with real people within culture history (Renfrew and Bahn, 2000: 38). Trigger (1989: 295) notes that 'the essential and enduring elements of the New Archaeology were the collective creation of a considerable number of American archaeologists during the 1950s'.

The emphasis shifted to explaining variation in the archaeological record in terms of general law-like relationships. 'Logico-deductive positivism', as expounded by Hempel (1962, 1965), was seen by some as an underpinning philosophical approach, 'eliminating subjective elements and establishing a basis for the objective scientific interpretation of archaeological data' (Trigger, 1989: 301). Johnson (1999: 21) states that 'the New Archaeology must be understood as a movement or mood of dissatisfaction rather than a specific set of beliefs' that was marked with 'revolutionary fervour'. Processual archaeology is described by Johnson (1999: 193) as a more mature form of the New Archaeology, with an emphasis on the idea of process, a tendency to generalize, and as adopting a broadly positivist approach. Leading figures included Lewis Binford in America and David Clarke in Britain, though the movement found more fertile soil in America.

Post-processual, or interpretive, archaeology is not a single theory but rather, as Johnson (1999: 101) stresses, 'a very diverse set of concerns and ideas that coalesced around certain slogans'. To highlight just one of the features of post-processual thought identified by Johnson (1999: 102–8), a positivist view of science is rejected, and data are regarded as theory-laden, so that confronting theory with data is chimerical, since data are not objectively defined and thus cannot provide an objective test of theory. In an extreme form this leads to the rejection of statistics, and quantification generally, as useful tools for interpreting the archaeological record (Shanks and Tilley, 1992: 56–9). Ringrose (1993a: 122), a statistician, commented that the 'bizarre claims' of Shanks and Tilley, who viewed the use of statistics in archaeology as 'stultifying and leading to thought being limited and non-creative', could be discounted. Some flavor of what Ringrose was reacting against can be seen in this passage from Shanks and Tilley (1992: 58):

the new archaeology has always looked to the construction of formal symbolic logic as an ultimate goal (e.g., Clarke, 1968: 62). This goal, when reached, destroys archaeology because it is not ultimately the data that matter any more but the internal coherence of the statistics to which they are fitted. It is the development of the statistics that provides the key to future work, not the conceptualization of the data. At least, the latter is placed in very much a subsidiary and peripheral role. Mathematical coherence replaces archaeological knowledge. Mathematization results in the dissolving of the physicality of the objects of archaeological knowledge in terms of the logical or mathematical relations. The very notion of objective substance opposed to subjectivity disappears.

Shanks and Tilley (1992: 59) also criticized what they characterize as Orton's (1980) view of 'mathematics as a cognitive instrument and a universal tool'. Orton (1999: 32) later riposted

what does statistics still have to offer archaeology in a post-processual era? Even post-processual archaeologists use statistical techniques, but it seems to me they have a problem with method. Archaeological analysis is primarily about *data* – we may argue about which data, and how to collect or measure them, or from which theoretical perspective we view them, but ultimately they are data like any other. The view that archaeologists create data is an attack of *hubris*: the data are there, and our role is to select and record. If you create your data and I create mine, then we are in the position of the witnesses to a traffic accident that consisted of a collision between two stationary vehicles.

Whatever the merits of these arguments, the important point is that those who bought into Shanks and Tilley's views have been disinclined to consider the merits of quantification and the use of statistical methods, as a matter of principle. Thus, to the extent that post-processual archaeologies have replaced processual archaeology as the dominant ideology, the development of statistical methodology within archaeology has been stifled. Shanks and Tilley (1992) was the second edition of their earlier, and controversial, 1987 book, and in their preface, in explaining why the book was largely unchanged, they suggested that 'there are still many, and particularly in the United States, who consider archaeology a scientific exercise'. This points to a difference in practice between (some) archaeologists in the United States and Europe, with Europe having been more penetrated by post-processual thought. This has obvious implications for the way statistics is routinely used on the two continents.

The growth of statistics in archaeology

To repeat an earlier caveat, this is a very simplified picture. Many archaeologists carry out their trade without undue reference to any particular theoretical (or philosophical or ideological) stance and some of these make routine use of statistical methods when it seems appropriate to do so. However, it is possible to see the period from the 1960s to the 1980s as a golden age for statistics in archaeology, nurtured and cherished within a climate of opinion that valued what was on offer.

One might associate the birth of statistics in archaeology with the work of Robinson (1951) and Spaulding (1953); its heady and carefree adolescence occurring in the 1960s (e.g., Binford, 1964; Binford and Binford, 1966; Hodson *et al.*, 1966; Clarke, 1968; Cowgill, 1968; Hodson, 1969); a coming of age in the early 1970s (Hodson *et al.*, 1971); culminating in early maturity in the mid-1970s with the publication of the books by Doran and Hodson (1975) and Hodder and Orton (1976). With increasing age the marriage between statistics and archaeology (to change analogies in midstream) fell into a more routine pattern, with statistics accepted as a suitable topic to be taught to archaeologists (e.g., see Fieller, 1993: 281–2) and texts eventually being produced for this purpose (Shennan, 1988). As in some marriages, the uncritical adoration of young love began to be replaced by a more critical awareness of the limitations, as well as merits, of the statistics partner (e.g., Thomas, 1978). This resulted in a growing concern for concordance in archaeological data analysis between archaeological theory and statistical method that preoccupied some scholars throughout the 1980s and beyond (Carr, 1985a; Aldenderfer, 1987a). Brandon, in the preface to Westcott and Brandon's (2000) book on geographical information systems (GIS) in archaeology, suggests that 'during its heyday, statistics had been waved above archaeologists' heads as an "answer" to dealing with a multitude of archaeological problems' but 'after much yelling and arm-waving, most agreed that statistics were not an answer in themselves but, like GIS, an extremely important tool available for archaeological use'.

Not all contributors to the development of statistics in archaeology viewed themselves as 'New Archaeologists'. Doran and Hodson (1975: 5) explicitly distanced themselves from New Archaeology, finding its claims 'greatly exaggerated and therefore dangerous' and 'a bizarre mixture of naivety and dogmatism'.

This highlights a difference in approach between some American (New Archaeology) and British quantitative archaeologists (Doran and Hodson) that recurs, which is that the former tend to pay more concern to 'theoretical' matters than the latter, who tend to be more pragmatic in their application of statistics to archaeological problems. A particular danger of embedding statistical use within a theoretical framework, positivism in this case, is that rejection of the framework also results in a rejection of the use of statistical methodology. Statistics in archaeology has suffered from this (Baxter, 1994a: 7; Shennan, 1997: 3; Orton, 1999: 31–2). There is, in fact, no necessary connection between believing statistical methods to be useful for archaeological data analysis, and adhering to a positivist view of how science should be conducted (unconsciously or otherwise).

Another important point concerns the role that archaeologists have conceived for statistics. Thus, Read (1989: 5) distinguishes between statistics as a collection of methods and 'statistics as a means to represent and reason about [archaeological] ideas and concepts'. He cites Clarke's (1968) *Analytical Archaeology* as 'a grandiose programmatic statement' promoting 'reasoning about, and representation of, archaeological statements using the conceptual framework provided by statistics and mathematics'. A rejection of such a conceptual framework, which many would not accept, does not entail rejecting the usefulness of statistics as methods.

Early statistical use in archaeology involved both inferential (or confirmatory) and exploratory methods of analysis, with the rigor of the Neyman–Pearson approach to inference attractive to some. With experience of this approach some archaeologists began to doubt if it was the best way of applying statistics to archaeological problems, and suggested that Bayesian methodology (Cowgill,

1977a) or exploratory data analysis (EDA) in the fashion of Tukey (1977) might provide a more appropriate framework (Clark and Stafford, 1982; Carr, 1987). Computational practicalities limited the use of Bayesian methodology until the 1990s, and even now its use outside the realm of dating problems is limited, though increasing (Chapter 14). As for EDA, Aldenderfer (1998: 109) suggested that archaeologists had 'elevated EDA to something of a philosophy of data analysis that is often seen as contradictory or opposed in some sense to classical, or confirmatory (or inferential), methods'.

The elevation of EDA to the status of a 'philosophy' is unhelpful and it is not accorded separate or special status in this book. I prefer to group the methods commonly labeled as EDA techniques with other data-analytic or exploratory methods, of the kind that dominate much statistical analysis in archaeology (Baxter, 1994a). In this I am thinking of techniques that look for structure or patterning in data, without making probabilistic assumptions about the nature of the population from which the data are drawn (and without necessarily having to conceive of the data as a sample from a population). Such exploratory methods can be contrasted with model-based methods, exemplified by but not exclusive to Bayesian analysis, where quite explicit probabilistic assumptions are made about the nature of the population generating the data. Whether model-based or exploratory analyses, or both, are appropriate is dependent on the data to hand, the questions asked of the data, and the assumptions one is prepared to make.

It is not always easy to pigeon-hole a method as 'exploratory' or 'model-based'. For example, some popular methods of cluster analysis, widely used in archaeology, such as Ward's method or the spatial k-means clustering of Kintigh and Ammerman (1982), while usually used in an exploratory or heuristic spirit, are equivalent to a model-based method that assumes clusters will be spherical and of equal volume (or area). The results from such analyses typically reflect these, often implicit, assumptions. Statistical theory helps explain why methods may impose structure on data, not always in accordance with what archaeological expectations might predict.

This brings us on to the subject of 'concordance' or 'congruence' between archaeological theory and statistical method (Carr, 1985a; Aldenderfer, 1987a). The essence of the problem, as it seems to me, can be simply stated. If the assumptions of an archaeological model (or theory) about the process(es) generating a set of data are incompatible with the assumptions of the statistical model or method used to analyze the data, then unsatisfactory results from the point of view of archaeological interpretation may be obtained. This is both an unexceptionable, and in some ways obvious, point of view that could be seen as a restatement of the 'garbage in, garbage out' principle, but the language in which the discussion about concordance has been conducted has possibly tended to obfuscate (a favorite word in the debate, incidentally) rather than enlighten.

As a result of a concern for concordance, the view has sometimes been expressed that statistical techniques for archaeological data analysis are best devised by archaeologists who understand their theories and data, rather than statisticians who don't. There is an element of Orwell's (1945) 'four legs good, two legs bad' here that, with particular reference to intrasite spatial analysis, Orton (1992: 137) has characterized as 'the Audrey syndrome ("An ill-favoured thing, sir, but mine own", *As You Like It*, Act V Scene iv)'. Orton notes that papers such as those of Kintigh and Ammerman (1982) and Whallon (1984), often cited as good examples of 'concordant' methodology by quantitative archaeologists, ignore developments in the theoretical statistical literature of direct relevance to the problems being addressed. Attempts by archaeologists to develop concordant methods have not always been judged successful. Ammerman (1992: 249), an archaeologist, describes Carr's (1985b) attempt to develop 'concordant' methods that deal with the complication that spatial patterns result from multiple processes as 'one of the more prolix and less fortunate chapters in the recent history of intrasite spatial analysis'.

There is no intention here of implying that statisticians are usually right and archaeologists wrong, or vice versa. I have described elsewhere (Baxter, 1994a: 219) an analysis of mine undertaken initially without archaeological input (fortunately never published) that has been used in undergraduate teaching at a leading British university archaeology department as a salutary example of how not to do things. Rather, the problem seems to me to be a lack of communication or contact between archaeologists and statisticians. This is exacerbated by the lack of statisticians interested in archaeological data analysis. I find it difficult to believe that statistical problems arising in archaeology are inherently more difficult than statistical problems in other disciplines, to whose solution statisticians have usefully contributed.

1.2.2   Landmark and other papers

It is possible to discern elements of statistical thinking in archaeological work in the late 19th century (e.g., Petrie, 1899), but most commentators date the serious emergence of quantitative methods in archaeology to the early 1950s. The papers (and one book) discussed below have been selected because of their importance in the development of statistics in archaeology and/or because they provide a convenient peg on which to hang further discussion of some of the points noted in the previous section. Discussion is limited to papers published before 1990; books, apart from Clarke (1968), are covered in the next subsection.

Robinson (1951)
Several commentators cite this as an important early paper (Aldenderfer, 1998; Orton, 1999). Robinson was concerned with the problem of seriation (Chapter 16), which involves the sequential ordering of archaeological contexts on the basis of their similarity in terms of the artifacts found within them. The hope is that such an ordering corresponds to a relative chronological ordering, not necessarily recoverable by other means. Given the percentage occurrence of p types of pottery in each of n contexts, Robinson defined a measure of similarity, or 'agreement coefficient', between pairs of contexts, and an ordering principle based on the $n \times n$ matrix of these coefficients. It is an example of a specific archaeological problem that admits of a clear formulation as a mathematical one as well, and subsequently attracted considerable attention from mathematicians and statisticians in the 1960s and 1970s (Chapter 16). This is in contrast to many archaeological problems, susceptible to statistical treatment, that do not lend themselves to such a tidy formulation in terms of a statistical or mathematical model.

Spaulding (1953)
Spaulding's (1953) paper, 'Statistical techniques for the discovery of artifact types', has variously been described as 'seminal' (Ammerman, 1992: 233; Dunnell, 1986: 191), a 'landmark paper' (Dunnell, 1986: 178), and as a 'classic

paper' that 'introduced modern statistical thinking to archaeology' (Aldenderfer, 1998: 91). It is difficult from a purely statistical point of view to appreciate the importance of the paper, which was devoted to typology, a subject central to archaeological practice. Technically, what was involved was the use of chi- squared statistics, applied to low-dimensional contingency tables, to identify cells that deviated significantly from their expectation under the null hypothesis of no association (Chapter 11). The categorical variables used described different aspects of an artifact, such as the nature of the fabric used in making a pot, type of surface, and type of decoration on the shoulder of a pot. Cells departing significantly from expectation under the hypothesis of no association – indicative of inter-action – identified combinations of attributes (levels) that Spaulding associated with particular types within the data. The much quoted, and disputed, definition of Spaulding was that an 'artifact type is here viewed as a group of artifacts exhibiting a consistent assemblage of attributes whose combined properties give a characteristic pattern' (Spaulding, 1953: 305), and chi-squared analysis was seen as a means of identifying such consistent assemblages of attributes.

Although the importance of the paper was recognized, Dunnell (1986: 182) noted that there had 'been few attempts to use Spaulding's initial formulation'. With the development of statistical methodology in the 1960s and 1970s others, such as Read (1974) and Spaulding (1976, 1977, 1982) himself, advocated the use of log-linear modeling for the purposes identified in the 1953 paper.

Given this, it might be asked why Spaulding (1953) was regarded as a seminal paper. The answer is to be found from the archaeological and not the statistical per-spective. Spaulding challenged thinking about the construction of archaeological typologies in a fundamental way and, moreover, did so in a 'rigorous quantitative manner' designed to eliminate the 'trial and error element' in the construction of typologies (Dunnell, 1986: 178). As Adams and Adams (1991: 291) put it, 'Spaulding was apparently the first archaeologist to learn something of the meth-ods of the professional statistician' and had argued with 'singular constancy' that 'statistical procedures hold the answers to many of the problems of classification' discussed in Adams and Adams's book.

Several points can be made here. One is that what is archaeologically important, stimulating, or useful, need not be statistically innovative. A second is that some important statistical developments in archaeological thinking and practice have either not gone very far, have gone in the wrong direction, or not gone in the direction intended by their proponents. What has been important, however, is not so much the success or otherwise of specific statistical endeavors, as the climate of opinion generated. The specific techniques promoted by Spaulding and others may not have borne much direct fruit, but they did encourage others to embrace statistical ideas and develop approaches to archaeological data analysis that, in some cases, are now routine. A third point is that ideas for statistical analysis in archaeology have sometimes been in advance of the available technology, both statistical and computational.

Thom (1955)
This is not an obvious choice. Thom presented his paper on the statistical examina-tion of megalithic sites in Britain to members of the Royal Statistical Society. He was, apparently, grateful for this forum that others were unwilling to provide him with (Aitchison, 1976). Among other things, Thom controversially proposed the hypothesis of the megalithic yard, a unit of measurement supposedly ubiquitous in

Britain from the Late Neolithic to the Middle Bronze Age over a period spanning several millennia (Chapter 19). This idea was anathema to many archaeologists.

The problem was that statistical technology at the time was unable to provide a rigorous test of Thom's ideas. About 20 years elapsed before the work of statisticians, Kendall (1974) and Freeman (1976), convinced most interested parties that the ubiquity of the megalithic yard was seriously to be doubted, any evidence for a common unit of measurement residing in a geographically circumscribed subset of data. This is an unusual instance of a specific archaeological problem being initially posed in a statistical setting and requiring the development of novel statistical methods to achieve some sort of resolution. Some archaeologists took the view that, on archaeological grounds alone, the hypothesis should never have been entertained (see the discussion of Freeman, 1976), but having statistical evidence to confirm their views did no harm.

Clarke (1962, 1968)
Most of the early influential papers in quantitative archaeology were published in the American literature. Clarke's (1962) use of matrix analysis to investigate the grouping of British 'Beaker' pottery was one of the earliest papers to alert British archaeologists to the existence of quantitative methodology. This drew on Robinson (1951). References to EDSAC II, 'one of the finest [electronic brains] in Europe' (Clarke, 1962: 375), and descriptions of data entry and programming using punched tape remind one of how difficult statistical analysis could be without modern computing facilities.

Jumping ahead in time a little, Clarke's (1968) book *Analytical Archaeology* has been described as a 'difficult but essential read' whose influence 'is hard to overestimate' (Johnson, 1999: 199). This influence was in the wider field of archaeological theory, but the book contains substantial sections on statistical methodology. Numerical taxonomy, in particular, was promoted as a means of providing a more objective definition of conceptual and actual archaeological entities, of allowing the assessment of affinity between such entities, and of allowing the ranking of such entities in hierarchical systems (Clarke, 1968: 518–19). Clarke drew on the book *Principles of Numerical Taxonomy*, by Sokal and Sneath (1963), a text that had a considerable influence on archaeological uses of cluster analysis. Later writers such as Thomas (1978) were to regard the dominance of numerical taxonomic methods as malign and were highly critical of Clarke's use of it.

Binford (1964)
Although Orton (2000a: 68) credits Vescelius (1960) with the idea of using probabilistic sampling methods in archaeology, he notes that it is only after Binford's (1964) influential paper that such ideas began to be acted on (Chapter 4). Hole (1980: 218) went so far as to say that 'no other single article has altered more radically the recent course of archaeology'. Binford argued that the region should be the primary unit of archaeological research and, given the impossibility of complete coverage, promoted the use of sampling theory in conjunction with properly designed sample surveys. The interest and activity this generated is reflected in the conference proceedings edited by Mueller (1975) and Cherry *et al.* (1978), based on events in 1973 in San Francisco in the USA and 1977 in Southampton in the UK. Binford had an enormous influence on the way quantitative methodology developed in archaeology, particularly in America.

Binford and Binford (1966)
This, along with Spaulding (1953), is one of the papers most frequently cited as important in the development of quantitative archaeology. Aldenderfer (1998: 91–2) describes it as a 'stimulating, if inaccurate, multivariate analysis of Mousterian assemblages from southern France and the Near East' that 'was almost mystical in its complexity', but which, with Spaulding (1953), 'provided exemplars of what could be done with quantitative methods'. Orton (1999: 25) has observed that the Binfords' paper was among those that put quantitative methods 'on the map', while Read (1989: 6–7) states that 'the paper seemed to show that the potential of what could be achieved with quantitative methods was limited only by the creativeness of the archaeologist', but also that it 'could serve as an example of incorrectly applied statistical methods'.

The paper introduced factor analysis (Chapter 7) to a wide archaeological audience, and Lewis Binford's influence was such that the technique was enthusiastically adopted for a period of time. As later commentators were to agree, this was not necessarily a good thing. The Binfords did not give full technical details and what they called factor analysis was, in reality, principal component analysis (PCA) with varimax rotation (Kimball, 1987: 119). Confusion between PCA and factor analysis, and misuse of what was called the latter, ensued and was not cleared up for some time. In this case a paper that was a mountainous landmark in the development of quantitative archaeology gave birth to numerous unsightly molehills that did not enhance the appearance of the emerging landscape.

Hodson et al. (1966) and Hodson (1969, 1970)
Much of the development in Britain took place under the influence of Hodson, whose early work on cluster analysis (Chapter 8) developed, among other things, an approach to artifact typology different from that initiated by Spaulding (1953). These differences were later to occasion a considerable amount of debate (e.g., Whallon and Brown, 1982), though there are those who have observed that such methodological controversy has had little impact on the way practical typologies are constructed (Dunnell, 1986; Adams and Adams, 1991).

Cowgill (1968, 1977a)
Some archaeologists with quantitative training became concerned, at an early stage, about the potential for the misuse of 'complex' statistical methodology in archaeology. A genre of what might be called constructive criticism that attempted to elucidate the assumptions underlying the 'new' statistical methods being used, and the circumstances under which they were appropriate for archaeological data, developed. Cowgill's papers are good examples of these. The 1968 paper contains a critique of cluster analysis, factor analysis and proximity analysis, methods that had been little used at that stage.

One attraction of statistics for quantitatively minded archaeologists was the apparent rigor of statistical procedures, as embodied in the Neyman–Pearson approach to statistical inference and significance testing. Cowgill (1977a: 351) was written with the intention of correcting 'serious errors in the interpretation of statistical results which are often committed by archaeologists'. The paper is largely an exposition of the Neyman–Pearson approach, but interesting as identifying the Bayesian approach as an alternative (Cowgill, 1977a: 361–2), a theme to which he returned in Cowgill (1993), little having been accomplished in the intervening years. The first published Bayesian analysis of archaeological data was Freeman (1976), and Bayesian methods only began to be used regularly for archaeological problems in the 1990s (Chapter 14).

Whallon (1973, 1974)

Wandsnider (1996: 324) cites Whallon's papers as having 'pioneered the subdiscipline of quantitative archaeological spatial analysis' (Chapter 13). Blankholm (1991: 39) defines the general operational approach in intrasite spatial analysis that was pioneered, as a three-stage procedure involving testing for randomness, identifying and delineating non-random concentrations of artifacts, and testing for spatial associations among artifact types. The 1973 and 1974 papers introduced techniques, dimensional analysis of variance and nearest neighbor analysis, that had been developed in the ecological literature. Blankholm (1991: 142) suggested that the former technique had largely fallen out of use by the end of the 1970s. Nearest neighbor analysis became much more widely used, and is discussed in Section 13.4.1.

Thomas (1978)

Having suggested the existence of a genre of papers involving 'constructive criticism' in the literature, one can also identify papers that can be classified as 'destructive criticism'. Thomas's graphically entitled 'The awful truth about statistics in archaeology' is an example. Such papers represented a backlash against the misuse of statistics rather than statistical usage *per se*. Thomas attacked 'archaeological bandwagons' including numerical taxonomy, probability sampling and multivariate statistical analysis, and was dismissive about the work of several of the authors whose papers have been discussed above. Other works of this kind include Hole (1980) and Scheps (1982) whose equally graphic title, 'Statistical blight', is indicative of its content. In some papers the target is more specific, as with Christenson and Read (1977) on cluster analysis, and Vierra and Carlson (1981) on factor analysis. I have noted elsewhere (Baxter, 1994a) that the effect of the onslaught appears to have been the disappearance of 'complex' statistics from the journal *American Antiquity* for a period of years.

Kintigh and Ammerman (1982)

This paper treats two statistical themes of importance that began to be explored seriously in the 1970s, the spatial analysis of data, and the use of simulation. The statistical content is unexceptional, with k-means cluster analysis used to try and identify structure in two-dimensional scatters of artifacts or sites (Chapters 8 and 13). The paper has been seen as one devised by, and for, archaeologists concerned to have models and methods tailored to the problems and data of their own discipline, rather than borrowing inappropriate methods from other disciplines. That it resulted in the use of a 'standard' statistical approach to clustering has been seen as incidental. This concern with 'congruency' (Whallon, 1984) or 'concordance' (Carr, 1985a) was discussed earlier. Whallon's (1984) paper on 'unconstrained clustering' of spatial data is another paper also often cited with approval in this context.

Bølviken et al. (1982)

Bølviken *et al.* is the paper that many regard as having introduced correspondence analysis into the archaeological literature (Chapter 11). Correspondence analysis is now widely used routinely for data analysis in archaeology. There are somewhat earlier archaeological applications in the statistical literature (Hill, 1974) and in the French archaeological literature (e.g., Djindjian and de Croisset, 1976a,b). The fact that Bølviken *et al.* often get the credit, and the story of correspondence analysis in archaeology, is indicative of the linguistic, geographical, disciplinary and computational barriers that can hamper the diffusion of useful statistical techniques in archaeology. Scandinavian scholars having introduced correspondence analysis to archaeologists in the English speaking world, it was German, South African and Australian scholars who made the method accessible by developing appropriate software, at a time when correspondence analysis was not available in most of the larger software packages used for statistical analysis by archaeologists (Scollar *et al.*, 1993; Wright, 1985, 1989). The technique 'took off' in the European literature from the late 1980s but does not seem to have been much used or noted in the American literature till the mid-1990s, when it was described as 'not well established in Americanist literature' (Duff, 1996: 90).

Naylor and Smith (1988)
Published in a statistics journal, Naylor and Smith (1988) is a Bayesian analysis that investigated how to incorporate knowledge of archaeological phasing into the calibration of a series of radiocarbon dates. It contains some archaeological misunderstandings, later corrected (Buck *et al.*, 1992). Litton and Leese (1991: 105) observed that Naylor and Smith 'were not interested in problems associated with radiocarbon dates *per se*, but wanted a complex problem, with a high dimensional, multimodal likelihood with constraints on the parameter values, with which to demonstrate the capabilities of their recently developed computer package, BAYES4'.

This might sound unpromising, however the basic method proposed was developed and extended by an intensive and fruitful collaboration between archaeologists and statisticians then working at Nottingham University, UK, in the early 1990s (Buck *et al.*, 1996). This led to the development of an approach for the calibration of multiple radiocarbon dates, capable of taking into account archaeological knowledge of stratigraphic relationships, phasing and so on (Chapters 14 and 15). As well as demonstrating the potential of Bayesian analysis for archaeology, the methods developed have also been implemented in the widely used OxCal calibration software (Bronk Ramsey, 1995) that is routinely used by archaeologists who may have only a limited appreciation of the statistical underpinnings. The story of this development, for which Naylor and Smith (1988) was one of the intellectual foundations, is a model of what can be achieved by close collaboration between archaeologists and statisticians.

1.2.3 Other literature

Texts
The book *Some Applications of Statistics to Archaeology* by Myers (1950) is the earliest text devoted to statistics in archaeology covering, among other things, the use of regression and correlation. Read (1989: 6) observed that the book was remarkable 'for its early insights into the manner in which statistical methods can address archaeological problems', but also that it had no impact. Orton (1999: 25) is the only other paper I have seen that refers to Myers' work, and it was developments in the American literature, and specifically papers published in *American Antiquity*, that proved more influential.

It was not until the mid-1970s that there was any attempt to emulate Myers, with *Mathematics and Computers in Archaeology* by Doran and Hodson (1975) the landmark publication. It brought together, and discussed in a critical way, many of the techniques that had been explored in the previous decade or so, particularly those of multivariate analysis. In the next year Hodder and Orton's (1976) *Spatial Analysis in Archaeology* did a similar job for statistical methods of spatial analysis. Thomas's (1976) *Figuring Anthropology* is a traditional introductory statistics textbook, covering topics from data description and probability through to hypothesis testing and regression analysis.

The 1980s saw the publication of Orton's (1980) introductory text *Mathematics in Archaeology*, the specialized work of Grayson (1984), *Quantitative Zooarchaeology*, and the second edition of Thomas's book (1986), now entitled *Refiguring Anthropology*. Shennan's (1988) *Quantifying Archaeology* rapidly established itself as the most widely used text on statistics for archaeology undergraduates. This placed some emphasis on regression methodology, probabilistic sampling and the multivariate methods of cluster analysis and PCA.

Other texts that complement rather than compete with Shennan appeared in the 1990s. *Digging Numbers* by Fletcher and Lock (1991) covers, at a basic level, some of the inferential statistics omitted by Shennan, while Drennan's (1996) *Statistics for Archaeologists* is at a more introductory level and emphasizes some of the techniques of exploratory data analysis. The second edition of *Quantifying Archaeology* (Shennan, 1997) included material on inferential statistics omitted in the first edition, and also reflected the growing popularity of correspondence analysis by devoting a new chapter to it. Banning's (2000) *The Archaeologist's Laboratory*, while not primarily a statistics text, contains useful sections, at an introductory level, on sampling, quantification, seriation and the interpretation of radiocarbon dates.

On the specialized front, the titles of the books *Intrasite Spatial Analysis in Theory and Practice* (Blankholm, 1991), *Exploratory Multivariate Analysis in Archaeology* (Baxter, 1994a), *Bayesian Approach to Interpreting Archaeological Data* (Buck *et al.*, 1996) and *Sampling in Archaeology* (Orton, 2000a) are all accurately descriptive of their contents. These are English-language texts. In other languages, I am aware of, but cannot claim to be familiar with, Ihm's (1978) *Statistik in der Archäologie* and Djindjian's (1991) *Méthodes pour l'Archéologie*.

Edited collections
Edited collections range from those themed round a single topic containing many papers with a statistical or quantitative content, through to collections where several themes may be covered and where most papers have a limited statistical input. In the former category, in chronological order of publication, are *Sampling in Archaeology* (Mueller, 1975), *Simulation Studies in Archaeology* (Hodder, 1978), *Sampling in Contemporary British Archaeology* (Cherry *et al.*, 1978), *Essays on Archaeological Typology* (Whallon and Brown, 1982), *Intrasite Spatial Analysis in Archaeology* (Hietala, 1984), *For Concordance in Archaeological Analysis* (Carr, 1985a), *Quantitative Research in Archaeology* (Aldenderfer, 1987a), *Multivariate Archaeology* (Madsen, 1988a), *Quantifying Diversity in Archaeology* (Leonard and Jones, 1989) and *The Interpretation of Archaeological Spatial Patterning* (Kroll and Price, 1991). It is noteworthy that nearly all these publications derive

from the 1970s and 1980s. The titles are mostly self-explanatory. The articles in Madsen (1988a) mostly, though not exclusively, make some use of correspondence analysis.

Heizer and Cook (1960) is the proceedings of a 1959 conference held at Burg Wartenstein, Austria, that Orton (1999) identifies as the first quantitative methods conference in archaeology. He notes that most of the contributions dealt with archaeometric themes. *Archéologie et Calculateurs* (Gardin, 1970) is the proceedings of another early conference on mathematics in archaeology with some papers of statistical interest (Read, 1989: 12–14).

Mathematics in the Archaeological and Historical Sciences (Hodson *et al.*, 1971) is the proceedings of a conference held in Mamaia, Romania, in 1970, and is regarded by Orton (1999: 25) as 'formal international recognition of the status of quantitative methods in archaeology, or as a rite of passage from a phase of experimentation to one of acceptance and optimism'. Many of the contributors were mathematicians and statisticians. Writing almost 30 years later, one of the participants (Wilcock and Sanie, 2000: 157) remembered that 'papers were excessively statistical in nature, however, and Clive Orton ... was applauded for what was almost the only slide of an archaeological site to be seen'. Topics of particular interest included taxonomy, multidimensional scaling and seriation.

The Union Internationale des Sciences Préhistoriques et Protohistoriques (UISPP) has a Scientific Commission 4 that organizes (sometimes jointly) meetings on 'Data Management and Mathematical Methods in Archaeology' that have resulted in publications such as Voorrips and Loving (1985), Aldenderfer (1987a) and Voorrips and Ottaway (1990). Voorrips (1990) also edited *Mathematics and Information Science in Archaeology: A Flexible Framework*, which includes overviews of topics such as sampling, seriation, predictive modeling of site location, classification and intrasite spatial analysis.

More recently, in 2000, Commission 4 of the UISPP held a joint meeting with the Computer Applications and Quantitative Methods in Archaeology (CAA) conference. The CAA conference has been held annually since 1973, and from being a small and largely British concern, has grown into an international conference that typically attracts several hundred participants. Proceedings are published and, from 1987 when publication has mostly been in the *British Archaeological Reports* International Series, are to be found (in the order in which conferences were held) in Ruggles and Rahtz (1987), Rahtz (1988), Rahtz and Richards (1989), Lockyear and Rahtz (1991), Lock and Moffett (1992), Andresen *et al.* (1993), Wilcock and Lockyear (1995), Huggett and Ryan (1995), Kamermans and Fennema (1996), Lockyear *et al.* (2000), Dingwall *et al.* (1999), Barceló *et al.* (1999), Stančič and Veljanovski (2001) and Burenhult (2002). Buck *et al.* (2000), which contains papers from a 1999 meeting of the British chapter of CAA, and Wheatley *et al.* (2002), which contains some papers from the 1998 British meeting, may also be regarded as part of this series. Content is varied, but there is usually a significant statistical presence.

There is also an 'International Symposium on Computing and Archaeology' that has similar concerns to CAA, but with a more southern European flavor, and publishing in several languages. Proceedings of the first two conferences are to be found in Valdés *et al.* (1995) and the two issues of volume 7 of the journal *Archaeologia e Calcolatori*.

Reviews
A useful way of tracking the development of statistical use in archaeology is through the reviews that appear fairly regularly in the literature.

Orton (1999) identifies Heizer and Cook (1956) as the first review of quantitative methods in archaeology. It pre-dates the period when quantitative methods began to be widely used by archaeologists, and some of it makes curious reading now. Subsequent reviews, in chronological order, include Clark (1982), Clark and Stafford (1982), Aldenderfer (1987b), Djindjian (1989), Read (1989), Barceló (1991), Ammerman (1992), Orton (1992), Fieller (1993), Aldenderfer (1998), Delicado (1999), Orton (1999), Wilcock (1999) and Cowgill (2001). The most comprehensive of these is Read (1989), while the more recent paper by Aldenderfer (1998) is also extensive. The range and content of these papers is highly variable. Some, such as Clark and Stafford (1982), concentrate primarily on developments in the American literature. Others are of a more technical statistical nature. Delicado (1999), for example, identifies statistical methods not currently used much in archaeology but of potential interest.

1.3   Simple statistics?

Whallon (1987: 135) argued that, to his knowledge, 'the vast majority of statistical analyses of archaeological data, published or unpublished [had] been done without adequate preliminary scrutiny of the data' and with what he termed 'simple statistics', in advance of, and often instead of, more 'complex' methods, among which PCA and factor analysis were named. In the same volume Kintigh (1987: 128) listed other methods, such as regression, discriminant analysis, cluster analysis, log-linear models and multidimensional scaling, that were similarly categorized as 'complex'.

No statistician would take issue with the idea that data analysis should begin with simple exploratory analysis. It can be argued that conceptions of 'complexity' and 'simplicity' in archaeological statistics need revision in the light of computational developments that have occurred since the mid-1980s. The simplicity, or otherwise, of a statistical method can be judged across several dimensions, such as the concept underlying the method, the mathematical underpinning, and ease of execution. When archaeologists use the term 'simple' to describe a method I suspect they often have in mind methods that are simple in all three senses. This includes graphical methods such as bivariate plots, histograms and box-plots, and the computation of statistics such as the mean or median of a set of data.

What is simple in terms of ease of execution depends on what software an investigator has access to, and how easily that software can be used. As a working definition, one could classify as computationally simple those techniques that are available in widely distributed software packages, such as SPSS and MINITAB, commonly used in teaching statistics to non-statisticians. On this definition methods such as PCA, factor analysis, linear regression, some variants of cluster analysis, discriminant analysis and, increasingly, correspondence analysis are computationally simple. Of other methods discussed in this book, kernel density estimation, projection pursuit, classification trees, Bayesian methods that involve MCMC simulation, shape analysis, various forms of non-linear regression, and, possibly, log-linear modeling are among those that are not. The categorization of a method can change with time. For example, it was only in the

1990s that correspondence analysis became available in some of the more popular software packages.

Most of the methods listed are, for a non-mathematician, mathematically complex, and this has undoubtedly acted as a deterrent to their archaeological use. It might be argued that, given suitable software to implement a method, such complexity is irrelevant, provided a method is conceptually simple, and given an understanding of its limitations and practical considerations in application.

Whether a method is to be regarded as conceptually simple or not is not always an easy judgment to make. One possible distinction is between methods that tend to be descriptive or exploratory in use on the one hand, and model-based methods on the other. Thus PCA is a straightforward exploratory method, the aim of which can be understood without a deep knowledge of the underlying mathematics. Factor analysis, which is often confused with PCA, is a model-based method the proper use of which involves many more considerations than the use of PCA. Similarly, correspondence analysis could be regarded as a simple method for investigating structure in a table of counted data, whereas log-linear modeling of such data is a more complex method of analysis. As a general rule, model-based methods tend to be more complex than exploratory ones, though this is not invariably so. Some methods of cluster analysis, for example, are widely used in an exploratory fashion and their intent is simple to understand, but they can also be viewed as model-based techniques and some of their limitations can be explained from this perspective.

As software develops I anticipate that some of the methods discussed in this book, that may seem complex, will find increasing use because they are so obviously useful. This has already happened with Bayesian approaches to radiocarbon calibration (Chapters 14 and 15) because of the development of software that hides the complexity from the user. Kernel density estimation, non-parametric curve fitting and classification trees are all useful methods whose use should expand if they are made more readily available. Correspondence analysis is now one of the most widely used multivariate methods in archaeology, not something that could have been said 10–15 years ago, and this is largely attributable to its ready availability as well as obvious utility.

In the next chapter some data sets, used for illustration in the book, are described, along with the kinds of archaeological problems that generate the collection of such data. Chapter 3 discusses kernel density estimation, a conceptually simple but mathematically complex method that is beginning to enjoy wider application in archaeology. Chapters 4 and 5 deal with sampling and regression. The treatment of the former topic is relatively short, in view of the recent publication of Orton's (2000a) excellent text on sampling in archaeology. Chapters 6 to 10 deal with a variety of methods of multivariate analysis. Techniques such as PCA, cluster analysis and discriminant analysis are among the staples of statistical use in archaeology and are discussed, but I have made some effort to highlight newer or neglected methods. Correspondence analysis might equally well have been included here, but is postponed to Chapter 11 which deals with the analysis of tabular data, including the use of log-linear models. I anticipate that the greatest developments in the use of statistics in archaeology over the next few years will involve the increased use of computer-intensive methods of analysis. Aspects of this are discussed in Chapter 12, but such methods are used in many other places in the book. Chapter 13 discusses methods of spatial analysis in archaeology. Archaeologists might expect a treatment of geographical information systems here; however, for reasons that are explained, discussion of GIS is cursory. Chapter 14 deals with aspects of the use of Bayesian methods in archaeology, perhaps the most important area to develop in the 1990s and illustrated in other chapters. Chapters 15 to 20 treat a number of specific topics that have attracted statistical attention, including absolute dating, relative dating, quantification, lead isotope analysis, the megalithic yard and the measurement of assemblage diversity. Chapter 21 contains a number of shorter studies, many of which would merit a chapter of their own in a longer book.

The analyses and figures in the book, unless otherwise stated, have been produced using S-Plus 2000 under Windows (Venables and Ripley, 1999). I have made a lot of use of libraries of functions written for S-Plus 2000, including the MASS library of Venables and Ripley (1999), and my general debt is acknowledged here. Specific details are provided in the Appendix.

2

Data sets and problems

2.1 Introduction

In this chapter a number of data sets, used elsewhere in the text for illustration, are described. In some cases these exemplify data of a kind commonly occurring in the literature. Some of the general problems that such data sets are collected to answer are discussed. Other data sets will be introduced throughout the text, as needed.

2.2 Data

2.2.1 Lead isotope ratio data

Ore bodies, mined in antiquity for copper, may be characterized by the proportions of four lead isotopes, ^{204}Pb, ^{206}Pb, ^{207}Pb and ^{208}Pb. These are constrained to sum to 100%, so that a sample from an ore body defines a point in three-dimensional space. In archaeological applications it is conventional to measure three ratios, ^{208}Pb/^{206}Pb, ^{207}Pb/^{206}Pb and ^{206}Pb/^{204}Pb. Table 2.1, based on Table 5 in Stos-Gale *et al.* (1996), shows the ratios for a subset of the 37 specimens sampled from an ore body on the island of Seriphos in the Cyclades.

A three-dimensional plot based on these ratios represents a sample from the *lead isotope field* for the ore body. One statistical issue that has arisen in practice concerns how to estimate the extent of the lead isotope field from the sample. If lead isotope fields for different ore bodies are distinct, this holds out the hope that if artifacts found in the archaeological record were made using ore from a single source their origin or *provenance* can be determined with reasonable confidence from their lead isotope signature. More cautiously, in the absence of good samples from all possible sources, one may at least hope to rule out a large number of ore bodies as possible sources.

This essentially simple idea has attracted controversy that can be followed in papers by Sayre *et al.* (1992a, 2001), Budd *et al.* (1993, 1995), Scaife *et al.* (1996, 1999), Stos-Gale *et al.* (1997), Baxter and Gale (1998), Baxter (1999a,b), Baxter *et al.* (2000), Knapp (2000) and Gale (2001), and a special section in the 1992

Table 2.1 Lead isotope data for galenas from the site of Moutalas on Seriphos, Cyclades (Stos-Gale *et al.*, 1996: 387–8)

$^{208}Pb/^{206}Pb$	$^{207}Pb/^{206}Pb$	$^{206}Pb/^{204}Pb$
2.06377	0.83069	18.8877
2.06431	0.83079	18.8978
2.06432	0.83087	18.9006
2.06323	0.83061	18.8880
2.06370	0.83066	18.8930
⋮	⋮	⋮
2.06384	0.83075	18.8788

volume of the journal *Archaeometry*. Reviews of the use of lead isotope ratio analysis in archaeology are to be found in Pollard and Heron (1996) and Gale and Stos-Gale (2000).

As far as statistical issues are concerned, opinion has polarized around the issue of whether graphical methods suffice to establish whether lead isotope fields are distinct, and whether fields could reasonably be the source of an artifact, or whether there is profit in using more 'sophisticated' statistical methods to answer these questions. Advocates of the latter view use the assumption that lead isotope fields have a trivariate normal distribution, as the basis for probability calculations based on the Mahalanobis distance between an artifact and a source. This has raised questions about whether or not normality is a reasonable assumption, the identification and treatment of outliers, data transformation, the adequacy of sample sizes that are typically available, and alternative methods of data presentation. These issues are discussed in Section 6.4.3 and Chapter 18.

2.2.2 Artifact compositional data

Table 2.2 shows the chemical composition of 48 specimens of Romano-British pottery, determined by atomic absorption spectrophotometry, for nine oxides (Tubb *et al.*, 1980). The specimens come from five kiln sites from three regions 1, (2,3) and (4,5). Tubb *et al.* demonstrate a clear chemical difference in the compositions for the three regions, arguing that only the oxides associated with the elements Fe, Mg, Ca and K are needed to discriminate between the products of kilns from the three regions. In Table 2.2 three values from the original publication, which may be typographical errors, are highlighted. These are retained for illustrative purposes.

The form of the data is typical of many data sets discussed in the archaeometric literature. Using multivariate methods, such as cluster analysis and principal component analysis, chemical compositional groups are sought within the data. If distinct compositional groups are found it is then of interest to see whether they are also associated with archaeological variables. Often, association with the site from which a specimen originated or was found will be of interest. If, as an example, it can be demonstrated that the products from different kiln sites are distinct it may then be possible to identify the origin of specimens, whose

Table 2.2 Results of chemical analyses of Romano-British pottery from Tubb *et al.* (1980). Numbers are percentage metal oxide; sample identifications are those from the original paper; 'Kiln' is the kiln site at which the pottery was found. Three probable typographical errors from the original publication, retained here for illustrative purposes, are highlighted

No.	Id.	Kiln	Al_2O_3	Fe_2O_3	MgO	CaO	Na_2O	K_2O	TiO_2	MnO	BaO
1	GA1	1	18.8	9.52	2.00	0.79	0.40	3.20	1.01	0.077	0.015
2	GA2	1	16.9	7.33	1.65	0.84	0.40	3.05	0.99	0.067	0.018
3	GA3	1	18.2	7.64	1.82	0.77	0.40	3.07	0.98	0.087	0.014
4	GA4	1	17.4	7.48	1.71	1.01	0.40	3.16	**0.03**	0.084	0.017
5	GA5	1	16.9	7.29	1.56	0.76	0.40	3.05	1.00	0.063	0.019
6	GB1	1	17.8	7.24	1.83	0.92	0.43	3.12	0.93	0.061	0.019
7	GB2	1	18.8	7.45	2.06	0.87	0.25	3.26	0.98	0.072	0.017
8	GB3	1	16.5	7.05	1.81	1.73	0.33	3.20	0.95	0.066	0.019
9	GB4	1	18.0	7.42	2.06	1.00	0.28	3.37	0.96	0.072	0.017
10	GB5	1	15.8	7.15	1.62	0.71	0.38	3.25	0.93	0.062	0.017
11	GC1	1	14.6	6.87	1.67	0.76	0.33	3.06	0.91	0.055	0.012
12	GC2	1	13.7	5.83	1.50	0.66	0.13	2.25	0.75	0.034	0.012
13	GC3	1	14.6	6.76	1.63	1.48	0.20	3.02	0.87	0.055	0.016
14	GC4	1	14.8	7.07	1.62	1.44	0.24	3.03	0.86	0.080	0.016
15	GD1	1	17.1	7.79	1.99	0.83	0.46	3.13	0.93	0.090	0.020
16	GD2	1	16.8	7.86	1.86	0.84	0.46	2.93	0.94	0.094	0.020
17	GD3	1	15.8	7.65	1.94	0.81	0.83	3.33	0.96	0.112	0.019
18	GD4	1	18.6	7.85	2.33	0.87	0.38	3.17	0.98	0.081	0.018
19	GD5	1	16.9	7.87	1.83	1.31	0.53	3.09	0.95	0.092	0.023
20	GE1	1	18.9	7.58	2.05	0.83	0.13	3.29	0.98	0.072	0.015
21	GE2	1	18.0	7.50	1.94	0.69	0.12	3.14	0.93	0.035	0.017
22	GE3	1	17.8	7.28	1.92	0.81	0.18	3.15	0.90	0.067	0.017
23	C01	2	14.4	7.00	4.30	0.15	0.51	4.25	0.79	0.160	0.019
24	C02	2	13.8	7.08	3.43	0.12	0.17	4.14	0.77	0.144	0.020
25	C03	2	14.6	7.09	3.88	0.13	0.20	4.36	0.81	0.124	0.019
26	C04	2	11.5	6.37	5.64	0.16	0.14	3.89	0.69	0.087	0.009
27	C05	2	13.8	7.06	5.34	0.20	0.20	4.31	0.71	0.101	0.021
28	C06	2	10.9	6.26	3.47	0.17	0.22	3.40	0.66	0.109	0.010
29	C07	2	10.1	4.26	4.26	0.20	0.18	3.32	0.59	0.149	0.017
30	C08	2	11.6	5.78	5.91	0.18	0.16	3.70	0.65	0.082	0.015
31	C09	2	11.1	5.49	4.52	0.29	0.30	4.03	0.63	0.080	0.016
32	C10	2	13.4	6.92	7.23	0.28	0.20	4.54	0.69	0.163	0.017
33	C11	2	12.4	6.13	5.69	0.22	0.54	4.65	0.70	0.159	0.015
34	C12	2	13.1	6.64	5.51	0.31	0.24	4.89	0.72	0.094	0.017
35	C13	2	12.7	6.69	4.45	0.20	0.22	4.70	0.73	**0.394**	0.024
36	C14	2	12.5	6.44	3.94	0.22	0.23	**0.81**	0.75	0.177	0.019
37	G01	3	11.6	5.39	3.77	0.29	0.06	4.51	0.56	0.110	0.015
38	G02	3	11.8	5.44	3.94	0.30	0.04	4.64	0.59	0.085	0.013
39	T11	4	18.3	1.28	0.67	0.03	0.03	1.96	0.65	0.001	0.014
40	T12	4	15.8	2.39	0.63	0.01	0.04	1.94	1.29	0.001	0.014
41	T13	4	18.0	1.50	0.67	0.01	0.06	2.11	0.92	0.001	0.016
42	T14	4	18.0	1.88	0.68	0.01	0.04	2.00	1.11	0.006	0.022
43	T16	4	20.8	1.51	0.72	0.07	0.10	2.37	1.26	0.002	0.016
44	A11	5	17.7	1.12	0.56	0.06	0.06	2.06	0.79	0.001	0.013
45	A12	5	18.3	1.14	0.67	0.06	0.05	2.11	0.89	0.006	0.019
46	A16	5	16.7	0.92	0.53	0.01	0.05	1.76	0.91	0.004	0.013
47	A18	5	14.8	2.74	0.67	0.03	0.05	2.15	1.34	0.003	0.015
48	A26	5	19.1	1.64	0.60	0.10	0.03	1.75	1.04	0.007	0.018

provenance is initially unknown, on the basis of their chemistry. This in turn may allow trade relationships in antiquity to be investigated.

In the form of analysis just described, archaeological information is used to interpret the output of groupings based on the chemical data but does not inform that grouping. An alternative approach to analysis is to take the archaeological groups as given and investigate whether they are indeed chemically distinct using methods such as discriminant analysis and related techniques.

There are a variety of different ways in which the chemical composition of a specimen may be determined. Among them are X-ray fluorescence analysis, neutron activation analysis, and inductively coupled plasma (ICP) emission spectroscopy. These and other methods are discussed in Pollard and Heron (1996) and Ciliberto and Spoto (2000). Methods differ in the range of elements they can measure and analytical precision. Modern analytical methods can typically measure up to about 30 elements, and this number can be increased by using different methods in combination. It is widely believed that the more elements that are measured the better, as one does not know in advance which elements may have discriminatory power. This raises interesting problems of variable selection, since the use of all measured variables in a statistical analysis can potentially obscure as well as reveal structure in the data (Baxter and Jackson, 2001).

Published data sets and analyses of them are readily found in journals such as *Archaeometry* and the *Journal of Archaeological Science*. The methods of statistical analysis used often reflect practice that has evolved in different laboratories with access to the necessary instrumentation (e.g., Bieber *et al.*, 1976; Pollard, 1986; Glascock, 1992; Beier and Mommsen, 1994). This evolution, rooted in a period when powerful and flexible statistical software was not as widespread as it is now, has meant that much of what is common practice does not take advantage of statistical methods developed over the last 15 or 20 years. Some possibilities are explored in later chapters. A review of common practice at the end of the 20th century is given in Baxter and Buck (2000). The data of Table 2.2 are analyzed in Sections 7.2.3 and 8.1.

2.2.3   Artifact typology

The data in Table 2.3 are taken from O'Hare (1990) and refer to measurements taken on a sample of 209 polished stone axes from southern Italy, relating to the size and shape of the axes. Analyses of these data are given in Sections 7.2.4 and 7.2.5.

More generally, data of these kind, on the dimensions and other qualities of artifacts, have been collected as input into studies of artifact typology. Methods such as cluster analysis have been proposed for deriving classifications from such data, Hodson *et al.* (1966) being an early application. This approach to deriving classifications, object clustering, has been contrasted with the attribute clustering approach pioneered by Spaulding (1953) (Section 1.2.2). The collection of Whallon and Brown (1982) contains a number of papers that discuss this issue, while Cowgill (1990) provides an overview. Despite the large amount of effort that has been devoted to statistical approaches to deriving classifications, there seems to be quite widespread agreement that it has had relatively little influence on archaeological practice (Cowgill, 1990: 63; Ammerman, 1992; Forsyth, 2000: 35).

Table 2.3 A subset of data from O'Hare (1990) on the dimensions of polished stone axes from southern Italy. The data refer to the dimensions, in millimeters, of the following variables: L1 = total length, L2 = length from the cutting edge to the point of maximum breadth, B1 = total breadth, B2 = breadth at distance $\frac{1}{5}$L1 from the butt, B3 = breadth at $\frac{1}{5}$L1 from the cutting edge, WC = maximum breadth of cutting edge, DC = depth of cutting edge, TH = total thickness, L3 = length from cutting edge to point of maximum thickness, T1 = thickness at $\frac{1}{5}$L1 from the butt, T2 = thickness at $\frac{1}{5}$L1 from the cutting edge. The final column is an indicator of butt shape: 1 = pointed, 2 = pointed/rounded, 3 = rounded, 4 = rounded/square, 5 = square

L1	L2	B1	B2	B3	WC	DC	TH	L3	T1	T2	Type
164.0	53.0	54.7	40.5	53.4	43.7	12.6	36.2	43.2	32.9	33.3	3
42.3	1.0	34.3	20.6	33.0	34.3	1.0	10.4	12.9	6.5	7.7	5
48.1	5.2	36.7	26.6	36.3	36.7	5.2	13.4	12.2	10.7	10.6	5
40.6	10.1	23.0	17.0	22.8	19.8	2.8	9.5	14.0	6.9	8.0	5
65.6	2.7	43.7	26.1	42.2	43.7	2.7	18.0	30.5	15.2	12.3	5
105.7	7.7	47.9	29.3	46.9	47.9	7.7	22.1	50.9	20.0	17.3	5
105.0	43.1	76.0	63.3	71.4	63.4	13.3	38.4	54.3	35.1	29.3	5
75.1	27.1	38.0	28.4	36.9	31.2	5.3	28.4	35.4	25.5	22.4	1
56.4	16.7	36.0	26.0	35.6	31.9	3.9	21.3	28.8	18.4	15.6	1
55.0	6.1	32.3	23.4	32.2	32.3	6.1	19.0	26.0	15.9	13.7	2
⋮	⋮	⋮	⋮	⋮	⋮	⋮	⋮	⋮	⋮	⋮	⋮

2.2.4 Assemblage comparison I

The data in Table 2.4 were originally discussed by Conkey (1980), and have been analyzed numerous times since (see Chapter 20). They are published in Kaufman (1998). Columns correspond to five different prehistoric hunter-gatherer sites in Cantabrian Spain, and rows to design elements on engraved bone artifacts. Conkey (1980: 611–12) was particularly concerned to examine the hypothesis that one site, Altamira, was an aggregation locale of otherwise dispersed hunter-gatherers, defining an aggregation site to be 'a place in which affiliated groups and individuals come together'. Under this hypothesis, the diversity of the design elements at Altamira was expected to be greater than that at other presumed dispersion sites.

'Diversity' can be defined in different ways, one of the simplest being *richness* defined as the number of different design elements *in the population*. The sample for Altamira is the richest, having 38 different design elements, with Cueto de la Mina the second richest with 27 elements. The statistical problem is to determine whether the difference in richness is a true reflection of population differences, or is simply a function of the fact that the sample for Altamira is larger (152) than that for Cueto de la Mina (69). Attempts to compare richness in the face of this sample-size effect are reviewed in Chapter 20.

Table 2.4 is an example of data collected to effect a comparison between assemblages. In this instance the assemblages for each site are used to investigate whether or not the populations they are drawn from have similar richness. Other forms of analysis for this and similarly structured data sets might be used. For example, a correspondence analysis of the data (Section 12.2.2 and Figure 12.2) suggests that the assemblage structures for Altamira and Cueto de la Mina differ markedly from each other and from the other three sites, these last having similar structure.

Table 2.4   Counts of 44 engraved bone design elements at five prehistoric hunter-gatherer sites in Cantabrian Spain. The sites are Altamira (A), Cueto de la Mina (CM), El Juyo (EJ), El Cierro (EC) and La Paloma (LP). These data appear in Kaufman (1998) but were originally analyzed by Conkey (1980)

Id.	A	CM	EJ	EC	LP	n_i
1	2	1	0	0	0	3
2	12	12	8	5	4	41
3	7	2	2	1	0	12
4	1	0	1	2	0	4
5	0	1	0	0	0	1
6	3	0	0	0	0	3
7	12	0	0	0	0	12
8	15	3	12	7	1	38
9	0	1	3	3	2	9
10	3	5	9	2	2	21
11	1	0	0	1	0	2
12	1	1	0	0	0	2
13	12	4	2	4	3	25
14	7	3	1	0	0	11
15	3	1	1	2	0	7
16	11	0	0	0	0	11
17	3	0	0	0	0	3
18	1	1	1	0	1	4
19	7	2	1	2	0	12
20	2	4	0	0	0	6
21	4	0	1	0	0	5
22	3	1	0	0	0	4
23	3	1	2	1	0	7
24	1	0	0	0	0	1
25	5	1	1	0	1	8
26	1	0	0	0	0	1
27	1	0	1	0	0	2
28	1	2	1	0	0	4
29	0	2	0	0	0	2
30	2	0	0	0	0	2
31	0	1	0	0	0	1
32	1	0	0	0	0	1
33	0	7	0	0	0	7
34	1	0	0	1	0	2
35	1	0	0	0	0	1
36	1	0	0	0	0	1
37	3	0	0	0	0	3
38	0	1	0	0	0	1
39	2	1	1	1	1	6
40	1	2	0	0	0	3
41	4	2	2	0	1	9
42	5	6	0	2	4	17
43	4	1	3	1	1	10
44	5	0	0	0	2	7
n_j	152	69	53	35	23	332

2.2.5 Assemblage comparison II

Table 2.5, based on data given in McClellan (1979), is similar in structure to Table 2.4 but was used to address different questions. The 16 assemblages

Type	100	200B	200C	201	229	500N	532	542	552	562	600	800	900B	900L	900S	900U
2	0	0	0	0	0	0	2	2	1	1	0	0	0	1	5	3
4	0	0	0	0	0	0	1	0	0	1	1	7	0	0	0	0
10	0	2	0	0	0	1	0	1	0	0	0	0	0	0	0	0
27	0	8	0	2	3	0	0	0	0	0	0	0	0	0	0	0
30	0	0	0	1	0	0	0	0	0	0	1	0	0	0	0	0
37	0	0	0	0	0	1	0	1	0	0	1	0	0	0	0	0
41	0	4	1	0	2	0	0	0	0	0	0	1	1	0	2	0
43	0	1	6	1	0	0	0	0	0	1	0	3	0	0	0	0
45	3	7	1	1	2	0	0	6	0	2	3	5	2	1	1	1
47	0	6	0	1	3	3	0	0	2	0	12	0	0	1	2	1
68	1	8	0	4	4	0	2	2	0	0	0	0	0	0	0	0
70	0	1	12	0	0	0	0	0	0	0	1	0	0	0	0	0
79	0	2	0	0	0	0	0	2	0	0	1	0	0	0	0	0
85	1	1	0	0	0	0	1	1	1	1	2	0	7	3	4	0
87	0	0	0	0	0	1	0	0	1	0	1	1	9	3	0	2
90	1	1	0	0	0	0	1	0	0	0	4	0	0	4	1	0
91	2	0	0	0	0	1	2	0	0	0	7	0	3	0	0	0
94	2	2	0	0	0	0	0	1	3	1	3	2	0	0	7	0
102	0	0	0	0	0	0	0	0	0	0	0	0	1	2	0	0
103	0	5	0	0	2	0	0	0	0	0	1	2	0	2	4	0
105	1	0	0	0	0	0	0	0	0	1	2	0	0	0	0	0
106	0	0	0	0	0	2	1	0	0	0	0	1	0	0	0	0
109	0	1	0	3	0	0	0	0	0	0	0	0	0	0	0	0
110	1	3	0	0	3	0	0	0	1	0	0	0	0	0	0	0
111	1	8	26	3	1	0	0	0	0	2	0	0	0	0	4	1
124	0	0	0	0	0	0	0	2	1	2	0	0	13	1	4	4
125	1	4	0	0	0	2	0	1	2	4	14	2	1	3	0	0

(continued)

Type | Assemblage | | | | | | | | | | | | | | |
---|---|---|---|---|---|---|---|---|---|---|---|---|---|---|---|---
 | 100 | 200B | 200C | 201 | 229 | 500N | 532 | 542 | 552 | 562 | 600 | 800 | 900B | 900L | 900S | 900U
126 | 1 | 1 | 13 | 1 | 0 | 2 | 0 | 0 | 0 | 0 | 0 | 0 | 0 | 0 | 0 | 0
127 | 0 | 3 | 14 | 0 | 0 | 0 | 1 | 0 | 0 | 0 | 1 | 0 | 0 | 0 | 0 | 0
128 | 1 | 6 | 0 | 0 | 0 | 1 | 1 | 1 | 3 | 5 | 4 | 27 | 4 | 5 | 1 | 1
132 | 10 | 14 | 2 | 0 | 0 | 11 | 6 | 16 | 3 | 5 | 24 | 1 | 0 | 5 | 2 | 2
133 | 1 | 0 | 0 | 0 | 0 | 3 | 0 | 0 | 0 | 0 | 0 | 2 | 0 | 0 | 0 | 0
134 | 0 | 0 | 0 | 0 | 0 | 0 | 0 | 0 | 0 | 1 | 5 | 1 | 2 | 0 | 0 | 0
138 | 0 | 4 | 0 | 0 | 0 | 1 | 0 | 0 | 0 | 0 | 4 | 0 | 0 | 1 | 7 | 0
139 | 4 | 5 | 0 | 0 | 0 | 1 | 1 | 1 | 1 | 0 | 3 | 0 | 1 | 2 | 0 | 1
159 | 0 | 0 | 0 | 0 | 0 | 0 | 0 | 0 | 3 | 0 | 1 | 1 | 0 | 4 | 2 | 1
163 | 0 | 0 | 0 | 0 | 0 | 0 | 0 | 0 | 2 | 0 | 0 | 0 | 0 | 0 | 0 | 0
165 | 0 | 0 | 0 | 2 | 2 | 5 | 7 | 5 | 3 | 5 | 0 | 0 | 0 | 4 | 0 | 0
174 | 2 | 13 | 0 | 4 | 0 | 0 | 0 | 0 | 0 | 9 | 15 | 5 | 0 | 4 | 6 | 3
182 | 0 | 0 | 0 | 0 | 0 | 5 | 7 | 7 | 0 | 1 | 0 | 0 | 0 | 8 | 3 | 1
186 | 5 | 9 | 3 | 4 | 0 | 0 | 7 | 7 | 11 | 9 | 13 | 13 | 14 | 8 | 11 | 12
210 | 0 | 1 | 0 | 0 | 0 | 0 | 0 | 0 | 0 | 1 | 0 | 0 | 8 | 0 | 1 | 1
213 | 0 | 0 | 0 | 0 | 0 | 0 | 1 | 2 | 1 | 2 | 0 | 0 | 0 | 0 | 0 | 0
223 | 0 | 0 | 0 | 0 | 0 | 0 | 1 | 5 | 0 | 1 | 0 | 3 | 0 | 0 | 0 | 0
245 | 0 | 0 | 1 | 0 | 0 | 1 | 0 | 2 | 1 | 1 | 2 | 2 | 0 | 0 | 0 | 0
247 | 0 | 0 | 0 | 0 | 0 | 1 | 0 | 0 | 0 | 0 | 1 | 0 | 0 | 0 | 0 | 0
258 | 0 | 5 | 0 | 0 | 1 | 2 | 0 | 0 | 0 | 0 | 2 | 2 | 0 | 0 | 2 | 0
267 | 1 | 3 | 1 | 0 | 1 | 5 | 2 | 5 | 0 | 0 | 9 | 2 | 1 | 0 | 0 | 1
271 | 0 | 0 | 0 | 1 | 0 | 0 | 1 | 0 | 3 | 3 | 0 | 0 | 0 | 0 | 0 | 0
301 | 1 | 7 | 0 | 1 | 2 | 3 | 1 | 4 | 3 | 1 | 9 | 2 | 1 | 2 | 1 | 0
316 | 3 | 12 | 0 | 0 | 1 | 1 | 1 | 7 | 1 | 0 | 1 | 2 | 0 | 0 | 4 | 4
317 | 0 | 1 | 0 | 0 | 0 | 1 | 1 | 2 | 0 | 0 | 3 | 0 | 0 | 0 | 0 | 0

correspond to individual tombs (201, 229, 532, 542, 552, 562) or groups of graves and tombs assumed to be contemporary, from Tell el-Far'ah (South), Palestine, from the Early Iron Age. The 52 rows correspond to different pottery types found in association with the burials.

One aim in collecting the data was to produce an ordering of the assemblages based on the relative similarity of the profiles of types, in order to obtain what is assumed to be a relative chronological ordering. This is a problem of *seriation*, discussed in detail in Chapter 16. Given the data, producing such an ordering is a mathematical or statistical problem, whereas the interpretation as a chronological sequence and identification of the early and late end points requires the application of archaeological expertise. McClellan applied a variety of methods, including a seriation technique due to Gelfand (1971a,b), cluster analysis and factor analysis, to investigate possible orderings. The sequencing of the tombs 532, 542, 552, 562, and their relationship to the assemblages in the 900 series was of particular interest because of the bearing this had on questions of absolute dating for some of the pottery types. Cluster analysis was also used to group the pottery types into five broad groups.

This type of data occurs very commonly in the archaeological literature and today correspondence analysis (Chapter 11), a method not widely known to archaeologists in the 1970s, would often be the method of choice for analysis. Its application to the data in Table 2.5 is illustrated in Section 11.4. In contrast to the problem involved in the analysis of the data in Table 2.4, relationships between the rows as well as the columns of the data matrix are of interest. A variant of the data structure exemplified in Table 2.5 arises when find types (not necessarily pottery) are recorded simply as present or absent, so that the table consists of zeros and ones. Such tables are called *incidence matrices*, and have also formed the basis of studies of seriation in archaeology.

2.2.6 Spatial data

Table 2.6 shows the location of a subset of 494 artifacts from the Mask Site, an Inuit camp site studied ethnoarchaeologically by Binford (1978). The data have been extensively used in the literature as a test-bed for different methods of spatial analysis (e.g., Whallon, 1984; Kintigh, 1990; Blankholm, 1991). The full data set is given in Blankholm.

The most numerous category of artifact is bone splinters. A plot of their locations, in Figure 2.1, shows evidence of some quite clear spatial patterning. In this case the distribution is associated with the positions of known hearths. A range of statistical methods can be used to try and identify patterning, or its lack, in the distribution for a single artifact type, and for examining the relationship betwen artifact distributions. Some of these methods are discussed in Sections 3.3.2 and 8.4, where analysis of the full data set is undertaken, and Chapter 13. With ethnoarchaeological data, where observation has established the way in which a site is used, it is of interest to ask whether statistical analysis can identify patterns in the artifact scatters, and whether the *interpretation* of such scatters is accurately informative about the way the site was used. This becomes a much more difficult problem with archaeological data, because of the processes that intervene between the deposition and eventual recovery of material (in extreme cases a site may disappear entirely!).

Table 2.6   A subset of the Mask Site data of Binford (1978), based on Blankholm (1991: 371–4). The data are the coordinates, x and y, of the locations of artifacts classified into five types, of which subsets for the first and fifth type are shown. In the data set published in Blankholm (1991) there are 494 artifacts. The artifact classifications are 1 = tools, 2 = projectiles, 3 = wood shavings, 4 = bone splinters, 5 = large bones

x	y	Type		x	y	Type
5.22	3.10	1	⋯	8.74	4.85	5
5.55	3.22	1	⋯	8.74	6.62	5
6.33	3.11	1	⋯	9.37	6.22	5
5.11	4.11	1	⋯	9.48	6.14	5
6.33	4.55	1	⋯	9.60	4.68	5
4.33	5.66	1	⋯	9.88	4.74	5
10.55	3.44	1	⋯	10.11	4.80	5
11.00	3.77	1	⋯	11.71	4.57	5
11.33	3.77	1	⋯	10.80	7.31	5
11.66	3.77	1	⋯	10.91	7.48	5
⋮	⋮	⋮	⋮	⋮	⋮	⋮

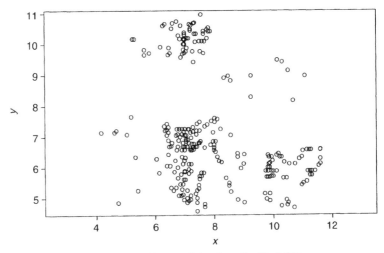

Figure 2.1   The spatial distribution of bone splinters from the Mask Site.

For such data, which are formed through the operation of multiple processes, formal statistical analysis may be of limited help. For example, the hypothesis that a scatter of data is the outcome of a Poisson process having a random distribution, or is the outcome of some other process that can be modeled mathematically, may often be obviously untenable. Possibly for such reasons statistical spatial analysis has not had the impact on archaeological data analysis that might once have been expected (see Chapter 13).

3
Kernel density estimates

3.1 Introduction

Kernel density estimates (KDEs) are conceptually simple. Their mathematical complexity and, at the time of writing, lack of availability in the software packages most likely to be used by archaeologists means they have not yet been widely used in archaeology, though use is increasing. At its simplest a KDE can be thought of as a smoothed histogram, that avoids some of the limitations of the histogram, and has advantages in terms of data presentation and generalization to two and three dimensions.

A histogram is appropriate for continuous data, and consists of a series of contiguous bars, the *areas* of which are proportional to the counts of observations that fall within the intervals covered by the bars. Anyone who has drawn a histogram will realize that the appearance is strongly affected by the (anchor) point at which one chooses to start the histogram, and the width of the intervals used. Whallon (1987: 146) illustrates this well.

It is relatively simple to produce a histogram using modern software, though users should be aware that some packages in widespread use do not produce a true histogram (Orton, 1999: 30). The appearance of computer-generated histograms is frequently not satisfactory, over-smoothing the data for example, and is not always straightforward to modify. Typically, intervals are of equal width so that height, as well as area, is proportional to frequency; however, instances occur when one might wish to use intervals of different width (using wider intervals in less dense regions of the data, for example). One use of the histogram is to compare the distribution of some variable for different sets of data. This can become unwieldy, even for a small number of data sets.

A KDE avoids many of these problems. The appearance of a KDE depends on the choice of a window width or bandwidth, analogous to the choice of interval width in a histogram. Theory exists to guide the choice of this, particularly in the univariate case, but it is quite legitimate, as with the histogram, to make the choice subjectively. KDEs are used for data presentation and comparison at various points in this book, including the applications section in this chapter. Readers who find the mathematics of the next section forbidding should bear in mind that, once software does become available, for simple applications it can be forgotten, and producing a KDE will be as simple as producing a histogram.

3.2 Kernel density estimates

3.2.1 Univariate KDEs

The simplicity of the idea of a KDE can be seen in Figure 3.1. Each data point, x_i, is associated with a 'bump' or (scaled) kernel, and the bumps are summed to get the KDE, represented by the solid line. The KDE, $\hat{f}(x)$, is an estimate of the unknown density, $f(x)$, of the population from which the sample is drawn. It should be emphasized that a KDE would not usually be calculated for a data set this size, which is used simply for clarity of illustration.

Mathematically, the KDE is estimated by

$$\hat{f}(x) = \frac{1}{nh} \sum_{i=1}^{n} K\left(\frac{x - x_i}{h}\right) = \frac{1}{n} \sum_{i=1}^{n} K_h(x - x_i), \tag{3.1}$$

where n is the sample size, h is the window width or bandwidth, the kernel $K(\cdot)$ is a probability density function so that

$$\int_{-\infty}^{\infty} K(t)\,dt = 1,$$

and

$$K_h(x - x_i) = h^{-1} K\left(\frac{x - x_i}{h}\right). \tag{3.2}$$

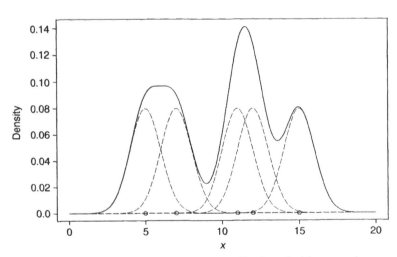

Figure 3.1 The five data points are each associated with a kernel, of the same shape, represented by the dashed lines. The kernel density estimate, represented by the solid line, is obtained by summing the heights of the kernels at each point on the line.

The KDE is relatively insensitive to the choice of kernel $K(t)$ and the normal or Gaussian kernel

$$K(t) = \frac{1}{\sqrt{2\pi}} \exp(-t^2/2),$$

equivalent to

$$K_h(t) = \frac{1}{h\sqrt{2\pi}} \exp(-t^2/2h^2),$$

will be used throughout. This is equivalent to taking $K_h \sim N(0, h^2)$, the normal distribution with standard deviation h, in equation (3.2). The 'bumps' in Figure 3.1 are of the form $n^{-1}K_h$, and the kernel can be thought of as spreading a 'probability mass' of size $1/n$ associated with each data point about its neighborhood (Wand and Jones, 1995: 12).

The spread, determined by the bandwidth h, is crucial for the appearance of the KDE. Values that are too large will tend to over-smooth the KDE, while values that are too small will tend to under-smooth. Most commentators are agreed that there is merit in examining KDEs for several choices of h. These might be determined subjectively; however, a number of rules have been developed for automatic bandwidth selection and these can be used, possibly as a guide to subjective choice.

Wand and Jones (1995) devote a chapter to the topic of bandwidth selection, and only a brief outline of some key ideas is given here. If one chooses h to make the KDE as 'close' as possible to the unknown $f(x)$, one measure of closeness is the asymptotic mean integrated squared error (AMISE) which can be shown to have the form

$$AMISE[\hat{f}(x)] = \frac{A}{nh} + \frac{h^4 B}{4} \tag{3.3}$$

that is minimized by

$$h_{AMISE} = \left[\frac{A}{nB}\right]^{1/5}$$

The second term in equation (3.3) is the asymptotic integrated squared bias and is proportional to h^4. This can be made small by making h small, but this has the effect of increasing the size of the first term in equation (3.3) and producing a rougher and more variable estimate of the KDE. Thus, in choosing h there is a variance–bias tradeoff that needs to be considered, with less variable, or smoother, estimates subject to greater bias.

In equation (3.3), A depends only on properties of the known kernel; B, however, depends on $\int f''(x) \, dx$, which can be interpreted as the 'roughness' of the unknown density $f(x)$. If it is *assumed* that $f(x)$ is normal with standard deviation σ, it can be shown (e.g., Silverman, 1986: 45; Wand and Jones, 1995: 60) that the so-called normal scale bandwidth estimate is obtained as

$$h_{NS} = 1.06\sigma n^{-1/5},$$

where σ has to be estimated. Either the usual sample estimate of the standard deviation, s, may be used, or a more robust estimate $IQR/1.34$, where IQR is the inter-quartile range. Silverman (1986: 47) recommends using the minimum of these. Unless the true density is close to normal this will typically produce an over-smoothed estimate; however, it can provide a starting point for more subjective choices.

Sheather and Jones (1991) developed a 'plug-in' method of estimating h that used an estimate $\hat{f}''(x)$ in estimating $\int f''(x) \, dx$, based on a different bandwidth from that used to estimate $\hat{f}(x)$. Details are given in Section 3.6 of Wand and Jones (1995) and in summary form in Baxter and Beardah (1996). The resultant estimate, h_{SJ}, is generally agreed to be effective in the univariate case (Bowman and Azzalini, 1997: 94; Simonoff, 1996: 46), and is that used in some of the applications to follow.

Problems can occur with bounded data when data points lie close to the boundary, for example, with strictly positive data and observations close to zero. The estimate at the boundary can 'spill over' into regions that are not feasible. One approach to this difficulty is to transform the data so that it is unbounded (by taking logarithms, for example), estimate a KDE, and transform back (Bowman and Azzalini, 1997: 14). Another approach is to modify the kernel function near the boundary (Jones, 1993).

Rather than using a fixed bandwidth, h, the possibility also exists of using variable bandwidths, h_i, that are larger in regions where the data are sparse, and produce a smoother *adaptive* KDE. Silverman (1986: 101) describes an approach for constructing an adaptive KDE in which the h_i are determined from an initial pilot estimate, based on fixed h, that provides an initial estimate of the density at the data points x_i. Some applications are given in Baxter and Gale (1998).

3.2.2 Multivariate KDEs

Apart from the advantages that KDEs have over the histogram in the univariate case, another potential attraction is that they generalize easily to the multidimensional case in a way that the histogram does not. Apart from discussion in the texts by Silverman (1986), Wand and Jones (1995), Simonoff (1996) and Bowman and Azzalini (1997), that of Scott (1992) has a particular focus on multivariate visualization.

For practical purposes the bivariate case is of most interest. The generalization of equation (3.1) for two-dimensional data, $\mathbf{x}_i = (x_{i1} \ x_{i2})$, is

$$\hat{f}(\mathbf{x}) = \frac{1}{n|\mathbf{H}|^{1/2}} \sum_{i=1}^{n} K(\mathbf{H}^{-1/2}(\mathbf{x} - \mathbf{x}_i)) = \frac{1}{n} \sum_{i=1}^{n} K_{\mathbf{H}}(\mathbf{x} - \mathbf{x}_i), \qquad (3.4)$$

where

$$K_{\mathbf{H}}(\mathbf{x} - \mathbf{x}_i) = |\mathbf{H}|^{-1/2} K(\mathbf{H}^{-1/2}(\mathbf{x} - \mathbf{x}_i))$$

and $\mathbf{H}$ is a symmetric positive definite bandwidth matrix of the form

$$\mathbf{H} = \begin{bmatrix} h_1^2 & h_3 \\ h_3 & h_2^2 \end{bmatrix}.$$

If $\mathbf{H}$ is diagonal, with $h_3 = 0$, then the KDE of equation (3.4) can be written as

$$\hat{f}(x_1, x_2) = \frac{1}{n h_1 h_2} \sum_{i=1}^{n} K\left(\frac{x_1 - x_{i1}}{h_1}, \frac{x_2 - x_{i2}}{h_2} \right). \qquad (3.5)$$

Further simplification is possible by taking $h_1 = h_2$, although in general this is not recommended (Wand and Jones, 1995). For all the examples in this book a bivariate normal kernel is used so that $K_H(\mathbf{x} - \mathbf{x}_i) \sim MVN(\mathbf{0}, \mathbf{H})$ where $MVN(\cdot)$ is the multivariate normal distribution. Generalization of equation (3.4) to three dimensions is direct.

Although multivariate KDEs provide richer possibilities for data visualization than univariate KDEs, there are also additional difficulties. One is that methods of automatic bandwidth selection are much less well developed. Wand and Jones (1995) devote just over one page to this, as opposed to 30 for the univariate case. It seems to be common to use the estimate given by equation (3.5), for which $h_3 = 0$, and common to use univariate rules for determining h_1 and h_2, though plug-in rules are available. This is discussed in the second of the examples to follow.

Sample size may also present a problem. Silverman (1986: 93–4) gives some guidance. For $f(\mathbf{x}) \sim MVN(\mathbf{0}, \mathbf{I})$, and a normal kernel, to obtain a relative mean square error of less than 0.1 at $\mathbf{0}$, sample sizes of about 19 and 67 are needed in two and three dimensions, respectively. For multimodal distributions much larger sample sizes than these may be needed to obtain satisfactory results (Baxter *et al.*, 2000).

Different methods of visualization, given a KDE, are available and will be illustrated in the examples. In the bivariate case, contouring in the manner suggested by Bowman and Foster (1993a) is available. After estimation, each data point is associated with a density height that may be ranked from largest to smallest. The first 75% ranked observations, for example, may be used to define contours that enclose the densest 75% of the data. This percentage may be varied, and several contours superimposed on the same plot, with the data if desired. Bowman and Foster (1993a: 173) note that, in some ways, this provides a two-dimensional analog of the univariate box-plot and may be useful for identifying clusters in the data. This idea extends to the use of contour shells for three-dimensional KDEs, and some examples are given in Baxter *et al.* (2000) and below.

3.3 Applications

Apart from an isolated application by Aitchison *et al.* (1991) to a dating problem, the work of Baxter, Beardah and colleagues was largely responsible for introducing kernel density estimation into the archaeological literature (Baxter and Beardah, 1995, 1996; Beardah and Baxter, 1996a,b; Baxter *et al.*, 1997) in the mid-1990s. The gap between this and the popularizing work of Silverman (1986) in the statistical literature is typical of the time-lag that can occur before potentially useful techniques developed in the statistical literature are investigated in archaeology. Beardah and Baxter (1996a) include details of a MATLAB package for kernel density estimation, developed by the former author, that is freely available. Subsequent to this Bowman and Azzalini (1997) made available a library of routines for the S-Plus package, and it is this that has been used in the examples in this book unless otherwise stated.

3.3.1 Italian Bronze Age cups

Lukesh and Howe (1978) published data on the dimensions of cups dated to the Middle Bronze Age from the Apennine culture of the central and southern

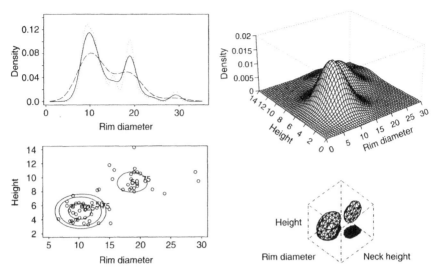

Figure 3.2   Univariate (upper left), bivariate (upper right and lower left) and trivariate (lower right) KDEs for dimensional measurements on 60 Italian Bronze Age cups. See the text for full details.

peninsula of Italy. A subset of the data, for 60 cups for which there are no missing data, is given in Baxter (1994a). This subset will be used to illustrate various aspects of the use of KDEs.

Figure 3.2 shows a variety of KDEs based on the data. The upper-left panel shows three univariate KDEs for the rim diameter of the cups. The dashed line is the KDE obtained using $h_{NS} = 2.46$; the solid line is the KDE using a considerably reduced $h_{SJ} = 1.17$; and the KDE represented by the dotted line uses $h = 0.70$. The data are clearly multimodal, with two main peaks at about 10 and 20 cm, and there is a much smaller mode associated with just two cups at about 30 cm. The pattern is most clearly revealed by the KDE using h_{SJ}. Although the KDE using h_{NS} does show the two main modes, they are much less apparent and the data are clearly being over-smoothed. The KDE with $h = 0.70$ shows the main features of the data and some spurious detail which, as Silverman (1986: 43) suggests, can be further 'smoothed' by eye.

The upper-right and lower-left panels of Figure 3.2 show two ways of visualizing the bivariate KDE for rim diameter and height, in which the KDE of equation (3.5) has been used, with h_1 and h_2 separately determined using the normal scale rule. Bimodality is evident in both cases. The lower-left panel illustrates the use of contouring as previously discussed. Beardah (1999) has analyzed these data in more detail and shown that using non-zero values of h_3 in the more general KDE (3.4) does not critically affect the results obtained. Examples where the choice of h_3 is crucial are discussed in Chapter 18.

There are five variables in the original data set. Three of these, rim, neck and shoulder diameters, are very highly correlated (Baxter, 1994a: 91) so that the data are effectively three-dimensional. The final panel shows a trivariate KDE based on the variables rim diameter, height and neck height, and using a 75% contour. Notwithstanding the fact that the sample size is rather smaller than we would like, the bimodality of the data is again evident.

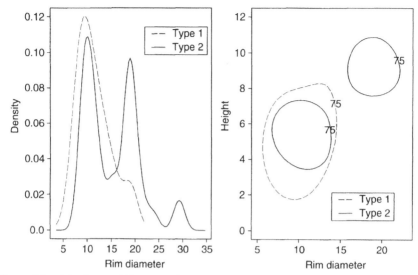

Figure 3.3 Univariate and bivariate KDEs based on dimensional measurements of 60 Italian Bronze Age cups, contrasting the distributions for two types.

The cups are classified into two different types, and it is of interest to see if and how the types differ in terms of their dimensions. This is investigated via KDEs in Figure 3.3. The left-hand panel shows separate KDEs for the two types for rim diameter, using h_{SJ} as the bandwidth estimator. Whereas type 2 reflects, and is clearly responsible for, the multimodality seen in Figure 3.2, type 1 cups show a unimodal distribution at the smaller end of the size range. The right-hand panel, based on bivariate KDEs for rim diameter and height for each type separately, and using 75% contours, shows the same picture. For comparative purposes such as this KDEs are less cumbersome than the use of multiple histograms, as is common, and are easier to digest.

3.3.2 The Mask Site bone splinter data

The background to the Mask Site data collected by Binford (1978) is discussed in Section 2.2.6. Spatial data are naturally two-dimensional and an obvious candidate for display using bivariate KDEs. Figure 3.4 shows two such estimates for the 276 bone splinters. The left-hand panel is based on separate estimation of $h_1 = 0.20$ and $h_2 = 0.28$ using the Sheather–Jones method; the right-hand panel uses $h_1 = 0.55$ and $h_2 = 0.61$ obtained from the normal scale rule. The contours are not labeled but encompass the 10% most dense points up to the 90% most dense in steps of 10%. Both methods identify two concentrations to the left of the figure, while to the right the left-hand panel suggests more structure than the right-hand panel. There are reasons for believing that the left-hand panel provides an archaeologically more interpretable result (Baxter *et al.*, 1997), and that the right-hand panel over-smooths the data.

This example raises a number of issues about the choice of smoothing parameters for bivariate KDEs since, unlike the previous example, the choice affects the interpretation. Beardah (1999) has examined this question using some of the

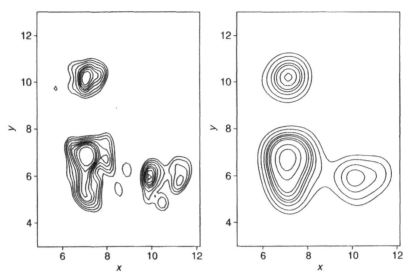

Figure 3.4 Two bivariate KDEs for the Mask Site bone splinter data using different smoothing parameters. See the text for full details.

methods discussed in Jones *et al.* (1996). The estimate in the right-hand panel of Figure 3.4 produces results that are very similar to those obtained using a bivariate direct plug-in rule, and Beardah (1999) observed that this rule tends to over-smooth, particularly if there are well-separated modes in the data. The same was found to be true of the biased cross-validation selection method of Sain *et al.* (1994) with this data set. Beardah (1999) argues that the use of univariate plug-in estimates for the h_i used in the left-hand panel of the figure may compensate for this over-smoothing. The argument is that, data dependent constants aside, the h_i in the univariate approach are proportional to $n^{-1/5}$, whereas in the bivariate case they are proportional to $n^{-1/6}$, and since $n^{-1/5} < n^{-1/6}$, the univariate rule will result in smaller smoothing parameters and smooth less.

Another application of KDEs applied to spatial data is given in Mameli *et al.* (2002).

3.3.3   Other applications

The range of application of KDEs can be extended by using them with derived variables, such as principal components (Chapter 7). Examples of such analyses, based on artifact compositional data, are reported in Baxter and Beardah (1995), Beardah and Baxter (1996a), Beardah (1999), Baxter *et al.* (1997), Hall *et al.* (1999) and Lockyear (1999).

Another relatively early application of KDEs in archaeology is Williams-Thorpe *et al.* (1996: 31) who exploited a univariate KDE to demonstrate bimodality in the logarithms of magnetic susceptibility measurements based on individual measurements of Mons Claudianus columns in Rome, Mons Claudianus being the quarry area from the Eastern Desert of Egypt that was the source of the material for the columns. This differed noticeably from the uni-modal distribution of similar readings from the Mons Claudianus quarries and led the authors to suggest that columns were taken preferentially from different areas of the quarries.

Where data are naturally low-dimensional, KDEs are a potentially attractive tool for exploring the data. This is the case for lead isotope ratio data, and Baxter and Gale (1998), Baxter (1999a,b), Baxter *et al.* (2000), Baxter and Buck (2000), Beardah (1999), Beardah and Baxter (1999), Scaife (1998) and Scaife *et al.* (1999) all exploit KDEs for the analysis of this kind of data. A fuller discussion of this topic is provided in Section 2.2.1 and Chapter 18.

There is some evidence that KDEs are increasingly being regarded as a 'standard' tool, available to analysts, that does not require detailed explanation in applications. Barnett (2000: 446), for example, used a univariate KDE, though he did not call it this, to summarize a set of inhomogeneous luminescence dates of pottery from Fornham St. Genevieve, Suffolk, UK, in order to establish a probable date range. A feature of his presentation was that the kernels used reflected the errors associated with different dates, and differ in width. Other recent applications, noted elsewhere in this book, include Pakkanen (2002) (Section 19.3), Steel and Weaver (2002) (Section 12.2.2) and a series of papers by Aykroyd *et al.* (1996, 1999) and Lucy *et al.* (2002) (Section 21.3).

4
Sampling

I think that there was also a practical element in the differential impact of New Archaeology. Stand in the middle of the Arizona desert and the need for sampling theory, understanding of variability, and so on is all too clear. By contrast, the British landscape is cramped; it has been intensively settled for millennia, and intensively studied for centuries. Most of its basic units (administrative boundaries, patches of woodland) are irregular in shape and are themselves hundreds if not thousands of years old. As a result, many of the techniques of New Archaeology such as sampling theory make clear sense as practical strategy in the Arizona desert, but are counter-intuitive in the densely settled palimpsest that is Wessex. (Johnson, 1999: 30–1)

4.1 Introduction

Sampling pervades archaeology. At a basic level what is found in excavation, even when excavation of a site is 'total', is only a sample of what once existed. From such samples archaeologists attempt to draw inferences about past human behavior. The terms 'sample' and 'inference' are used rather loosely here. This chapter examines the somewhat narrower topic of the use of *probabilistic sampling* and associated methods of *statistical inference* in archaeology.

While Vescelius (1960) has been credited with introducing ideas of probabilistic sampling to archaeology (Hole, 1980; Orton, 2000a: 5), it was Binford's (1964) advocacy of probabilistic sampling in regional survey that had the greatest initial impact. Hole (1980: 218) said of this paper that 'no other single article has altered more radically the recent course of archaeology'. Interest in probabilistic sampling in archaeology was intense in the 1970s, resulting in two collections of conference proceedings on the topic, one American (Mueller, 1975) and one British (Cherry *et al.*, 1978). Ideas and procedures formulated then have, to some extent, become 'fossilized' in archaeological practice (Orton, 2000a: 11). One concern in Orton's (2000a) overview was to evaluate the relevance of more recent statistical developments for archaeological practice.

In the period between Cherry *et al.* and Orton's books, review articles devoted wholly or partly to the practice of sampling in archaeology include Hole (1980), Read (1989: 25–30) and Nance (1990, 1994). Cowgill's (1975) contribution to Mueller (1975) is a useful overview of thinking at that time.

Orton (2000a: 206) emphasized the 'centrality of sampling to both archaeological theory and practice' and illustrated how sampling theory could 'make a useful contribution to debates about archaeological inference in the face of distorting factors imposed by site formation processes'. He followed this by noting that the main value of sampling theory 'is in providing a language and frame of reference within which such problems can be discussed, rather than in definitive answers'. This last point is important. As with the use of other statistical methods in archaeology, attacks on, and rejection of, the value of probabilistic sampling in archaeology (e.g., Hole, 1980) have been provoked more by misuse of the available methods, and unrealistic expectations of what can be achieved, than by any fundamental problems with the statistical methodology itself. Shennan (1997: 361), for example, noted that

> the belief that forms of probability sampling could provide solutions to such problems [of drawing inferences about the whole from a part – the sample] was one of the reasons for the popularity of this subject in the 1960s and 1970s, and the disillusion when it was realised that such techniques did not provide any sort of panacea is one of the reasons why it has subsequently faded from attention.

The central part of Orton's (2000a) book distinguishes between sampling at four scales: the regional, the site, the feature or deposit, and the artifactual or ecofactual. Although different scales of analysis pose their own problems, the natural sampling unit (given aims and practicalities) is often an area of land or volume (of soil) distributed over space. Much of the material in this chapter is pitched in general terms with the aim of highlighting some concerns specific to sampling in archaeology. For illustration, examples at the regional or site scale of survey are largely, though not exclusively, used.

Some of the more basic sampling methods available are outlined below. Mathematical details are limited and reference may be made to Orton (2000a) and, particularly, Cochran (1977) for more detail.

4.2 Sampling methods

4.2.1 Notation

We follow the notational conventions used in Cochran (1977) and Orton (2000a), in which n units are selected from a population of size N, so the sampling fraction is $f = n/N$. Variable values are x_i, with the sample mean $\bar{x}$ and population mean $\bar{X}$. The population variance is $S^2 = \sum_{i=1}^{N} (x_i - \bar{X})^2/(N - 1)$, defined this way by convention (Cochran, 1977: 23). The 'hat' notation is used for an estimate so that, for example, $\hat{\bar{X}}$ is an estimate of the population mean.

4.2.2 Simple random sampling

In simple random sampling any set of n units has an equal probability of being selected. The population mean is estimated as

$$\hat{\bar{X}} = \bar{x} = \sum_{i=1}^{n} x_i/n \tag{4.1}$$

with variance

$$\sigma_{\bar{x}}^2 = (1 - f)S^2/n \tag{4.2}$$

estimated by

$$s_{\bar{x}}^2 = (1 - f)s^2/n, \tag{4.3}$$

where

$$s^2 = \sum_{i=1}^{n} (x_i - \bar{x})^2/(n - 1)$$

is an unbiased estimate of S^2. For population totals, multiply equations (4.1) and (4.3) by N and N^2, respectively. Note that is necessary to know N for this.

For estimating proportions, the estimate is just $\hat{p}$, the observed proportion of units in the class of interest, with estimated variance

$$\frac{(1 - f)\hat{p}\hat{q}}{n - 1}$$

where $\hat{q} = (1 - \hat{p})$.

An example of simple random sampling would be a regional survey in which the units are equal sized parcels of land, and the variable of interest is the number of sites in a unit, used as the basis for estimating the number of sites in a region.

Simple random sampling (srs) is conceptually easy to understand and, as results in sampling theory go, estimates take a straightforward form. It is important as the basis for other procedures but possibly of less use in its own right than one might think. One reason is that archaeological sampling often takes place in a spatial context, and it may be desirable to ensure more even spatial coverage than can be guaranteed by simple random sampling. A related reason is that the units sampled may be defined areally, such as a quadrat or transect, whereas the focus of interest is on *elements* within the unit. The theory of cluster sampling is relevant here and is discussed later. Simple random sampling, particularly over a large area, can also be costly and administratively inconvenient.

Simple random sampling can be used to illustrate a point that has been much labored, and equally often ignored, in the literature. Suppose we are prepared to specify the precision required for the estimate of, say, the population mean by specifying a value for $\sigma_{\bar{x}}^2$. Assuming f is small enough to ignore, equation (4.2) can be solved for n to get

$$n = S^2/\sigma_{\bar{x}}^2,$$

the sample size needed to obtain the required precision. The important point of principle here is that (for large enough populations) the precision depends only on the sample size, and not on the proportion of the population sampled.

It follows (again for large populations), that it may be possible to achieve good results with small sampling fractions, and that there is no such thing as an ideal sampling fraction. In practice, to use this result, a good estimate of S^2 is needed and obtaining this may not be easy. Two-stage sampling, in which an estimate is obtained from the first stage and used to determine the sample size at the second stage, or a pilot survey are two possibilities, but these may not always be practical in applications.

4.2.3 Stratified random sampling

Stratified random sampling occurs if the population is first divided into subgroups, and then simple random sampling is applied within each subgroup. If there are G groups, and sample sizes and estimated means and variances within group g are subscripted with g, then

$$\hat{\bar{X}} = \sum_{g=1}^{G} N_g \bar{x}_g / N$$

with estimated variance

$$s_{\hat{\bar{x}}}^2 = \frac{\sum_{g=1}^{G} N_g (N_g - n_g) s_g^2 / n_g}{N^2}.$$

An example might be a regional survey in which the sampling units are classified by type of terrain. Proportional allocation occurs if the same sampling fraction, $f_g = n_g / N_g$, is used within each group, and this could be used if it was considered important that each type of terrain was equally well sampled. If the S_g^2 are known, or good approximations are available, results exist for optimally allocating the sample to strata, which can take into account differential costs of sampling units within different strata (Cochran, 1977: 96–101). This can result in different sampling fractions for different strata, since the smaller S_g^2 is the smaller n_g can be. An extreme, and hypothetical, example would be for types of terrain where it was judged that that there could be no archaeology of interest, resulting in sample sizes of zero for such strata. For a fixed sample size n, stratified random sampling will generally be more efficient than simple random sampling (i.e., have smaller variance) if the strata can be designed such that the variances within strata, S_g^2, are smaller than S^2. This greater efficiency is not guaranteed if the strata are defined for reasons of convenience.

4.2.4 Cluster sampling

For reasons of practicality much archaeological sampling can be viewed as cluster sampling. For example, in regional survey the units sampled may be grid squares on a map, whereas the elements of interest may be sites, for which some variable such as size is to be measured. Here the unit sampled is a *cluster* of smaller units (sites) or elements. The elements of interest are unknown in advance of survey so the population of interest is unknown and cannot be labeled for sampling purposes. Note that this differs from the examples previously used, since we are no longer interested in the number of sites (a property of the unit) but in size of site (a property of elements within the unit).

Orton (2000a) emphasizes the prevalence of cluster sampling in archaeological applications, and the fact that it is not always recognized that cluster sampling is being used. Failure to appreciate that cluster sampling is being used, and application of results for some other sampling method, can result in misleading inferences being drawn. Examples of cluster sampling in archaeology cited by Orton include the following.

1. Intrasite sampling, in the absence of much in the way of surface features, where the objective may be to estimate the proportion of a particular class of artifact in an assemblage (Orton, 2000a: 128).
2. Soil sampling, in which the cluster is a volume of soil, and the elements of interest might be botanical remains, such as seeds, classified by type (Orton, 2000a: 153).
3. Ceramic thin-section analysis, from which the proportions of different types of inclusion in a vessel are estimated from a single section (Orton, 2000a: 184).

The problem identified by Orton may most simply be illustrated by considering the problem of estimating a proportion from a cluster sampling strategy. For illustration, suppose a soil sample has been taken and is found to contain m_i seeds of which a proportion, $\hat{p}_i$, are of a particular type. This is a single number derived from a single cluster sample, and it is impossible to attach a standard error to it. If n soil (cluster) samples are taken an estimate of the proportion of the type is

$$\hat{p} = \sum_{i=1}^{n} m_i \hat{p}_i \bigg/ \sum_{i=1}^{n} m_i$$

with an approximate variance estimate of

$$\frac{1-f}{n\bar{m}^2} \frac{\sum_{i=1}^{n} (m_i^2 \hat{p}_i^2 - 2\hat{p}\hat{p}_i m_i^2 + \hat{p}^2 m_i^2)}{n-1},$$

where $\bar{m} = \sum_i m_i/n$ (Orton, 2000a: 213).

What is sometimes incorrectly done is to treat the sample of $n\bar{m}$ as a simple random sample in which a proportion, $\hat{p}$, of seeds (elements) are of a particular type and estimate the variance, assuming f is negligible, as $\hat{p}\hat{q}/n\bar{m}$. This covers the case of a single cluster sample ($n = 1$) or several cluster samples for which counts are amalgamated. This approach is potentially misleading unless the elements and type of interest are homogeneously distributed so that a single cluster sample can be treated as 'representative' (Orton, 2000a: 167).

This discussion only touches on the topic of cluster sampling, and has highlighted Orton's argument that cluster samples are sometimes mistakenly treated as simple random samples at the smaller scales of sampling used by archaeologists. Single-stage cluster sampling has been described, but multistage cluster sampling is also possible. Multistage designs can also incorporate more than one sampling method. Another issue not discussed here is the allocation of the sample between clusters, on which there is a large literature. For the theory of this and other matters Cochran (1977) may be consulted, while Orton (2000a) discusses a number of archaeological case studies. Discussion of the method of adaptive cluster sampling, of which Orton is an enthusiastic advocate, is deferred to Section 4.4.

4.2.5 Systematic sampling

In regional or site sampling, conducted in an initial state of relative ignorance, random sampling can give rise to concerns that areas of potential interest remain unsurveyed. Where this is a major concern systematic sampling is a possibility. This may involve sampling square or rectangular (i.e., transects) units regularly spaced over the region of interest.

The theory of systematic sampling is discussed in Chapter 8 of Cochran (1977) and he concluded, on the basis of analyses of both artificial and natural populations, that systematic sampling often compared favorably in precision with stratified random sampling (Cochran, 1977: 229). Limitations included problems with reliably estimating the variance of the estimated mean, and poor precision in the presence of periodicity in the data. This last disadvantage is potentially important for archaeological spatial sampling. For instance, if an area is surveyed using regularly spaced transects it would be easy to miss linear features, such as ditches, aligned with, but not within, the transects. This raises other issues concerning the optimal size and spacing of sampling units, touched on later, that are not easily resolved.

4.3 Archaeological considerations

4.3.1 Random versus purposive sampling

Probabilistic or random sampling methods are frequently contrasted favorably with non-random sampling methods (variously described as 'opportunistic', 'self-selecting', 'informal', 'haphazard', 'grab', 'quota', 'non-probability', 'judgment' or 'purposive' sampling etc.). The variety of terminology used to characterize non-random sampling reflects, in part, the differing degrees of disapprobation that different methods attract. Properly designed quota sampling, as used in marketing research or opinion polling can, for example, be quite effective, whereas grab sampling – selecting what happens to be convenient – may produce very unrepresentative results. It should therefore be emphasized that some of the most informed commentators on sampling methods in archaeology have been at pains to stress that in many circumstances non-random or purposive sampling may be the best approach. (By 'informed' I mean those commentators familiar with both archaeological and statistical considerations who are not, by temperament, training, philosophy, or whatever, predisposed *against* the use of statistical methods in archaeology.)

It is worth citing some of these observations at length. Cowgill (1975: 260), in reviewing the papers in Mueller (1975), distinguishes between *selection* and *sampling* strategies (emphasis in the original). He observes that sometimes

> our resources, the nature of our data, and the nature of critical test implications derived from competing hypotheses are all such that we can define obvious criteria of relevance, and use these criteria as a basis for picking a manageable number of intrinsically important observations. In these cases there are other observations we might make but do not make because we are satisfied that they are relatively unimportant for our purposes.... This strategy I propose to call *purposive selection*.... There is little question that

purposive selection is preferable to sampling whenever selection is feasible, sufficient for one's research objectives, and not wasteful.

There is often a good deal of prior knowledge about the populations of interest in archaeological sampling, though defining populations can be problematic. Even when knowledge is limited it may quickly be accumulated in the course of sampling. Cowgill (1975: 259) makes the useful point that, when little is known about the population of interest, what he calls *probing* may be the most economical strategy, which he defines as a 'sort of preliminary exploration to get some notion of the gross characteristics of the population' which might include such things as 'judgmental or systematic or opportunistic pits or borings in a site', but 'can also include such things as exploratory surface reconnaissance of a region'. In other words, even when some form of random sampling strategy may be considered desirable, there is an important role for non-random sampling informed by archaeological common sense, judgment and expertise. Sometimes this may be all that is needed.

Bellhouse (1980: 123), in a quite technical paper on sampling methods in archaeology, makes similar observations in his introduction, to the effect that the

> bias in purposive selection, if large enough, could result in faulty conclusions. This leads to a body of erroneous literature if the same bias is carried through a number of research projects. However a slavish adherence to randomization could result in important items, known to exist, being ignored. . . . One solution . . . is to take a sample which is a hybrid of randomization techniques and other methods.

In similar vein, Shennan (1997: 363) notes that

> probability sampling isn't the only approach available. Judgmental or purposive sampling involves investigators making their choice about, for example, which part of a site to excavate or which part of a region to survey on the basis of their academic judgment without regard to any statistical criteria.

While warning of the potential pitfalls in this approach, Shennan concludes that 'there are many archaeological questions to which the techniques of probability sampling are not relevant'.

Finally, in the following, Orton's (2000a: 2–3), use of the terms 'formal' and 'informal' may be equated with probabilistic and non-probabilistic sampling.

> It might seem that I am making the equation: formal = good, informal = bad, but that would be a gross over-simplification. . . . formal methods are to some extent a rational response to a state of ignorance and the more we know about a situation, the less necessary they may be. . . . to ignore evidence from aerial photographs or geophysical surveys in designing a purely random sample of (for example) test-pits would be wasteful and unproductive, and a more targeted approach would be likely to give more useful results. . . . there needs to be a balance between statistical rigour and the use of what statisticians call *prior information*, in order to make best use of resources and to achieve reliable outcomes.

The desire to incorporate prior information in an analysis leads Orton to suggest Bayesian methods as a possible tool. Some of these ideas are described more fully in Chapter 14, though at the time of writing I am not aware that they have been applied in anger to archaeological sampling problems.

It should also be emphasized that all these commentators are firmly convinced of the value of probabilistic sampling in archaeology. The point, as the quotation that heads this chapter suggests, is that archaeologists are often in possession of a considerable amount of knowledge in advance of any sample survey they might consider undertaking, and this needs to be accounted for in the design of any survey. Bayesian methodology notwithstanding, it is not necessarily obvious how such prior knowledge should be incorporated into a sampling design, and perhaps impossible to generalize. Some appear to have seen the existence of such prior knowledge as a good reason for not using probabilistic sampling in archaeology (Hole, 1980: 232). This position may be perfectly defensible in specific instances, but seems rather extreme as a generalization.

Even when one is initially in a state of relative ignorance about the specific characteristics of the population being sampled, and a random sampling strategy is employed, evidence may quickly become available that, from a common-sense point of view, ought to be able to modify the sampling strategy. In regional or site survey, for example, one would usually wish to pursue evidence for an important and previously unsuspected site or feature, even if the evidence for this lay largely in areas not destined to be sampled according to the original scheme. One approach to sampling that allows this flexibility, while maintaining statistical rigor, is adaptive cluster sampling. This is discussed in Section 4.4.

4.3.2 What is being sampled?

Orton's (2000a) Chapter 3, entitled 'If this is the sample, what was the population?', raises some thorny issues about the relation between statistical sampling and *archaeological* inference. Hypothetically, suppose a regional sample survey is to be undertaken, with the aim of investigating the number of sites in a region and their type (e.g., size or period). Suppose, further, that this is to be based on surface survey, so that sites not visible on the surface will not be recorded. Assume that a random sampling survey is commissioned, to avoid biases inherent in existing records that may have accumulated in a haphazard fashion, and assume that this is carried out in a rigorous and satisfactory manner. What will emerge is an estimate of the number of sites in the region, visible on the surface, broken down by type, and with associated confidence intervals.

Of what use are these estimates? If the population of interest is indeed all archaeological sites visible on the surface then there is no problem. If the population of interest is all extant archaeological sites (regardless of visibility) then assumptions need to be made about the relationship between visible remains and extant remains. Is it probable that most extant sites are visible? If not, are some kinds of site more likely to be visible than others? If the population of interest is all the sites that once existed, for some of which there is no longer any detectable evidence, the sample may be of limited use without very strong assumptions about the nature of survival.

The main point being made here is that random sampling procedures, however well executed, may only be informative about populations that are of limited interest from the point of view of archaeological inference. In the hypothetical example above, if the interest is in all the sites that once existed, an ability to draw rigorous statistical inferences about existing surface remains may be of limited interest.

Further discussion is provided in Section 17.2.1, concerning the quantification of assemblages of bones found in archaeological survey or excavation. There, it is noted that what is recovered, whether by random sampling or not, can be conceived of as the end product of a sequence of sampling mechanisms, mostly non-random. The real interest often centers on the population sampled at one of the early stages of the sequence. Making a connection between what is actually recovered and the population of interest typically requires the use of very strong assumptions, which may not be satisfied, and/or an understanding of the processes leading to the sample eventually obtained, which may not exist.

Orton (2000a: 65) is realistic about the problems these issues pose for the application of sampling theory in archaeology. He regards sampling theory as providing a useful language within which some problems of archaeological inference can be discussed and understood. It is not a panacea.

4.4 Adaptive sampling

Orton (2000a: 16) attributes the development of adaptive sampling to a series of papers by Thompson in the late 1980s and early 1990s, culminating in the book by Thompson and Seber (1996). In adaptive sampling the sample design is modified in the course of sampling in the light of what is discovered. Another way of putting this is that the units to be sampled are not fixed 'once and for all' in advance of carrying out a survey. The advantage of this is that it accords with what many archaeologists do in practice while retaining statistical rigor.

To the extent that archaeological sampling practice is locked into methodologies evolved in the 1970s and earlier, adaptive sampling, consciously adopted as a rigorous random sampling procedure, has had limited use in archaeology, and Orton (2000a) has been its major proponent. It is worth noting, however, that the ideas have been around a lot longer. For example, Cowgill (1975: 266–7) observed that, as 'numerous other experiences have demonstrated, test pits are not the way to find out about structures larger than the test pit. In some cases pits may be the best way to *locate* features, but they should then be expanded to reveal more or less the entire feature'. This, without the term being used, is the essence of adaptive sampling.

The following more formal account of adaptive cluster sampling is based on Orton (2000a). Assume for definiteness that the survey involves an area divided into sampling units of quadrats. Define a neighborhood for each unit, such as quadrats with an edge in common, or quadrats with an edge or corner in common. (These are the Rook's and Queen's cases familiar from studies of spatial autocorrelation – Section 13.5.) Select a cluster sample of n_0 units, and define some criterion that a unit must meet to be archaeologically 'interesting'. If a unit is surveyed and found to be interesting, all the units in its neighborhood are also sampled, and this process is repeated. This results eventually in the sampling of a cluster of interesting units bounded by edge units that are not interesting. The network for the unit initially sampled consists of the interesting units in the same cluster. A unit in the original sample found to be uninteresting is defined as a network of size 1.

Let the number of networks in the sample be K, and note that this may differ from n_0 because units in the original sample may form part of the same network. Let the number of units in the kth network be n_k, so that the final number of units sampled is $n = \sum_{k=1}^{K} n_k$. In contrast to more conventional sampling procedures, n is a variable rather than fixed quantity that cannot be predicted in advance of survey. Let $x_k^* = \sum_{i \in k} x_i$ be the sum of the variable of interest in the kth network. With this notation in place, an unbiased estimate of the mean per unit is

$$\frac{1}{N} \sum_{k=1}^{K} \frac{x_k^*}{\alpha_k},$$

where

$$\alpha_k = 1 - \left[\binom{N - n_k}{n_0} \middle/ \binom{N}{n_0} \right],$$

and the estimated variance is

$$1 - \frac{1}{N^2} \left[\sum_{j=1}^{K} \sum_{k=1}^{K} \frac{x_j^* x_k^*}{\alpha_{jk}} \left(\frac{\alpha_{jk}}{\alpha_j \alpha_k} - 1 \right) \right],$$

where

$$\alpha_{jk} = \alpha_j + \alpha_k - (1 - p_{jk})$$

with

$$p_{jk} = \binom{N - n_j - n_k}{n_0} \middle/ \binom{N}{n_0}.$$

Orton (2000a: 93–8, 133–5) reported on some sampling experiments at the regional and site level, where adaptive sampling was applied to a region with a known number of sites, and sites with known number of features, in order to estimate the number of sites and features, respectively. The results from adaptive sampling were compared with those from previous analyses (Plog, 1976) and simple random sampling. Results were equivocal. In both analyses more sites/features were located than with the other methods, but the efficiency was generally poorer. In the regional experiment some experimentation was needed, with the number of initial units sampled, and with the criterion used to determine if neighboring units were to be sampled, to obtain the best results. Such experimentation would not be possible when using the technique in anger. Against this, Orton sets the facts that adaptive cluster sampling is likely to be most efficient when sites/features are strongly clustered, whereas in his experiments only weak clustering was observed, and that the cost of conducting the adaptive survey as opposed to other probabilistic sampling schemes is competitive. It is clear that adaptive sampling has great intuitive appeal at these scales of survey, but more experience applying the method 'in the field' is also clearly needed.

4.5 Sampling for discovery/rare features

4.5.1 Sampling for discovery

A sampling problem that has attracted some attention is that of sampling for the discovery of sites. In line with Krakker *et al.* (1983: 471) and Nance and

Ball (1986: 458), a site is viewed here as a well-defined cluster of artifacts, not necessarily visible on the surface. Test-pit sampling, or shovel-test sampling, involves sampling a region of interest using test-pits spread over the region in order to discover sites. These test-pits, and the total excavated area, would typically be small in relation to the size of the region and square in shape.

Mathematical and statistical investigations of test-pit sampling have typically idealized sites as circular, or possibly elliptical in shape (e.g., Sundstrom, 1993). For a site to be discovered it has to be intersected by a test-pit *and* recognized as a site, from, for example, the presence of at least one artifact in the excavated area. The probability that an intersected site will be recognized as such will depend, at the very least, on the density of artifacts within the site, and their distribution.

The probability of intersection depends on the size (area) of the site, the size of a test-pit, and the number and configuration of the test-pits. In formal investigations the configuration is often taken to be a square grid, staggered square grid, hexagonal, or variants thereof (Kintigh, 1988). A square grid is one in which there are t equally spaced transects, of distance s apart, and test-pits within transects are aligned and distance $i = s$ apart. For a square grid the maximum size of a circular site that can go undetected is $d = 1.414i$, where d is the diameter of the site. Thus, as $d/i \rightarrow 1.414$, from below, the probability that a site is intersected tends to 1.

A (regularly) staggered square grid is one for which $i = s$, with tests in adjacent rows offset by a distance of $s/2$ (Krakker *et al.*, 1983: 472). The maximum size of a circular site that can go undetected is $d = 1.25i$. This is a more efficient layout for detecting sites than a square grid, in the sense that the size of the largest site that can go undetected is smaller. A hexagonal grid, where $s = i\sqrt{3}/2$, is better still, with $d = 1.155i$ the size of the largest site that can go undetected (Kintigh, 1988: 688). Results for more complex situations, such as arbitrarily staggered grids and narrow survey regions, are discussed in Kintigh (1988) and Krakker *et al.* (1983).

Krakker *et al.* (1983: 472) provided a graph of $P(I)$ against d/i for square and staggered grids, where $P(I)$ is the probability that a site is intersected, that illustrated the superiority of the staggered grid, particularly over the range $0.4 < d/i < 0.9$. If $P(D)$ is the probability that a site is discovered, this may be modeled as $P(D) = P(I) \times P(R)$, where $P(R)$ is the probability that an intersected site is recognized. Krakker *et al.* (1983) noted that if the density (number of artifacts per unit area) in a site was ρ and the distribution was random (Poisson) then a test-pit of area a has a probability of $P(R) = 1 - e^{-a\rho}$ of detecting at least one artifact. This can be used, for example, to estimate the value of a needed to recognize a site at some specified level of probability, for assumed ρ. Such calculations assume that artifacts present in a test-pit would be recognized in excavation.

The assumption of a random distribution of artifacts within a site is, of course, questionable. Kintigh (1988) reported the results of an extensive simulation study that, among many other things, compared probabilities of discovery for sites with a random distribution, sites for which the distribution was clustered, and sites for which the distribution was centrally peaked with a lower density towards the edge of the site. Intuition suggests that in the last two situations the probability of discovery will be reduced, and the simulations confirmed this. Sites characterized by an artifact distribution consisting of many small and reasonably well-defined clusters were less likely to be discovered than sites where clusters tended to be larger and more diffuse.

Krakker *et al.* (1983), Kintigh (1988) and Shott (1985, 1989) are all pessimistic about the efficacy of test-pit sampling, if the intention is to discover the majority of sites in a region. The fundamental problem is that, to achieve this aim an impractical (uneconomic) amount of excavation is needed. Shott (1989: 396) goes so far as to suggest that 'shovel-test sampling is a survey method whose time, hopefully, has come and gone'. Shott (1989) is a robust critique of earlier work by Nance and Ball (1986) and Lightfoot (1986) which takes issue with them because they 'appear to endorse shovel-test sampling as a site-discovery technique'. The former paper is criticized because it focuses on probabilities of recognition, at the expense of probabilities of intersection, and the latter paper is criticized for the opposite reason. In their equally robust replies Nance and Ball (1989) and Lightfoot (1989) contend, among other things, that test-pit sampling has more of a role to play than Shott allows, in the absence of demonstrably better practical alternatives.

4.5.2 Sampling for 'nothing'

A problem that has been addressed by several authors is, if a sampling scheme reveals no archaeological remains, how large the sample should be to have confidence that there really is nothing there. In an idealized form this corresponds to a situation where there are n sampling points, where the true proportion of archaeological remains on the site is θ, and the observed proportion is $\hat{\theta} = 0$. An upper $100p\%$ confidence limit for θ is

$$\theta_p = 1 - (1 - p)^{1/n},$$

which can be solved to give

$$n = \frac{\log(1 - p)}{\log(1 - \theta_p)},$$

with the interpretation that if no archaeological remains are found with a sample of this size there is $100p\%$ confidence that θ is no greater than θ_p (Nicholson and Barry, 1995: 75; Read, 1986: 484; Shott, 1987: 365; Orton, 2000b). A Bayesian analysis of this problem, by Orton (2000b), is described in Section 14.3.1.

5

Regression and related models

5.1 Introduction

Read (1989) notes that regression methods were extensively used to address archaeological problems in Myers (1950), and Orton (1980) cites Harrington (1954) as an early use of regression to model the dates of clay tobacco pipes in terms of their bore size. However, given Spaulding's (1960: 82) comments to the effect that he was unaware of any application of regression methods in archaeology, it is clear that regression methods had little impact on archaeology before the 1960s. They are now a standard tool. Books on general quantitative archaeology typically contain material on regression. Shennan (1997) is the most thorough treatment, while Hodder and Orton (1976) and Orton (1980) have the widest range of examples. There are numerous statistical texts on regression methods.

This chapter documents some of the ways regression methods have been used in archaeology. It is useful at the outset to distinguish between the use of regression models for *descriptive*, *predictive* and *explanatory* purposes. The distinction may be illustrated with reference to Figure 5.1, using data reconstructed from Figure 2A of Morris (1994). This shows the frequency of occurrence of a particular type of Middle–Late Iron Age pottery, recorded as a percentage, plotted against distance from the source area, for 12 sites.

There is a clear non-linear distance decay relationship, with an obvious outlier at about 33–34 km where the occurrence is much greater than expected, given the pattern shown in the rest of the data. For the purposes of describing data such as these a simple linear regression model is sometimes (inappropriately) fitted. Such a model has the form

$$y = \beta_0 + \beta_1 x + \varepsilon \tag{5.1}$$

where y is the *dependent* (or *response*) variable, x is the *independent* (or *predictor* or *explanatory*) variable, ε is a random *error* term, and β_0 and β_1 are intercept and slope parameters to be estimated. Unless otherwise stated, the x are regarded as fixed, or measured without error, and are assumed to be uncorrelated with the ε. The fitted model, using the method of least squares, is

$$\hat{y} = \hat{\beta}_0 + \hat{\beta}_1 x$$

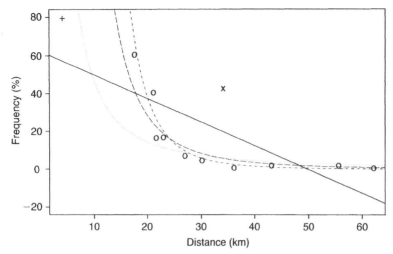

Figure 5.1 Plot of the percentage presence of a particular type of Iron Age pottery against distance from the source for 12 sites (based on Morris, 1994: Figure 2A). The fitted lines and curves are for different models discussed in the text, with sites labeled '+' and 'x' omitted in some analyses.

where the ˆ symbol indicates an estimated value, and residuals, which are estimates of the unknown errors, are given by

$$\hat{\varepsilon} = y - \hat{y}.$$

For the purposes of statistical inference it is commonly assumed that the errors are independent, identically distributed normal random variables with zero mean and constant variance,

$$\varepsilon \sim N(0, \sigma^2).$$

Model (5.1) may also be written as

$$E(y) = \beta_0 + \beta_1 x,$$

where $E(\cdot)$ is the expectation operator.

In the present case the fitted model is $\hat{y} = 62.4 - 1.25x$, shown by the solid line in Figure 5.1. The goodness of fit measured by the coefficient of determination, R^2, is 56%, which is not especially good. Quite apart from the outlier, a linear model is inappropriate for the data. Nevertheless, similar data do sometimes approximate to a linear scatter (e.g., Figure 2B of Morris, 1994), and a linear model might provide a reasonable descriptive model. It would not, however, be appropriate for prediction since, as is clear from Figure 5.1, at sufficiently large distances negative frequencies are predicted.

A better model for both descriptive and predictive purposes would respect the fact that the observed frequency cannot be negative. One such model is

$$y = \alpha x^\beta \delta, \tag{5.2}$$

which can be transformed to linearity by

$$\log y = \beta_0 + \beta_1 \log x + \varepsilon, \tag{5.3}$$

where $\beta_0 = \log \alpha$, $\beta_1 = \beta$, $\varepsilon = \log \delta$ and the error term, δ, enters into equation (5.2) in a multiplicative fashion. The dotted line in Figure 5.1 shows the fit obtained from this model on omitting the obvious outlier labeled by '×'. The dashed line is the fit omitting this point and the influential case labeled by '+'. The latter model fits the remaining data well, and model (5.3) has an R^2 of 88%.

Models such as equation (5.2) which can be transformed so that they are linear in the parameters are said to be *linearizable*. Another model which sometimes fits this kind of data, though not so well in this instance, is

$$y = \alpha \exp(\beta x)\delta, \tag{5.4}$$

which can be linearized as

$$\log y = \beta_0 + \beta_1 x + \varepsilon, \tag{5.5}$$

where natural logarithms are used.

The models just described provide a better description of the data than the linear model, and are certainly much better for prediction, though care must always be exercised in predicting beyond the range of the data used in the fitting process. It happens that model (5.3) is a simple form of a distance–decay model for which a variety of different theoretical justifications have been proposed. Where a theoretical justification is possible, and if the theory is correct, regression models have explanatory power and can be used for prediction outside the range of the data used to fit the model.

Several warning notes need to be sounded here. One is that a good fit to a regression model does not establish a causal relationship between the variables involved. Substantive theory is needed to determine whether or not a good fit is anything other than simply a good description of the data to hand. Even if a model has theoretical justification, a good statistical fit does not, in itself, establish that the theory is correct. It is possible for the same statistical model to be derived from more than one theoretical perspective, and statistics alone cannot determine which of the competing theories is to be preferred. Statistical analysis can suggest that data are compatible with a particular theory, but non-statistical arguments would also need to be adduced in favor of any theory so supported.

5.2 Simple linear regression

5.2.1 An example

Models (5.1), (5.3) and (5.5) are all examples of simple linear regression models, where just one independent variable, x (or $\log x$), is used to describe, predict or explain the variation in the dependent variable y (or $\log y$).

As another example, to be used in later developments, some data collected by Schreiber and Kintigh (1996), discussed also by Kvamme (1997) and given in Table 5.1, are used. It can be of interest to estimate prehistoric population sizes, and changes in population over time, but population size cannot be observed directly. Surrogate variables are frequently used, and the size of an archaeological site, as measured by its areal extent, is one possible index of the number of occupants on a site. The assumption is that site size is proportional to population size.

Table 5.1 Site population (number of tribute payers), size, and type (P = political center, R = regular village) for 11 sites surveyed in a 1540 census (Schreiber and Kintigh, 1996)

Population	30	66	76	163	19	56	13	19	36	16	30
Size (ha)	4.25	2.00	3.75	22.60	0.80	8.60	0.80	1.00	5.00	0.75	1.20
Type	P	R	R	P	R	P	R	R	P	R	R

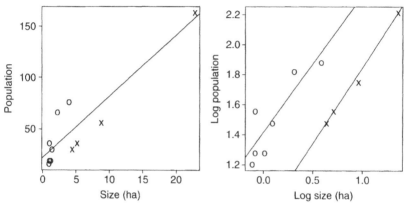

Figure 5.2 Estimated population in 1540 against site size for 11 settlements in the Peruvian Andes, classified as regular villages (o) and political settlements (×) (Schreiber and Kintigh, 1996). The left-hand plot shows a simple linear regression fitted to the original data; the right-hand plot shows separate fits to the two settlement types after a logarithmic transformation of both variables.

In order to test this assumption, Schreiber and Kintigh (1996) used population data from a 1540 census of 11 settlements in the Carhuarazo valley in the Peruvian Andes, undertaken after the Spanish conquest. The census was of tribute payers, likely to be male heads of households, and the 'population', y, so recorded was assumed to be proportional to the true population. The site size, in hectares, was measured in an archaeological survey of 1981. On the basis of population density a distinction was drawn between 'political centers' and 'regular villages', with the latter the more dense. The data are plotted in the left-hand panel of Figure 5.2, where the predicted line from a simple linear regression is shown.

Schreiber and Kintigh (1996) used the data to argue that there was no obvious support for the hypothesis of a strong relation between site size and population. Their argument was that the fit looked poor and that the high correlation between site size and population, of 0.91, was largely attributable to an 'outlier' at the top left of the plot. Removing this results in a less impressive correlation of 0.49. They then argued that if the regression was controlled for site type the expected relationship between size and population held. (It is recognized in the paper that small samples are involved.) The correlation for political centers is 0.99 however the 'outlier' is treated, and for villages 0.90, rising to 0.99 if two villages with constrained boundaries are excluded.

Kvamme (1997) commented that, as an analytical strategy, there was merit in transforming to logarithms before analysis. This downweights the impact of the unusual value. The right-hand panel of Figure 5.2 shows the log-transformed data (to base 10) and separate regressions for the two site types. Kvamme (1997: 720) further noted that this implied a model of type (5.2) within each site type, and that theoretical models of precisely this form had been proposed for relationships between settlement area and population size.

5.2.2 Models for data

Model (5.2) has a systematic component, αx^β, that has been given theoretical justification in other archaeological contexts. For example, using theories from structural engineering, Cavanagh and Laxton (1981) derive model (5.2) to describe the curvature in prehistoric corbeled domes from the Bronze Age Aegean and Sardinia, and Neolithic passage graves in France, where y is the radius of the tomb at a depth x below the apex. Model (5.2) is also used in studies of diversity that relate sample size to the richness of observed assemblages (Section 20.3).

The same formal model occurs in studies of rank–size relationships. The rank–size rule was derived from empirical observation by geographers and given a theoretical foundation by Zipf (1949). Let x_i be the size of a site (taken as a surrogate for population) ranked from largest, x_1, to smallest, and let y_i be the actual rank. The parameter α is taken to be equal to x_1 and in application $\beta = -1$ is often assumed. In this form the law expresses the idea that the size of the jth ranked place is just x_1/j. The model, and deviations from it, has been used to try and explain different processes related to urbanism and social complexity (Savage, 1997a). With $\beta = -1$ and α known, the model is linear on a log–log scale. This is used as a reference line against which a plot of log-size against log-rank is judged. A variety of methodological or culturally based explanations have been advanced to explain departures from the expected linearity (Savage, 1997a: 234). There are some subtle problems associated with the archaeological use of this model, to do with the fact that not all the sites relevant to an analysis will be known, and missing data may be biased towards the smaller sites (Savage, 1997a).

One problem in using some of the models described above is the quality of the data. Hodder and Orton (1976: 104–7) emphasize this in their discussion of distance–decay models in archaeology. The problem of missing data in rank–size models has already been noted. In analyzing shapes of corbeled domes measurements may be inaccurate if, for instance, the apex of the tomb is not known exactly because of the presence of a capstone. This does not mean that the models are wrong or fundamentally inappropriate, but it does mean that care may be needed in interpretation and analysis, or that a modification of the model is needed.

5.2.3 Flake size and flake-size distribution

Regression methods, of varying degrees of complexity, have been used in the analysis of flake debris, defined by Shott (1994: 70) as a collective term 'for all waste material generated by humans in lithic reduction'. Such debris is generated in the making of stone tools, and is far more abundant than the tools themselves, as well as being more likely to remain at the site of manufacture than the tool. Shott (1994: 71) argued that flake types or entire debris assemblages may be culturally or chronologically diagnostic, and that debris is a product of, and registers the kind and amount of, reduction and sharpening undertaken by artisans.

Patterson (1990) has argued that flake-size distribution patterns are different for different lithic reduction processes and, in particular, that model (5.4) is characteristic of bifacial reduction processes, where x is flake size in square millimeters and y is the percentage of the total in the size category. Patterson's (1990) experimental work provided some support for this, though he noted that, typically, the small percentage of data in the largest class size may fit the model less well. Shott's (1994: 91–4) analyses suggested that Patterson's model was reasonably robust, but that not all biface-reduction experiments conformed to his model, and that other reduction modes sometimes generated similar patterns.

Brown (2001) preferred model (5.2) for flake-size distributions, calling this a fractal or power-law distribution and relating it to fractal theories of fragmentation based in mathematical and physical theory. The value of $-\beta$ was interpreted as the fractal dimension of the data. Brown's Table 2, of the fractal dimension and R^2 for a variety of experimentally generated data sets using model (5.3), showed that the fit was often good with a wide range of different fractal dimensions. Applications to archaeological assemblages produced similarly good results. Brown (2001: 619) claimed that fractal analysis is a simpler and more effective method of analyzing debitage size distributions than other methods. This claim appears to rest on its theoretical merits as there are no empirical comparisons of his model (5.2) with Patterson's (1990) equally simple model (5.4).

Regression methods have also been used to investigate the relationship between the *original* size of a flake and the platform size, an attribute of the flake tool created in its manufacture. If valid predictive equations, of flake size from platform attributes, can be derived from experimental studies then the difference between a tool's original and discarded size can be estimated, which bears on a variety of interpretive issues in Paleolithic archaeology (Shott *et al.*, 2000: 877). In their study Shott *et al.* used regression and correlation to investigate the relationship between different measures of flake size, such as area and mass, and platform attributes such as the width, thickness and area, taking into account whether hard or soft hammers had been used in flake production, and using a variety of experimental assemblages. They found that better results were generally obtained after log-transforming both the dependent and independent variables. Typical correlations vary between about 0.5 and 0.8, so that while there is often a clearly significant relation between flake size and platform attributes the predictive relation is not strong, these correlations corresponding to R^2 values of 0.25, which is quite poor, and 0.64, which is only moderate.

5.3 Multiple linear regression

5.3.1 The basic model

The simple linear regression model generalizes to multiple linear regression, with p independent variables, where the model is

$$y = \beta_0 + \beta_1 x_1 + \beta_2 x_2 + \cdots + \beta_p x_p + \varepsilon = \mathbf{x}'\boldsymbol{\beta} + \varepsilon \tag{5.6}$$

$\mathbf{x}'$ is the $1 \times (p+1)$ vector $(1\ x_1\ x_2 \cdots x_p)$, and $\boldsymbol{\beta}$ is the $(p+1) \times 1$ parameter vector $(\beta_0\ \beta_1\ \cdots\ \beta_p)$. The model is linear in the sense that the unknown parameters enter into it in a linear fashion. If the independent variables are known non-linear functions of other variables, the model of equation (5.6) remains linear in the sense just described.

5.3.2   Polynomial regression and related models

A variant of multiple regression arises with polynomial regression, when the x_i are powers of a single independent variable, x. Liversidge and Molleson (1999), for example, developed regression equations to predict age (y) from tooth length (x) using data on permanent teeth from birth to maturity. Data were available from the skeletal remains of 76 individuals buried in the 18th century at Spitalfields, London, for which the age was known from parish records. Using variables of the form $x_i = x^i$, equations up to the fifth order were developed for predictive purposes.

In a rather different example, related to the prediction of flake mass from platform attributes discussed in the previous section, Shott *et al.* (2000) referenced models derived by Pelchin (1996) for predicting flake mass from platform thickness (x) using a third-order polynomial model and with $\beta_0 = 0$.

Some simple models for predicting population size were discussed in Section 5.2. Multiple regression models have also been exploited in this kind of context. Curet (1998), for example, developed formulas to estimate prehistoric populations for lowland South America and the Caribbean islands, at the household and community levels. Curet's models were based on ethnographic data, it being assumed that these were valid when applied to appropriate culture-specific or culture-area-specific contexts. Most of the analyses were simple linear regressions; however, some were multiple regressions, regressing, for example, the number of occupants in multifamily households against $\exp(-2x)$ and $\exp(-5x^2)$, where x is the floor area in square meters (Curet, 1998: Figure 4). While archaeological theory, or common sense, might dictate the choice of variables used, the functional form of the predictive equation in such applications seems usually to be determined by the quality of fit, as measured by R^2, after what may be considerable empirical exploration.

Given two variables, x_1 and x_2, in a regression model, the interaction term x_1x_2 might also be included in a regression analysis, possibly together with polynomial (quadratic) terms. Thus, in controlled experiments to investigate if the mass of flakes produced in flintknapping could be predicted from characteristics of the flake, Dibble and Pelchin (1995) proposed the model

$$\hat{y} = -0.361x_1 - 3.305x_2 + 1.663x_1x_2,$$

where y is flake mass, x_1 is the platform thickness, and x_2 is the tangent of the exterior platform angle. The model appears to be empirically rather than theoretically derived. Dibble and Pelchin reported that this yielded a correlation coefficient of 0.903, which presumably refers to the correlation between y and $\hat{y}$.

5.3.3   Trend-surface analysis

Trend-surface analysis is a form of polynomial regression used for fitting surfaces to spatially distributed data. Initially 'borrowed' from the quantitative geographical literature, Hodder and Orton (1976) treat the topic at some length. Wheatley and Gillings (2002: 187) note the potential of trend-surface analysis for interpolation in geographical information systems (see Chapter 13 for a discussion of GIS).

Given observed values of a dependent variable, y, at spatial coordinates, x_1 and x_2, a fitted linear trend-surface model has the form

$$E(y) = \beta_0 + \beta_1 x_1 + \beta_2 x_2,$$

a second-order model has the form

$$E(y) = \beta_0 + \beta_1 x_1 + \beta_2 x_2 + \beta_{11} x_1^2 + \beta_{12} x_1 x_2 + \beta_{22} x_2^2,$$

a third-order model has the form

$$E(y) = \beta_0 + \beta_1 x_1 + \beta_2 x_2 + \beta_{11} x_1^2 + \beta_{12} x_1 x_2 + \beta_{22} x_2^2 + \beta_{111} x_1^3 \\ + \beta_{112} x_1^2 x_2 + \beta_{122} x_1 x_2^2 + \beta_{222} x_2^3,$$

and so on.

Bove (1981) investigated the relationship between time and distance for 47 sites in the Lowland Maya region of Mesoamerica in the Late Classic period of about AD 750–900. The end of monument construction activity at these sites can be dated, at least approximately, by the dates associated with the most recent monuments, given in Bove's Table 1. Cessation of construction activity occurred over a relatively short timespan, and this is sometimes called the 'Classic Maya collapse'. A variety of archaeological hypotheses have been advanced to explain this collapse.

One question of relevance to these hypotheses is whether the dates exhibit a spatial pattern. Bove (1981) investigated this by fitting a series of trend-surface models to the data, mapping these, and inspecting residuals from the fits. It was concluded (Bove, 1981: 108–9) that there was some support for the hypothesis of a west to east sequential diffusion of the events leading to the cessation of construction activity, but that the trend was a weak one. The third-order model, for example, while significant at the 5% level, had a modest goodness of fit of $R^2 = 36\%$. Examination of residuals from the third-order model identified clusters of positive residuals that could be interpreted as being in accord with the hypothesis that the collapse involved the decay of a series of small city-states (Whitley and Clark, 1985: 386). Other analyses of these data are discussed in Section 13.5.2.

5.3.4 Regression with 0–1 dependent variables

A special case of multiple regression arises when the dependent variable is categorical, taking on only two levels, conventionally coded as *dummy* or *indicator* variables, 1 and 0. Examples include artifacts classified into one of two types, with an attempt being made to discriminate between types on the basis of measurements on the artifacts (e.g., Lukesh and Howe, 1978). This use of regression is equivalent to two-group discriminant analysis and is discussed more fully in Chapter 9. The two levels might also represent presence and absence of a site type in some region, which is to be modeled in terms of the characteristics of the region. This kind of use for predictive modeling has been superseded by the use of logistic regression models, discussed below, and geographical information systems, discussed in Chapter 13.

5.3.5   Regression with 0–1 independent variables

Forms of model
Dummy variables can also be used as independent variables, and the results in the right-hand panel of Figure 5.2 were generated using multiple regression with a dummy variable. The model actually used was

$$\log y = \beta_0 + \beta_1 \log x + \alpha_1 z_1 + \gamma_1 (z_1 \log x) + \varepsilon$$

where $z_1 = 1$ for a site characterized as a political center and 0 otherwise. For a political center

$$E(\log y) = (\beta_0 + \alpha_1) + (\beta_1 + \gamma_1) \log x,$$

and for a regular village

$$E(\log y) = \beta_0 + \beta_1 \log x.$$

In effect, the two single regressions are fitted using a multiple regression with a dummy variable and interaction term $z_1 \log x$. Regression lines with separate intercepts, $(\beta_0 + \alpha_1)$ and β_0, and slopes, $(\beta_1 + \gamma_1)$ and β_1, are produced. If the interaction term is omitted (i.e., $\gamma_1 = 0$) this fits parallel regressions.

It would be equally possible to define a dummy variable z_2 taking the value 1 for a regular village and 0 for a political center, and replace z_1 with z_2 in the above formulation with essentially the same result. Including both z_1 and z_2 in the model and dropping the interaction term would give

$$E(\log y) = \beta_0 1 + \beta_1 \log x + \alpha_1 z_1 + \alpha_2 z_2,$$

where the 1 is included as a 'variable' associated with the constant, β_0. Because $z_{1i} + z_{2i} = 1$ for all i there is a linear dependency in the data that will cause regression software either to fail, or report a warning and take corrective action. Such action usually involves omitting one of the dummy variables or, equivalently, setting α_1 or $\alpha_2 = 0$. More generally, imposing some linear constraint such as $\alpha_1 + \alpha_2 = 0$ will resolve the problem.

Suppose, hypothetically, that all settlements are of the same size, so that terms in x may be dropped from the model, and the only interest lies in whether or not the mean population differs between settlement types. Reverting to y rather than $\log y$ as the dependent variable, the regression model may be written as

$$E(y_i) = \beta_0 + \alpha_1 z_{1i} + \alpha_2 z_{2i}.$$

Adopting the constraint that $\alpha_2 = 0$, the test of the null hypothesis that settlement types do not differ in terms of their mean population amounts to testing the null hypothesis that $\alpha_1 = 0$. This is equivalent to an independent two-sample t-test.

If $I > 2$ settlement types are involved the model may be written as

$$E(y_i) = \beta_0 + \alpha_1 z_{1i} + \alpha_2 z_{2i} + \cdots + \alpha_I z_{Ii} \tag{5.7}$$

on defining dummy variables for each of the I settlement types. With the constraint that $\alpha_I = 0$, testing the hypothesis that settlement types do not differ in terms of mean population amounts to testing the null hypothesis that the remaining α_i are simultaneously zero, or $\alpha_1 = \alpha_2 = \cdots = \alpha_{I-1} = 0$. The test, in a regression context, is equivalent to a one-way analysis of variance (ANOVA) that tests for the equivalence of I population means using an F statistic.

A notational digression

In equation (5.7), $i = 1, \ldots, n$, where n is the total sample size. An alternative way of writing the model is to let $i = 1, \ldots, I$, the number of *levels* for the settlement type variable, and write the model more concisely as

$$E(y_{ij}) = \beta_0 + \alpha_i, \tag{5.8}$$

where $j = 1, \ldots, n_i$, the number of observations for level i. Settlement type is an example of a *factor* with I *levels*. Equation (5.8) is the form in which models for one-way classifications, analyzed by one-way ANOVA, are often written (e.g., Draper and Smith, 1998: 475).

If a second classifying factor with J levels is available, model (5.8) extends to

$$E(y_{ijk}) = \beta_0 + \alpha_i + \gamma_j, \tag{5.9}$$

where $k = 1, \ldots, n_{ij}$ is the number of observations at level i of the first factor and j of the second. This is the form of model used for two-way classifications leading to two-way ANOVA without interactions.

The above model might be written in a more cumbersome fashion by defining J dummy variables, w_1 to w_J say, for each level of the second factor and adding these to model (5.7). As with the first factor there are linear dependencies among the variables and one parameter, say γ_J, may be set to 0. This means that in model (5.9) there are $1 + (I - 1) + (J - 1) = I + J - 1$ parameters to be estimated.

Interactions may be included by the addition of $(I - 1)(J - 1)$ extra parameters of the form $(\alpha\gamma)_{ij}$ to get

$$E(y_{ijk}) = \beta_0 + \alpha_i + \gamma_j + (\alpha\gamma)_{ij}. \tag{5.10}$$

In the extended form of the model the $(\alpha\gamma)_{ij}$ are associated with variables of the form $z_i w_j$ and parameters associated with terms involving z_I or w_J are set to zero.

These ANOVA models are not widely used in archaeology, and one of the main purposes for developing the notation here is that similar models are used in the log-linear models for tabulated data, discussed in Chapter 11. A case of some importance is when $n_{ij} = 1$, so there is a single observation for each combination of levels. In this case, for model (5.10) there are $1 + (I - 1) + (J - 1) + (I - 1)(J - 1) = IJ$ parameters to be estimated. This is identical to the number of observations available and means that the model can be fitted exactly. In the context of the log-linear models discussed in Chapter 11, such a model is said to be *saturated*.

5.4 Generalized linear models

5.4.1 Basic ideas

Given the assumption of independent normally distributed errors, the multiple linear regression model can be written as

$$y_i \sim N(\beta_0 + \beta_1 x_{1i} + \beta_2 x_{2i} + \cdots + \beta_p x_{pi}, \sigma^2) \sim N(\mu_i, \sigma^2), \tag{5.11}$$

where $\mu_i = \beta_0 + \beta_1 x_{1i} + \beta_2 x_{2i} + \cdots + \beta_p x_{pi}$, and estimated using the method of *maximum likelihood* rather than least squares. The differences are minimal for this particular model, but maximum likelihood provides a more general approach to estimation. If y_i is distributed other than normal, but the distribution comes from the *exponential family of distributions*, and if the mean of the distribution, μ_i, is linearizable then a *generalized linear model* is obtained. Maximum likelihood estimates of parameters in such models can be obtained using iteratively weighted least squares methods, and many of the ideas and statistics used in connection with linear models can be extended to them (McCullagh and Nelder, 1989).

The transform that linearizes μ_i is called the *link* function. Model (5.11), for instance, is a generalized linear model with a normal distribution and identity link. In equation (5.2) the error term was assumed to be multiplicative and a simple linear regression model was obtained on log transformation. If, instead, an additive error had been assumed so that

$$y_i = \alpha x_i^\beta + \varepsilon, \tag{5.12}$$

no such simple transformation would be possible. However, assuming normal errors, we may write

$$y_i \sim N(\alpha x_i^\beta, \sigma^2)$$

and note that, as $\mu_i = \alpha x_i^\beta$ can be linearized by taking logarithms, we have a generalized linear model with normal distribution and log link. Using model (5.12) to fit the data of Morris (1994), omitting the two points noted, gives the dot-dashed line in Figure 5.1. This can be seen to fit the data better than the competing model derived from equation (5.3), particularly at the lower distances.

For counted data a standard generalized linear model is the log-linear model, which is multiplicative in the parameters and estimated assuming a Poisson distribution and log link. Archaeological applications of this model are discussed in Chapter 11. Another generalized linear model is the logistic regression model discussed below.

5.4.2 Logistic regression

The underlying probability model is that the observations are sampled from the binomial distribution, so that

$$y_i \sim B(n_i, \pi_i)$$

where y_i is the number of 'successes' in n_i independent trials, with π_i the probability of success in a single trial. An important special case is when the data are binary, so that $n_i = 1$ for all i and $y_i = 1$ or 0. This is the form of the model most widely used in archaeology and the one discussed here.

A common use is in predictive site location models (see Section 13.3). One possible sampling design for such models involves dividing an area into grid squares; selecting a sample of the squares; and noting whether an archaeological site is present in the square ($y_i = 1$) or not ($y_i = 0$). If biophysical variables, x_{ji}, descriptive of a location, are measured then the aim is to build a model that is a good predictor of the location of archaeological sites. Parker (1985) provides a review of the considerations that arise in using such models and an application.

In *logistic regression* a model for π_i, the probability that a site is present, is

$$\pi_i = \frac{\exp(\beta_0 + \beta_1 x_{1i} + \beta_2 x_{2i} + \cdots + \beta_p x_{pi})}{1 + \exp(\beta_0 + \beta_1 x_{1i} + \beta_2 x_{2i} + \cdots + \beta_p x_{pi})} = \frac{\exp(\mathbf{x}_i'\boldsymbol{\beta})}{1 + \exp(\mathbf{x}_i'\boldsymbol{\beta})}. \tag{5.13}$$

It follows that the probability that there is no site is $1/[1 + \exp(\mathbf{x}_i'\boldsymbol{\beta})]$. From equation (5.13) may be derived the *logit* transformation

$$\log\left(\frac{\pi_i}{1 - \pi_i}\right) = \beta_0 + \beta_1 x_{1i} + \beta_2 x_{2i} + \cdots + \beta_p x_{pi}. \tag{5.14}$$

Since the binomial distribution is a member of the exponential family of distributions, and since the mean can be linearized as in equation (5.14), the logistic regression model is a particular example of a generalized linear model.

An archaeological problem in using such models is defining and measuring the independent variables to be used. Statistical problems include selecting the most useful predictive variables, determining the fit and validating the model.

To assess whether one model improves significantly on another, or equivalently whether it is legitimate to simplify a model by omitting variables, likelihood ratio statistics may be used. Let L_s be the maximized likelihood for a saturated model that fits the data exactly and let L_A be the maximized likelihood for a model A with df_A degrees of freedom. The *deviance* of model A is defined as

$$G_A = -2\log(L_A/L_s).$$

If A is nested within a larger model, B, in the sense that all the variables in A are contained in B, then the difference in deviances, $G_A - G_B$, can be tested against the chi-squared distribution with $df_A - df_B$ degrees of freedom to assess whether the additional variables in model B make a significant contribution.

Site location in prehistoric Thailand
For an example the paper of Higham *et al.* (1982), concerned with site location in prehistoric Thailand, will be used. This was based on intensive field surveys in two areas, with eight variables being defined for all 1 km² squares in both study areas. Two variables were found to be suitable for predictive purposes. These were x_1, a soil index between 2 and 5, measuring the suitability of the soil for rice, and x_2, the sum of the distances to the nearest two water sources. Models for which the logit transformation took the form

$$\beta_0 + \beta_1 x_{1i} + \beta_2 x_{1i}^2 + \beta_3 x_{2i} + \beta_4 x_{2i}^2$$

were fitted, with the estimated parameter vectors for the two regions being $(-27.0$ 15.2 -2.24 -0.0303 $-0.242)$ and $(-41.3$ 23.7 -3.48 -1.20 $0.210)$. As only two variables were involved it was possible to contrast the different areas using contours of equal probability derived from these results, as shown in Figure 5.3.

It was possible to infer (Higham *et al.*, 1982: 10–12) that in the first area, represented by the solid contours in the figure, prehistoric sites tended to be located in squares where the soil index was between 3 and 4, and where summed distance to the nearest two water sources was never more than 1.7 km. The different pattern of contouring for the second area was explained by the presence of several prehistoric sites for which the summed distance to the nearest two water sources was somewhat in excess of 1.7 km. Some of these sites may have been closer to water in the past, but no longer are, owing to environmental change, or were close to their nearest water source but distant from the second nearest. The overall conclusion was that location on medium-quality rice lands, near a watercourse, characterized prehistoric settlement in both study areas.

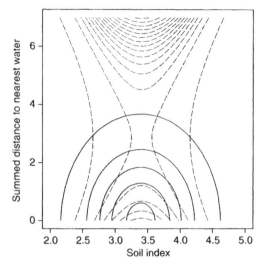

Figure 5.3 Contours of equal probability for two areas, represented by solid and by dashed lines, showing the probability of site location in relation to soil quality and the summed distance to the two nearest water sources (Higham *et al.*, 1982).

Other applications

Applications of logistic regression to problems other than predictive modeling include Tennessen *et al.* (2002) and McCorriston and Weisberg (2002). In the former paper logistic regression was used to show that modern samples of aspen and cottonwood (both species of the genus *Populus*), from northwestern New Mexico and southwestern Colorado, could be differentiated reasonably well by their mean pore size. The regression equation derived was applied to 50 archaeological wood samples from the Anasazi great house at Aztec Ruins National Monument, Aztec, New Mexico, dating to the early AD 1100s, to assess whether they were likely to be aspen or cottonwood.

McCorriston and Weisberg (2002) used a random effects logistic regression model to investigate spatial and temporal variation in Mesopotamian agricultural practices in the Khabur Basin, Syrian Jazira. On the basis of a large number of analyses, McCorriston and Weisberg (2002: 485) highlighted a 'significant rise in crop-processing wastes', probably associated with a focus on barley processing in the 3rd millennium BC, and 'probably linked with an emerging specialization in pastoral production and re-settlement in arid regions of the northern Mesopotamian steppe'.

5.5 Non-linear regression

A non-linear model is one that cannot be linearized in any of the senses discussed in the previous sections. A simple example arises in the study of the shape of corbeled domes discussed earlier in this chapter (Cavanagh and Laxton, 1981). Whereas previously *y* was the radius of the tomb and *x* the distance below the apex, the true position of the apex may not be known if the tomb is capped and an alternative model is

$$\log y = \log \alpha + \beta \log(x + \delta) + \varepsilon,$$

where x is now the distance below the cap and δ is the distance between the cap and the apex, a parameter to be estimated. Cavanagh and Laxton (1982) estimated this model using non-linear least squares.

Another non-linear model that has found some use is the growth model of the form

$$y_i = \beta_0[1 - \exp(-\beta_1 x_i)], \tag{5.15}$$

which has a limit of β_0 as $x \rightarrow \infty$ for $\beta_1 > 0$. The model used by Byrd (1997) for the relationship between assemblage size and richness is superficially similar, but the parameters are determined in an *ad hoc* fashion. The issues involved, including those of estimation, are deferred to the case study of Chapter 20. Trompetter and Coote (1993) use the same model for shellfish growth, using modern shells, with the ultimate aim of being able to predict age from shell length, and hence identify the season of collection, in archaeological specimens.

Non-linear models often arise in specialized applications where simple 'off-the-shelf' models may not be appropriate. As an example, Williams-Thorpe *et al.* (2000) were concerned to model the relationship between measures of magnetic susceptibility and the thickness of artifacts such as stone axes. Magnetic suscepti-bility has been found to be useful in the characterization of source rocks, but their instrumentation was designed for measurements on objects at least 50 mm thick, and they wished to provide correction factors for objects of smaller thickness. For their experimental data 26 readings were taken on prepared blocks of rock between 3 mm and 102 mm in thickness and a plot of susceptibility (y) against thickness (x) showed a non-linear curve, with initially low readings increasing as thickness increased and leveling off at about 50 mm. After some experimentation with different models, a subjective choice of a constrained double exponential model of the form

$$y = \beta_0 + \beta_1 \delta_1^x + \beta_2 \delta_2^x,$$

where $(\beta_0 + \beta_1 + \beta_2) = 0$ and $\delta_1, \delta_2 < 0$, was chosen. This was estimated using maximum likelihood and correction factors were then determined from the fitted model.

5.6 Non-parametric regression

Non-parametric regression has had relatively little use in archaeology but, given the widespread use of scatterplots in archaeology to summarize relationships, there is an obvious role for non-parametric summary methods and I would anticipate their wider use once they become more available in accessible software.

Such use as has occurred has often been in the context of radiocarbon calibration (Aitchison *et al.*, 1991; Clark, 1979, 1980; Bowman *et al.*, 1998; Bowman and Azzalini, 1997: 48–52). The left-hand panel of Figure 5.4 shows a section of the radiocarbon calibration curve established by Stuiver *et al.* (1998). Points correspond to exact dates, determined from tree-ring dating (in years BC for this portion of the curve) and uncalibrated radiocarbon dates (in years BP, defined conventionally as the number of years before AD 1950). For reasons discussed

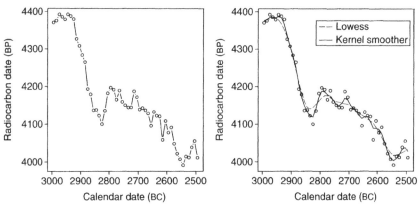

Figure 5.4 The left-hand panel shows a section of the radiocarbon calibration curve of Stuiver *et al.* (1998) with piecewise linear interpolation. The right-hand panel shows the same data and two alternative non-parametric smoothing regressions.

in Chapter 15 a radiocarbon date, which is what would normally be available in practice, does not correspond to the true date, which is what is of interest. The curve itself is 'wiggly' and only sampled at intervals.

Radiocarbon dates are reported with an error and converting the date plus its error range into a probability distribution for the true date is complex. This is discussed at length in Chapter 15, where it will be seen that for calibration purposes dates need to be interpolated at the points where they are not observed. The simplest approach is to use the piecewise linear function shown in the left-hand panel of Figure 5.4; however, in the references cited above smoother estimates based on non-parametric estimation have also been illustrated. The right-hand panel of Figure 5.4 shows two such non-parametric estimates, also sometimes called *scatterplot smoothers*.

The underlying model may be written as

$$y_i = m(x_i) + \varepsilon$$

for some function $m(x)$. Rather than specifying $m(x)$ as a function of a small number of parameters, the idea in non-parametric regression is to allow the data to determine its form, with the general aim of smoothing the data sufficiently to see any underlying structure without 'tracking' the data in too much detail. This is analogous to the use of kernel density estimates for representing univariate data, where a balance has to be struck between over-smoothing the data and missing structure, and under-smoothing and showing spurious detail. Indeed, kernel-like estimators play an important role in several of the available methods.

One general idea is to estimate $m(x)$ at a data point x by $\hat{m}(x) = \hat{\beta}_0$, where $\hat{\beta}_0$ is obtained from the weighted polynomial least squares regression that minimizes

$$\sum_{i=1}^{n} [y_i - \beta_0 - \beta_1(x_i - x) - \cdots - \beta_p(x_i - x)^p]^2 K_h(x_i - x)$$

for some p, and where the weights $K_h(x_i - x)$ are defined by a kernel density function. Explicit formulas exist for $p = 0$ and $p = 1$ (Wand and Jones, 1995: 119;

Simonoff, 1996: 139), which are called the local mean estimate and local linear estimate, respectively. The latter has good properties compared to the former (Bowman and Azzalini, 1997: 50) and is shown as the solid curve in the right-hand panel of Figure 5.4.

Similar issues concerning the choice of bandwidth arise as with univariate KDEs. The plug-in selector of Ruppert *et al.* (1995) was used here, estimated in S-Plus using a function from Wand's KernSmooth library of routines, which was also used to achieve the fit. The sm.regression function within the sm library of Bowman and Azzalini (1997) can also be used to obtain the regression. This does not have a facility for automatic bandwidth selection, but does allow for variable bandwidths that can be used to make allowance for the precision of the radiocarbon date estimate (Bowman and Azzalini, 1997: 67).

The dashed curve in the figure is obtained using the lowess method of Cleveland (1979). Weights are assigned to data within some neighborhood of a point x_i, and the predicted value at that point is obtained by weighted linear regression. This can be used to calculate residuals that, in an iterative procedure, are used to modify the weights, the effect being to downweight the influence of outlying cases. The smoothness of the curve is affected by the proportion of data, or local neighborhood, used to define the neighborhood in the fitting procedure, the proportion used in the example being 0.2. The lowess estimator in the figure produces a smoother fit to the data than that achieved by the kernel smoother, which possibly tracks the data too closely. This could, of course, be 'corrected for' by increasing the size of the smoothing parameter used. A slightly fuller discussion of technicalities, including choice of the size of local neighborhood, is included in the context of a specific application in Section 13.5.2.

The smoothing methods discussed here, while conceptually simple, are computationally demanding and, as yet, have been little used by archaeologists. Rogers (2000a) and Rogers and Broughton (2001) provide examples. Neiman (1997) made extensive use of the lowess smoother in his study of the Classic Maya collapse, discussed above in Section 5.3.3, and this is considered further in Section 13.5.2.

6

Multivariate methods – an introduction

6.1 Introduction

Multivariate methods such as *principal component analysis, factor analysis, correspondence analysis, discriminant analysis* and *cluster analysis* are widely used in archaeology. There are many good texts on multivariate analysis (e.g., Mardia *et al.*, 1979; Seber, 1984; Krzanowski, 1988; Everitt and Dunn, 2001; Krzanowski and Marriott, 1994, 1995; Manly, 1994). Texts written for an archaeological readership include Shennan (1997), who provides an introductory account of several of the main methods, and Baxter (1994a), who provides a comprehensive treatment of the methods listed above.

In this chapter notation and ideas used in several subsequent chapters are introduced, with the concept of *distance* being particularly important. Some readers may prefer to skim the more technical parts of this chapter, and refer back as necessary when reading later chapters, though the section on Mahalanobis distance is of practical importance.

6.2 Notation and terminology

Initial discussion is based on tables of continuous data. Let $X_1, X_2, \ldots, X_p$ be p *variables* measured for n cases, resulting in an $n \times p$ raw data matrix, $\mathbf{X}$, whose typical element is x_{ij}. The mean of X_j is $\bar{x}_j$, and $\bar{\mathbf{x}}$ is the $p \times 1$ vector of means, so

$$\bar{\mathbf{x}}' = (\bar{x}_1\ \bar{x}_2\ \cdots\ \bar{x}_p)$$

where the prime notation, $'$, denotes vector or matrix transpose.

The data matrix $\mathbf{X}$ is a sample from a population with mean $\boldsymbol{\mu}$ and covariance matrix $\boldsymbol{\Sigma}$, and $\bar{\mathbf{x}}$ is an estimate of $\boldsymbol{\mu}$. The $p \times p$ matrix, $\mathbf{S}_X$, defined as

$$\mathbf{S}_X = \sum_{j=1}^{n} (\mathbf{x}_j - \bar{\mathbf{x}})(\mathbf{x}_j - \bar{\mathbf{x}})'/(n-1), \tag{6.1}$$

where $\mathbf{x}_j$ is a $p \times 1$ vector of the observations for the jth case, is an unbiased estimate of $\mathbf{\Sigma}$. The off-diagonal elements of $\mathbf{S}_X$,

$$s_{ij} = s_{ji} = \sum_{k=1}^{n} (x_{ki} - \bar{x}_i)(x_{kj} - \bar{x}_j)/(n-1),$$

are the estimated covariances for X_i and X_j, and the diagonal elements,

$$s_{ii} = s_i^2 = \sum_{k=1}^{n} (x_{ki} - \bar{x}_i)^2/(n-1),$$

are the estimated variances of X_i, with s_i the estimated standard deviation.

Many methods of multivariate analysis operate on a data matrix, $\mathbf{Y}$, derived from $\mathbf{X}$, to produce linear transformations of the variables $Y_1, Y_2, \ldots, Y_p$ that result in new variables $Z_1, Z_2, \ldots, Z_q$, where $q \leq p$. The $n \times q$ matrix $\mathbf{Z}$ is used to investigate features of the original data $\mathbf{X}$, often by using plots based on the first few columns of $\mathbf{Z}$.

The analytical process described above can be represented as

$$\mathbf{Z} \leftarrow \mathbf{Y} \leftarrow \mathbf{X}$$

or

$$\mathbf{Z} = \mathbf{YA}$$

so

$$
\begin{aligned}
Z_1 &= a_{11} Y_1 + a_{12} Y_2 + \cdots + a_{1p} Y_p, \\
Z_2 &= a_{21} Y_1 + a_{22} Y_2 + \cdots + a_{2p} Y_p, \\
&\;\;\vdots \\
Z_q &= a_{q1} Y_1 + a_{q2} Y_2 + \cdots + a_{qp} Y_p,
\end{aligned}
\tag{6.2}
$$

where $\mathbf{A}$ is a $p \times q$ matrix of the coefficients a_{ij}.

A typical element of $\mathbf{Y}$ is y_{ij}, and common choices for y_{ij} are *centered* variables of the form

$$y_{ij} = x_{ij} - \bar{x}_j \tag{6.3}$$

or *standardized* variables, having zero mean and unit variance, of the form

$$y_{ij} = (x_{ij} - \bar{x}_j)/s_j. \tag{6.4}$$

Sometimes the x_{ij} are *transformed*, by taking logarithms for example, before centering or standardization, and in later chapters a distinction – not always drawn in the literature – will be made between transformation and standardization. Issues of standardization and transformation are discussed more fully in Section 7.2.

Henceforth the term *data matrix* will be used for both $\mathbf{X}$ and the derived data matrix $\mathbf{Y}$, with the context making it clear which is intended. Estimated means and covariance matrices may be defined for $\mathbf{Y}$ and $\mathbf{Z}$. If the data are centered, as in equation (6.3), then $\mathbf{S}_X = \mathbf{S}_Y = \mathbf{S}$. If the data are standardized as in equation (6.4) then $\mathbf{S}_Y = \mathbf{R}$, where $\mathbf{R}$ is the correlation matrix of $\mathbf{X}$. Unless there is a need for emphasis, subscripting will be dropped in what follows and the notation **S** and **R** used.

The *trace* operator, tr($\cdot$), is the sum of the diagonal elements of a square matrix so that, for example,

$$\text{tr}(\mathbf{S}) = \sum_{i=1}^{p} s_i^2 = s_1^2 + s_2^2 + \cdots + s_p^2$$

can be interpreted as the total variance in the data matrix **X**. Since all the diagonal elements of **R** are equal to 1, it follows that tr(**R**) $= p$.

6.3 The singular value decomposition

A data matrix **Y** can be factorized, using the singular value decomposition, as

$$\mathbf{Y} = \mathbf{UDV}' \tag{6.5}$$

where **U** is $n \times p$, **V** is $p \times p$, $\mathbf{U}'\mathbf{U} = \mathbf{V}'\mathbf{V} = \mathbf{I}$, the $p \times p$ identity matrix, and **D** is a diagonal matrix of *singular values*, σ_i. Equation (6.5) can also be written as $\mathbf{Y} = \mathbf{ZA}'$, where $\mathbf{Z} = \mathbf{UD} = \mathbf{YV} = \mathbf{YA}$ and $\mathbf{A} = \mathbf{V}$.
Define $y_{ij} = (x_{ij} - \bar{x}_j)/(n-1)^{-1/2}$ so that

$$\mathbf{Y}'\mathbf{Y} = \mathbf{S} = \mathbf{VAV}', \tag{6.6}$$

where $\mathbf{\Lambda} = \mathbf{D}^2$ is diagonal, with diagonal elements $\lambda_i = \sigma_i^2$. It follows from equation (6.6) that $\mathbf{SV} = \mathbf{V\Lambda}$ so that the ith column of $\mathbf{V} = \mathbf{A}$ is the ith eigenvector of the estimated covariance matrix **S** and λ_i is the ith eigenvalue. It also follows that

$$\mathbf{S}_Z = \mathbf{Z}'\mathbf{Z} = \mathbf{DU}'\mathbf{UD} = \mathbf{D}^2 = \mathbf{\Lambda}.$$

This shows that the variables Z_i are uncorrelated with each other, and that the variance of Z_i is λ_i, where we assume the λ_i are ordered by size with λ_1 the largest. Using the result that, for any two matrices **B** and **C**, tr(**BC**) $=$ tr(**CB**) if the matrix products are well defined,

$$\text{tr}(\mathbf{S}) = \text{tr}(\mathbf{V\Lambda V}') = \text{tr}(\mathbf{V}'\mathbf{V\Lambda}) = \text{tr}(\mathbf{\Lambda}) = \text{tr}(\mathbf{S}_Z),$$

which shows that the total variance of **Z** is the same as that of **Y**. If $y_{ij} = (x_{ij} - \bar{x}_j)/s_i(n-1)^{-1/2}$ is used then **S** may be replaced by **R** in the foregoing development and $\sum_{i=1}^{p} \lambda_i = p$.

6.4 Measures of distance

6.4.1 Manhattan distance

Manhattan (or city-block) distance, defined as

$$d_{ij} = \sum_{k=1}^{p} |y_{ik} - y_{jk}|, \tag{6.7}$$

is a particular case of the Minkowski metric

$$\sum_{k=1}^{p} |y_{ik} - y_{jk}|^r \qquad (6.8)$$

for $r = 1$.

It is occasionally used as a measure of dissimilarity in archaeological applications of cluster analysis, and is related to the similarity coefficient used by Robinson (1951) in his seminal paper on seriation methodology (Chapter 16).

6.4.2 Euclidean distance

The squared Euclidean distance between two cases, where continuous data are assumed, is defined as

$$d_{ij}^2 = \sum_{k=1}^{p} (y_{ik} - y_{jk})^2 = (\mathbf{y}_i - \mathbf{y}_j)'(\mathbf{y}_i - \mathbf{y}_j) \qquad (6.9)$$

and is a special case of equation (6.8) when $r = 2$. The square root, d_{ij}, is Euclidean distance, also referred to as Pythagorean distance in some treatments (Gower and Hand, 1996). It is straightforward to show that, for the transformation $\mathbf{Z} = \mathbf{YV} = \mathbf{UD}$, discussed in the previous section,

$$(\mathbf{z}_i - \mathbf{z}_j)'(\mathbf{z}_i - \mathbf{z}_j) = (\mathbf{y}_i - \mathbf{y}_j)'(\mathbf{y}_i - \mathbf{y}_j)$$

so that the Euclidean distance between cases is preserved under this transformation.

6.4.3 Mahalanobis distance

Theory
Euclidean distance arises naturally in the context of principal component analysis‧ (Chapter 7), and other forms of distance arise in connection with other multivariate methods. To anticipate the material of Chapter 9, linear discriminant analysis is a method in which membership of G groups within the data is given and linear functions that separate the groups in some optimal fashion are defined. With G groups $q = G - 1$ discriminant functions can be defined so that for $G = 2$ there is one discriminant function which, once calculated, can be used to determine means, $\bar{z}_1$ and $\bar{z}_2$, that are the means of Z_1 for the two groups. Maximizing the Euclidean distance between the group means on the transformed scale gives rise to the result that

$$(\bar{z}_1 - \bar{z}_2) = d_{12}^2 = (\bar{\mathbf{x}}_1 - \bar{\mathbf{x}}_2)'\mathbf{S}_w^{-1}(\bar{\mathbf{x}}_1 - \bar{\mathbf{x}}_2), \qquad (6.10)$$

where

$$\mathbf{S}_w = [(n_1 - 1)\mathbf{S}_1 + (n_2 - 1)\mathbf{S}_2]/(n_1 + n_2 - 2) \qquad (6.11)$$

is a pooled estimate of the covariance matrices of the two groups, $\mathbf{S}_1$ and $\mathbf{S}_2$, and $\bar{\mathbf{x}}_1$ and $\bar{\mathbf{x}}_2$ are the means of the two groups. Equation (6.10) is the Mahalanobis distance (MD) between the means of the two groups that constitute the data matrix $\mathbf{X}$.

To test whether the two groups are significantly different, assuming that they are sampled from two populations that have multivariate normal distributions with equal covariance matrices, the statistic

$$\frac{(n_1 + n_2 - p - 1)n_1 n_2 d_{12}^2}{(n_1 + n_2 - 2)(n_1 + n_2)p} \sim F_{p,n-p-1}$$

may be used, where $F_{p,n-p-1}$ is the F distribution with p and $n - p - 1$ degrees of freedom (Glascock, 1992).

For two cases, i and j, within a single well-defined group, the MD between the cases is defined as

$$d_{ij}^2 = (\mathbf{x}_i - \mathbf{x}_j)'\mathbf{S}^{-1}(\mathbf{x}_i - \mathbf{x}_j) \tag{6.12}$$

where $\mathbf{S}$ is given by equation (6.1). If $s_{ij} = 0$ for all $i \neq j$, so that $\mathbf{S}$ is diagonal, the MD reduces to Euclidean distance for standardized data.

The MD between a case, $\mathbf{w}_i$, and the mean of a group, $\bar{\mathbf{x}}$ is

$$d_i^2 = (\mathbf{w}_i - \bar{\mathbf{x}})'\mathbf{S}^{-1}(\mathbf{w}_i - \bar{\mathbf{x}}) \tag{6.13}$$

where $\mathbf{w}_i$ may or may not be a member of the group.

Unlike Euclidean distance, MD allows for the fact that variables are correlated. One use of distance measures in archaeology is for the purpose of allocating isolated cases to one of several predefined groups on the basis of distance from the group. For example, in lead isotope studies (Section 2.2.1; Chapter 18), it is of interest to see if an artifact plausibly originates from an ore source on the basis of its isotopic composition (Sayre *et al.*, 2001). Distance calculations based on Euclidean distance can be misleading, as the example to follow will illustrate.

MD forms the basis of some of the more interesting approaches to clustering data that have been developed in archaeological (more specifically, archaeometric) study, and some results whose use will be described later are collected here. If $\mathbf{X}$ consists of a single group that is assumed to be a sample from a multivariate normal distribution and $\mathbf{w}_i$ is not a member of the group, it can be shown that

$$n d_i^2/(n + 1) \sim T^2(p, n - 1)$$

where T is Hotelling's T statistic with p and $n - 1$ degrees of freedom (Mardia *et al.*, 1979: 77). It then follows that

$$\frac{(n - p)n d_i^2}{(n^2 - 1)p} \sim F_{p,n-p}. \tag{6.14}$$

When n is large in relation to p then, approximately, $d_i^2 \sim p F_{p,n-p}$. Under these conditions it is also the case that $F_{p,n-p} \sim \chi_p^2/p$, where χ_p^2 is the chi-squared distribution with p degrees of freedom (Johnson and Kotz, 1970: 75) and it follows that $d_i^2 \sim \chi_p^2$ for large n. These results allow the MD for $\mathbf{w}_i$ to be associated with a probability which may be used to assess whether $\mathbf{w}_i$ can plausibly be regarded as a member of the group.

If $\mathbf{w}_i = \mathbf{X}_i$ is a member of $\mathbf{X}$ some modification to the probability calculations is needed because of the influence it has on the estimate of $\bar{\mathbf{x}}$, $\mathbf{S}$ and hence d_i^2. Let the *leave-one-out* estimate of MD, based on equation (6.13) after omitting $\mathbf{x}_i$ from all calculations, be $d_{(i)}^2$. It can then be shown that

$$\frac{(n - p - 1)(n - 1)d_{(i)}^2}{n(n - 2)p} \sim F_{p, n-p-1} \tag{6.15}$$

(Campbell, 1985; Penny, 1996). As

$$d_{(i)}^2 = \frac{(n - 2)n^2}{(n - 1)^3} \frac{d_i^2}{1 - nd_i^2/(n - 1)^2}, \tag{6.16}$$

computations can be based on calculations of MD for the full data set (Atkinson and Mulira, 1993; Leese and Main, 1994; Penny, 1996).

Two examples of the use of MD in archaeometric applications, where MD has been used most, now follow. A third example is given in Section 8.3.

Example 1: The use of confidence ellipsoids
One way in which MD has been exploited is in the construction of confidence ellipsoids in bivariate plots, used to assess the degree of separation or otherwise of different groups. Given the mean and covariance matrix of an $n \times 2$ data matrix $\mathbf{X}$, pairs of values for which $d_i^2 = c$, where c is a constant, define elliptical contours of constant probability. The chosen value of c depends on the probability level required and whether the χ^2 or F distribution is used.

Figure 6.1 shows a bivariate plot for two of the lead isotope ratio fields discussed in Chapter 2. The 90% confidence ellipsoids for the two fields are shown. These are based on the questionable assumption that the data for each field are sampled

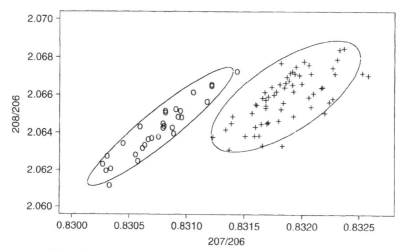

Figure 6.1 ^{208}Pb/^{206}Pb and ^{207}Pb/^{206}Pb ratios for the Seriphos (o) and Kea (+) lead isotope ratio data discussed in Chapter 2, with 90% confidence ellipsoids for each field.

from a bivariate normal distribution (Chapter 18); however, the plot is typical of those used in the literature to demonstrate field separation, and the fields are visually distinct. It can be seen that the case to the extreme right of the Seriphos field appears to be closer to the center of the Kea data than the Seriphos data. (Note that equal scaling of axes is not used here, in the interests of making the graph more readable, so some care is needed in making this judgment. Using equal scaling or exact calculation confirms the impression.) This extreme case is about three times closer to the centroid of the Kea field if Euclidean distance is used, but closer to the centroid of the Seriphos field using either d_i^2 or $d_{(i)}^2$.

This use of confidence ellipsoids to approximate regions occupied by a population from which a sample is drawn is now quite common in archaeology. Apart from the use in lead isotope studies illustrated here, the method is employed to highlight the similarity or differences between presumed groups in two-dimensional PCA plots (e.g., García-Heras *et al.*, 2001), and discriminant analysis plots (e.g., Attanasio *et al.*, 2000), as well as simple bivariate plots (e.g., Arnold *et al.*, 2000).

Example 2: Provenancing British stone axes
Jones and Williams-Thorpe (2001) develop an atypicality index, related to MD, for investigating whether two areas of South Wales were likely sources for 11 British prehistoric implements, mainly axes. Their development is based on equations (6.13) and (6.14), modified to allow for measurement errors in the components of $\mathbf{w}_i$ which is not a member of the source group to which it is compared. In their application

$$d_i^2 = (\mathbf{w}_i - \bar{\mathbf{x}})'(\mathbf{S} + k^{-1}\boldsymbol{\Sigma}_w)^{-1}(\mathbf{w}_i - \bar{\mathbf{x}}) \qquad (6.17)$$

where $\mathbf{w}_i$ is the *mean* of K independent measurements on an implement, the kth of which is associated with a diagonal matrix, $\boldsymbol{\Sigma}_{wk}$, whose diagonal elements are measurement error variances assumed to be known; and

$$\boldsymbol{\Sigma}_w = k^{-1}(\boldsymbol{\Sigma}_{w1} + \cdots + \boldsymbol{\Sigma}_{wK}).$$

Although the distribution theory associated with the use of (6.14) is no longer strictly valid, it is argued that it will be approximately so.

After careful selection of four of ten available elements, whose distribution, as far as can be judged, seemed to satisfy the assumption of multivariate normality, Jones and Williams-Thorpe (2001) used (6.17) to calculate atypicality probabilities for six dolerite implements related to one possible source, and five rhyolite implements in relation to a second source. They concluded that their results, which produced several very low atypicality probabilities that suggested the implements did not originate from the sources used, called into question previously accepted assignments.

An important feature of the use of MD in the paper was the incorporation of measurement error into the calculation of (6.17), through $\boldsymbol{\Sigma}_w$. Earlier work by Beier and Mommsen (1994), on the analysis of artifact compositional data obtained from neutron activation analysis, developed a similar modification. Beier and Mommsen (1994) based probability calculations on the chi-squared approximation given under equation (6.14), rather than equation (6.14) itself, and this is only strictly valid for large n (Jones and Williams-Thorpe, 2001: 12).

7
Principal component analysis and related methods

7.1 Introduction

Principal component analysis, as widely applied, is conceptually simple. The fundamental idea is that p correlated variables are linearly transformed to p new uncorrelated variables, as in equation (6.2), so that the properties of the data can be explored using simple graphical methods involving $r < p$ of the new variables, where typically $r = 2$ or 3. The mathematics of this is summarized in Section 7.2.1 below. An example, shown in Figure 7.2, is discussed in more detail later. It is based on a PCA of the standardized data of Table 2.2. The left-hand plot shows three groups in the data, corresponding to different regions, and suggests that pots from sites 1, (2, 3) and (4, 5) are chemically distinct. The outliers in the data blur the distinction between sites 1 and (2, 3).

PCA is scale-dependent, so that decisions concerning standardization and transformation of the data are important. These and other practical issues are discussed in the following sections. Although PCA is most often used to investigate the relationship between cases, it can be used to examine relationships between variables. Joint examination of the relationship between cases and variables involves the use of a biplot, and this approach, which 'caught on' in archaeology in the 1990s, is also examined.

After extracting components some practitioners choose to rotate them to enhance their interpretability. The issue of rotation of components is potentially confusing. Many early applications of factor analysis in archaeology were PCA with rotation. Not all commentators make a sharp distinction between PCA and factor analysis (Cowgill, 1977b: 129), but the view taken here is that they are distinct techniques that it is important not to confuse. I view PCA as a simple exploratory technique that can be readily applied and understood, and factor analysis as a more complex model-based method that has often been abused. The issues involved are outlined in Section 7.3.

The main use of PCA in practice is to obtain a low-dimensional representation of the data. Other techniques with the same aim include multidimensional scaling and projection pursuit methods, and discussions of these conclude the chapter. Correspondence analysis, which can be viewed as a form of PCA for discrete data, is considered in Chapter 11. General texts on multivariate analysis invariably include material on PCA. Two texts devoted to the topic are Jolliffe (2002) and Jackson (1991). Shennan (1997) and Baxter (1994a) provide treatments aimed at archaeologists.

7.2 Aspects of principal component analysis

7.2.1 The mathematics

It has been noted that PCA involves a linear transformation of variables, $Y_1, \ldots, Y_p$ to *uncorrelated* variables $Z_1, \ldots, Z_p$. The Z_i are defined so that Z_1 has maximum variance, Z_2 has second maximum variance and so on. To avoid indeterminacy a constraint has to be placed on the a_{ij} of equation (6.2) and

$$a_{i1}^2 + a_{i2}^2 + \cdots + a_{ip}^2 = 1 \tag{7.1}$$

is often used, though other constraints are possible.

Referring to Section 6.3, the a_{ij} are obtained from the eigenvectors defined by $\mathbf{V}$ in the singular value decomposition of $\mathbf{Y}$. The principal component *scores* are held in the columns of $\mathbf{Z}$. The eigenvectors and scores are scale-dependent, in that they depend on how the Y_i are obtained from the X_i, and in particular, if standardization has taken place or not, after any transformation.

There are p principal components. The result cited at the end of Section 6.4.2 shows that the Euclidean distance between cases is the same in the space defined by the principal components as in the original coordinate system. If only the first two or three components are used as the basis for plotting then, for an equally scaled plot, the observed distances between cases in the low-dimensional space approximate the Euclidean distances between cases in the full p-dimensional space.

There is no requirement that the original data have a multivariate normal distribution; indeed, PCA is often most useful when there are evident groups in the data, clearly violating the normality assumption. Where the distribution of variables is highly skew, or values differ by orders of magnitude, data transformation (often logarithmic) may be advantageous; however, this is because it makes the subsequent plots easier to read, and may downweight the visual impact of outliers, rather than because it transforms to normality. Some practitioners (e.g., Bieber *et al.*, 1976) assume that transformation will induce approximate multivariate normality, *within* subgroups of the full data set, but this requirement is desirable for subsequent analysis rather than PCA itself.

7.2.2 Transformation

Types of transformation
Some possibilities for transformation/standardization are:

1. $y_{ij} \leftarrow (x_{ij} - \bar{x}_j)$ – *centered* data;
2. $y_{ij} \leftarrow (x_{ij} - \bar{x}_j)/s_j$ – *standardized* data;
3. $y_{ij} \leftarrow \log x_{ij}$ – *logged* data;
4. $y_{ij} \leftarrow x_{ij}/x_{ip}$ for $j = 1, 2, \ldots, p - 1$ – *ratio* transformed data;
5. $y_{ij} \leftarrow \log(x_{ij}/x_{ip})$ for $j = 1, 2, \ldots, p - 1$ – *log-ratio* transformed data.

These can be combined, so that logged data may subsequently be standardized, for example.

A PCA will be dominated by variables with a large variance. If variables are measured in different units, standardization is usual. This scales variables to have equal variance so that they can potentially play an equal role in an analysis. The main exception to this generalization is in archaeometric studies, based on the chemical composition of a sample of artifacts, where variables are measured both in terms of percentage presence and in parts per million. In this case the use of logged data, to base 10, is common (Bieber *et al.*, 1976; Glascock, 1992). The reasoning is that the logged data will be of a similar order of magnitude with comparable, though not equal, variances.

If data are measured in the same units a wider choice is available. Use of centered data is subject to the problem that variables with large variances dominate, so that the use of standardized or logged data (if the data permit) is again usual. Some statisticians prefer the latter, on the grounds that standardization involves an arbitrary reweighting of the variables. After logging the data, a subset of variables with relatively large variances may still dominate the analysis. The use of centered unstandardized data will typically emphasize those variables exhibiting the largest difference in their absolute values, whereas logging the data will tend to emphasize those variables exhibiting the greatest difference in their relative values.

The use of ratios is less common, but can be motivated in various ways. If all the variables in an analysis are positively correlated then the first component will typically have an interpretation as a 'size' component. 'Size' can be interpreted literally in the case of analyses based on the measurement of artifact dimensions. If 'size' is not of interest one way to try and eliminate it is to divide $p - 1$ of the variables with a pth variable chosen as a general indicator of size (e.g., Madsen, 1988b).

Another context in which ratios are used is in artifact compositional studies where a *dilution effect* occurs. Adan-Bayewitz *et al.* (1999), in their study of pottery provenance in Roman Galilee, observe that varying amounts of calcareous material and quartz sand in the clays and tempers used in manufacture differentially dilute the abundance of most of the other elements measured by their instrumentation. They base statistical analysis on the ratios of the affected elements to one of the most precisely determined elements, Fe, so that the dilution effect is canceled out. Having taken ratios, they subsequently standardize these in their PCA. An alternative is to use log-ratio data, as Leese *et al.* (1989) do in their compositional analysis of medieval tiles.

The analysis of compositional data

Data for which the sum over the p variables is a constant (often 1 or 100%) will be described as *fully compositional*. Such data are common in archaeometric studies, when the chemical composition of artifacts is measured with respect to all the major and minor constituents (as well as trace elements). Data that can be viewed as a subset of fully compositional data will be described as *subcompositional*.

For fully compositional data, knowledge of any $p - 1$ components determines the final component, so that the data are $(p - 1)$-dimensional. This is most simply seen when $p = 3$, and what are variously referred to as ternary, triangular, tripolar or phase diagrams can be used to represent data exactly in two dimensions. This is illustrated in Figure 7.1 where data on the numbers of cores, blanks and tools from

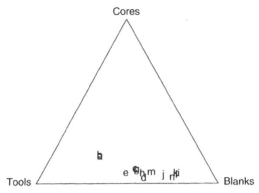

Figure 7.1 Ternary diagram representing the percentages of cores, blanks and tools from 14 levels from the Paleolithic site of Ksar Asil. Labels, a to n, are from earliest to latest levels. The data used are from Doran and Hodson (1975: 259).

the Paleolithic site of Ksar Asil (Doran and Hodson, 1975: 259) are represented as relative proportions in a ternary diagram. The perpendicular distance from a point to any two sides of the triangle is proportional to the proportions of the two types associated with the sides, and determines the distance, and hence proportion, for the third type. A side is associated with the type labeled on the apex opposite it. The labeling, a to n, is from the earliest to latest levels. For the earliest level the relative proportions of cores : blanks : tools are 17 : 26 : 57, while for the latest level they are 4 : 72 : 24. The figure suggests a decline in the proportion of cores over time and, more noticeably, a substantial increase in the proportion of blanks relative to tools.

The triangular plot of Figure 7.1 is an example of a simplex. For fully compositional data with $p = 4$, points would lie within a three-dimensional tetrahedron. Aitchison (1986, 2001) has argued that statistical procedures appropriate for an unconstrained space are inappropriate for the constrained space of the simplex. His solution is to transform the data using log-ratios, which gives rise to unconstrained data in $(p - 1)$-dimensional space for which 'standard' methodology is more appropriate.

Statistical analysis of fully compositional data in archaeology is most often undertaken using standardized data, or logged data to base 10, sometimes omitting the dominant variable. The arguments against this treatment are rehearsed, for an archaeological audience, in Aitchison *et al.* (2002: 296), the essence being that 'compositions provide information on relative rather than absolute values of the components of compositions, that relative values are characterized by ratios and that logarithms of ratios are simpler to handle mathematically and interpret statistically than ratios'.

Aitchison *et al.* also argue that the log-ratio approach is appropriate for subcompositional data. In terms of artifact compositional analysis this is equivalent to an analysis using logged data if only trace elements are used. The equivalence can be seen by assuming that variables 1 to $p - 1$ correspond to the trace elements, and equating the divisor x_{ip} in $\log(x_{ij}/x_{ip})$ with the sum of all other elements in a fully compositional analysis. Since x_{ip} is approximately 1 (in proportional terms) $\log(x_{ij}/x_{ip})$ is approximately equal to $\log x_{ij}$ (Aitchison *et al.*, 2002: 302).

Glass compositional data are often approximately constrained in the manner described above, often with one oxide, SiO_2, having the dominant presence at 60–70%. In this context Baxter (1989) suggested adopting Aitchison's methods. Later experience led to a waning of this enthusiasm (Baxter, 1993), and others also experienced problems with the method (Tangri and Wright, 1993). The particular form of log-ratio used was

$$\log[x_{ij}/g(\mathbf{x}_i)] \tag{7.2}$$

where $g(\mathbf{x}_i)$ is the geometric mean of values for the ith case. One problem is the undue influence that minor oxides (with a presence of less than 1%) can have on the analysis. In many cases two or three variables dominate an analysis, and these are typically those present at a low level and measured to just one leading digit (e.g., 0.01, 0.02). Results can be less interpretable than those obtained by using standardized data after omitting SiO_2. This measurement difficulty may have contributed to the unusual results sometimes obtained. Where comparisons have been made with ceramic data not suffering from these limitations it has been found that the use of log-ratios and logged data give similar results. For example, Buxeda's log-ratio analysis of ceramic compositional data using 14 variables can be reproduced almost exactly using logged data even though most variables are not trace elements (Baxter, 2001a). Aitchison *et al.* (2002) is a rebuttal of the arguments in Tangri and Wright (1993) against the use of log-ratio analysis. The theoretical reasons for preferring log-ratio analysis are well grounded. Doubts about the approach that have been expressed by archaeological users were, how- ever, founded on practical concerns because it was felt that 'standard' approaches produced more interpretable results.

Correspondence analysis
Correspondence analysis, treated at greater length in Chapter 11, has occasionally been used for the analysis of artifact compositional data (e.g., Underhill and Peisach, 1985; Bollong *et al.*, 1997). For fully compositional data Baxter *et al.* (1990) show that correspondence analysis of the raw data matrix will approximate a PCA of standardized data if $\bar{x}_j = \beta s_j^2$ holds approximately for some β, and this will sometimes be the case. Underhill and Peisach (1985) prefer to divide the x_{ij} by the variable means $\bar{x}_j$ before analysis and correspondence analysis then amounts to a PCA of

$$y_{ij} = (w_{ij} - \bar{w}_i)/\bar{w}_i^{1/2}, \tag{7.3}$$

where $w_{ij} = x_{ij}/\bar{x}_j$ and $\bar{w}_i = p^{-1} \sum_j w_{ij}$. Arguments are presented in Baxter *et al.* (1990) to show that this can be expected to produce similar results to a log-ratio analysis based on equation (7.2). If one wishes to undertake a log-ratio analysis with zero values in the data, for which equation (7.2) is not defined, a PCA based on equation (7.3) is an alternative method of analysis.

Summary
While some form of standardization and/or transformation is usually desirable, what is appropriate will depend on the aims of the analysis (e.g., whether 'size' is of interest) and the data structure (e.g., whether measurements are in the same units or fully compositional). Often a clear-cut choice will not be obvious and the use of more than one data treatment may be desirable. For instance, even if prior considerations suggest a preference for an unstandardized log-ratio or

logged analysis, problems may arise because of measurements that are close to zero (or zero) that are avoided using standardized data.

Baxter (1995) compared the performance of PCA using standardized data, standardized and unstandardized logged data, and ranked data, on 20 archaeometric data sets. It was found that analyses of standardized data, whether logged or not, and ranked data often gave rise to similar results except in the presence of outliers. Use of unstandardized logged data produced results that, while apparently different from other data treatments, often had a similar substantive interpretation. Baxter (1994b), also in an archaeometric context, compared the same four data treatments and the log-ratio transformation of equation (7.2), with and without subsequent standardization. One example was given where the unstandardized log-ratio analysis identified interpretable structure in the data less apparent in other methods.

7.2.3　Biplots

It was noted in Section 6.3 that the singular value decomposition of $Y = UDV'$ can be written as ZA', where $Z = UD = YV$ and $A = V$. More generally, Y may be factorized as $Y = GH'$, where $G = UD^\alpha$ and $H = VD^{1-\alpha}$ with $0 \le \alpha \le 1$.

The usual PCA score plot is based on the first two columns of G with $\alpha = 1$ and has the interpretation that distances between cases are two-dimensional approximations to the Euclidean distances in p-dimensional space. Occasionally plots are based on the first three columns of G (e.g., Hall, 2001). If $\alpha = 0$ the distances in the plot approximate Mahalanobis distance (see Section 6.4).

A graphical representation of the relationship between the variables can be based on the first two columns of H. If $\alpha = 1$ then the plot is of the coefficients of the first principal component against the second. Plots using $\alpha = 0$ have a particularly nice interpretation. The cosines of the angles between the vectors corresponding to different variables approximate the correlation between them, and the lengths of the vectors represent, approximately, the variances of the variables. Thus the plot can be viewed as an approximation to the covariance matrix of the data, or the correlation matrix if data are standardized.

A joint plot of both G and H constitutes a *biplot* and can be viewed as an approximation to the data matrix as a whole, including its covariance structure. Figure 7.2 shows an example arising from a PCA of the standardized data of Table 2.2. Here $\alpha = 1$ has been used, though the plot for $\alpha = 0$ is very similar. Cases are labeled according to the kiln site of origin, apart from three outliers, and the axes of the plot are scaled equally so that the plot can be read as a 'map'. Apart from the outliers, which obscure the picture a little, and one case from kiln site 1, it can be seen that there are three main groups in the data that correspond to the three regions in which the kilns are located. This example is characteristic of a great number of applications that have been published in the archaeological literature.

Several things can be inferred from the right-hand variable plot. It is based on the correlation matrix so, if the analysis is a good one, all variables have the same variance and the extremes of the vectors that represent them should lie approximately on a circle. This is so, with the exception of Ba which has a relatively short ray, so that most of the variables are reasonably well represented. Given this, it can be inferred, from the sharpness of the angles subtended by pairs of variables, that the subsets (Al, Ti), (Ca, Na, Fe) and (Mn, K, Mg) will tend to

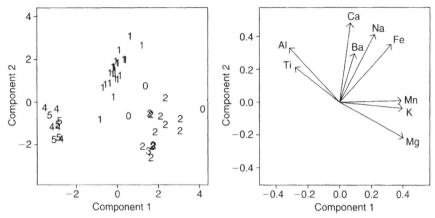

Figure 7.2 The left-hand figure is a plot of the scores based on the first two components from a principal component analysis of the standardized data of Table 2.2. Labels correspond to the kiln site of origin, except that outliers are labeled 0. The right-hand figure is a plot based on the coefficients of the leading two components.

exhibit positive correlations within the groups; the subsets (Al, Ti) and (Mn, K, Mg) will tend to be negatively correlated with each other; and both these groups will be weakly correlated with (Ca, Na, Fe). That these inferences are broadly true can be checked by direct inspection of the correlation matrix.

If one imagines the two plots to be superimposed with a common origin then it is also possible to infer something about the chemical composition of the different groups evident in the left-hand plot. For example, cases for kiln sites 2 and 3 lie in the same region of the plot as the markers for variables (Mn, K, Mg) and away from the markers for variables (Al, Ti), so can be expected to be relatively rich in the latter and poor in the former, while kiln sites 4 and 5 can be expected to have relatively low amounts of everything except (Al, Ti). Univariate analysis using multiple boxplots confirms these interpretations.

Baxter (1992) reviewed the use of biplots in archaeology and observed that they had been little used at that time. Neff (1994) discussed their use, under the heading of 'RQ-mode PCA' in the context of the analysis of ceramic compositional data. Since then the use of biplots in archaeological applications has noticeably increased (e.g., Arnold *et al.*, 2000; Lentfer *et al.*, 2002).

7.2.4 Choosing the number of components

In Section 6.3 the result that the variance of the principal components was equal to the variance of the data from which they were derived was given. Since the ith principal component has variance λ_i, and components are uncorrelated, this leads to the interpretation that

$$\frac{(\lambda_1 + \lambda_2 + \cdots + \lambda_r)}{\sum_{i=1}^{p} \lambda_i} \tag{7.4}$$

is the proportion of variation in **Y** 'explained' by the first r components. Equation (7.4) is commonly used as a measure of how well the first r principal components explain the variation in the data, and this begs the question of what constitutes a 'good' explanation.

If the data are centered then $\mathrm{tr}(\mathbf{S}_Y) = \mathrm{tr}(\mathbf{S})$. If data are standardized then $\mathbf{S}_Y = \mathbf{R}$, the correlation matrix of $\mathbf{X}$, and $\mathrm{tr}(\mathbf{S}_Y) = p$. For the example presented in Figure 7.2, where the data are standardized and have total variance 9, the first two components have variances of 3.70 and 2.34. Singly they account for 41% and 26% of the variance in the data, and cumulatively for 67%. The cumulative proportion of variance explained is used to measure the success of a PCA. Judging how successful a PCA is, or how many components are needed to approximate the data well, is often based on rules of thumb. Such rules may require that a certain percentage of the variance in the data be explained by the components, 70% or 80% for example. If the data are standardized then all components for which $\lambda_i = s_{Z_i}^2 > 1$ are sometimes considered to be important, though Jolliffe (2002: 115) suggests $\lambda_i > 0.7$ may be better.

For the data used to obtain Figure 7.2 the first three components account for 79% of the variance, all are associated with variance greater than 1, and no other component has a variance greater than 0.7. This suggests that the first three components provide a good approximation to the data. The main feature of plots involving the third component here is to highlight one of the outliers. For data sets such as that in Table 2.2 the first two components in a PCA typically explain 50–80% of the variance, with the lower figures usually associated with larger values of p. Even with explanation as low as 50% a plot based on the first two components will often be informative, though it is then advisable to inspect the higher-order components as well.

In many archaeological applications of PCA the question of how many components to extract, or examine, need not trouble the practitioner too much. If interpretation is based on plots of scores on the first few principal components it is easy enough to examine all possible pairwise plots, possibly labeled, without worrying about whether a 'significant' amount of variation is 'explained' by a given number of components. The question is more critical if one attempts to rotate the principal components, possibly in the guise of a factor analysis, to enhance interpretation, since the interpretation may be sensitive to the number of components extracted. This is discussed in more detail in the next two sections.

One device sometimes suggested for helping choose the important components is the scree plot. This is simply a plot of the variance explained by a component against its index. This is examined for 'elbows' – points at which the plot levels off – and the components before this are considered to be the important ones. For illustration the Italian stone axe data of O'Hare (1990), described in Section 2.2.3, will be used. The scree plot from a PCA of the standardized data is shown in Figure 7.3. There is a fairly clear elbow at 3, suggesting that two components are adequate to summarize the data. It must be remarked that it is frequently very difficult to detect an elbow in such plots, the pattern often resembling that of 'exponential decay'. In other circumstances more than one elbow may be apparent.

7.2.5   Rotation

In Section 6.3 the result that $\mathbf{Y} = \mathbf{Z}\mathbf{A}'$ was given, from which $\mathbf{Z} = \mathbf{Y}\mathbf{A}$, where $\mathbf{Z}$ is the $n \times p$ matrix of principal component scores and $\mathbf{A}$ is the $p \times p$ matrix of coefficients. The constraint (7.1) is assumed here. Score plots intended to reveal structure in the data are typically based on the first few columns of $\mathbf{Z}$. Though not an essential aspect of PCA, a desire to interpret the components (reification)

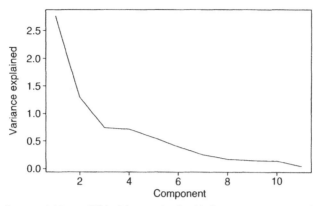

Figure 7.3 A scree plot from a PCA of the standardized Italian stone axe data of O'Hare (1990).

is sometimes manifested, and this is often not easy with the components as they stand.

Rotation of the principal components is sometimes used in an attempt to achieve this. Typically the first q columns of $\mathbf{A}$, say $\mathbf{A}_q$, are selected as being important. Without rotation, interpretation of the results is then often based on plots of the columns of $\mathbf{Z}_q = \mathbf{Y}\mathbf{A}_q$. Rotation involves post-multiplying $\mathbf{A}_q$ by a $q \times q$ matrix $\mathbf{T}$ to get a new coefficient matrix $\mathbf{B} = \mathbf{A}_q\mathbf{T}$ and new score matrix $\tilde{\mathbf{Z}}_q = \mathbf{Y}\mathbf{A}_q\mathbf{T}$. The rotation matrix $\mathbf{T}$ is chosen so that the structure of $\mathbf{B}$ is 'simple' in some sense and, it is hoped, more interpretable than $\mathbf{A}_q$. Commonly $\mathbf{T}$ is chosen to be an orthogonal matrix.

One immediate problem is that there are numerous criteria for simple structure that can be defined. In archaeology Kaiser's (1958) *varimax* criterion is by far the most widely used, almost certainly because it is the default in many of the software packages used for analysis. Mathematical definitions of this are given in Mardia *et al.* (1979: 269) and Krzanowski and Marriott (1995: 138). The essential idea is to produce rotated components in which only a few of the coefficients are 'large', and the rest are close to zero. Ideally a variable will have a large coefficient for only one component.

To illustrate the use of rotation and some potential limitations, the Italian stone axe data of O'Hare (1990) are used again. Figure 7.4 shows coefficient plots for these data after standardization. From the left-hand plot it can be inferred that the variables tend to be positively correlated, so that the first component has an obvious size interpretation. It can also be seen that the variables tend to split into two groups, one comprising the length and thickness measurements (L1, L2, L3, T1, T2, TH) and the other the breadth measurements (B1, B2, B3, WC) and the depth of cutting edge (DC). This last has a shorter ray than other variables, which suggests it is not well represented on the first two components, and the right-hand plot using the third component separates it out from the other two groups of variables.

The first four components account individually for 69.4%, 15.4%, 5.0% and 4.6% of the variation in the data, and cumulatively for 69%, 85%, 90% and 94% of the variation. The first two eigenvalues are in excess of 1, while the first four are in excess of 0.7. Taking these figures and the screeplot of Figure 7.3 into account,

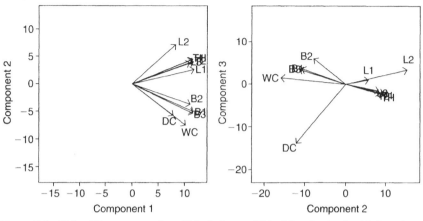

Figure 7.4   Plots of the component coefficients from a PCA of the standardized Italian stone axe data of O'Hare (1990).

Table 7.1   Component coefficients from various PCA analyses of the Italian stone axe data of O'Hare (1990), before and after varimax rotation. Rj indicates that the analysis is based on rotating the first j components. Coefficients are rounded to one decimal place and those of 0.1 or less in absolute value are left blank

	Method	B1	B2	B3	WC	TH	T1	T2	L1	L2	L3	DC
Component 1	PCA	0.3	0.3	0.3	0.3	0.3	0.3	0.3	0.3	0.2	0.3	0.2
	R2					0.4	0.4	0.4	0.4	0.5	0.4	
	R3					0.4	0.4	0.4	0.3	0.4	0.4	
	R4					0.6	0.6	0.5		−0.2	0.2	
Component 2	PCA	0.3	0.2	0.3	0.4	−0.3	−0.3	−0.2	−0.2	−0.4	−0.2	0.3
	R2	0.4	0.4	0.5	0.5					−0.2		0.4
	R3	0.5	0.5	0.5	0.5							
	R4	0.5	0.5	0.5	0.5							
Component 3	PCA	−0.2	−0.4	−0.2		0.2				−0.2		0.8
	R3		−0.2		0.2					−0.3		0.9
	R4											1

a case could be made for regarding two, three or four components as adequate to represent the data, according to the criterion used.

Table 7.1 shows the outcome from the original PCA and the results of rotating the first two, three and four components. Rotating two components separates the variables into the two groups evident from the left-hand plot of Figure 7.4, and these might be labeled length/thickness and breadth/depth of cutting-edge variables. Rotation of the first three components separates out depth of cutting edge from the breadth variables, giving the same interpretation that could be derived from Figure 7.4. Rotation of the first four components resulted in the first rotated component having an interpretation solely as a 'thickness' variable, while the fourth component was interpretable as a 'length' variable, being dominated by L2 with a coefficient of 0.8.

None of these results is 'wrong', but they are different and illustrate the point, which can be problematic, that results and interpretation may depend on the number of components one chooses to rotate. Rotation also destroys the perfectly good size interpretation that the first component in the PCA has, replacing it, in a sense, with separate size variables related to thickness, breadth and length when the four-component solution is rotated. Arguably rotation adds little to what can be deduced from the PCA in this example.

Other potential problems with rotation have been discussed in Jolliffe (1989, 1995). One is that the property that principal components successively have maximum variance no longer applies. Another is that the outcome of rotation depends on the normalization used for the coefficients, and substantively different interpretations can arise from different normalizations. In many implementations of PCA the normalization

$$a_{i1}^2 + a_{i2}^2 + \cdots + a_{ip}^2 = 1$$

is used, but

$$a_{i1}^2 + a_{i2}^2 + \cdots + a_{ip}^2 = \lambda_i$$

occurs in some software packages, with the interpretation that a_{ij} is the correlation between the ith component and jth variable if variables have been standardized. Jolliffe (1995) shows that, for the first of these normalizations, the rotated components will be correlated, and a consequence of this is that the appearance of plots based on the components will differ between the unrotated and rotated solutions, with possible differences in interpretation.

Rotation of principal components has been quite widely, but sometimes unwittingly, used in archaeology under the guise of factor analysis. The conscious description of an analysis as the rotation *of principal components* is rarer. The reason for wishing to rotate principle components is usually either not discussed at all or, if it is, is expressed as a desire to enhance interpretation of the components. What is meant by 'interpretation' is often not explained explicitly, though the suspicion often is that users want to assign some kind of 'meaning' to the rotated components. The merits of this are frequently unclear, since adequate interpretation is often available via biplots without the need for rotation.

7.3 Factor analysis

What was called 'factor analysis' was widely used, and misused, in the early days of quantification in archaeology. This was due, in part, to the influence of the paper by Binford and Binford (1966), discussed in Section 1.2.2. Although technical details were often omitted in early applications, many were really PCA with rotation. The relative merits of PCA and factor analysis were discussed in Doran and Hodson (1975). They were very much in favor of the former technique, though some commentators, such as Cowgill (1977b), thought they exaggerated the distinction between the two techniques. The term 'factor analysis' is sometimes used generically to encompass true factor analysis, PCA and correspondence analysis. Close inspection of applications described as factor analysis will sometimes reveal that what has actually been carried out is a PCA or correspondence analysis, and these last two techniques are now much more commonly used in archaeology than true factor analysis.

Modifying notation for the purposes of this section, in PCA the data, $\mathbf{Y}$, are transformed to $\mathbf{Z}_p = \mathbf{YA}$ and can be expressed as $\mathbf{Y} = \mathbf{Z}_p \mathbf{\Lambda}_p$, or

$$Y_1 = \lambda_{11} Z_1 + \lambda_{12} Z_2 + \cdots + \lambda_{1p} Z_p,$$
$$Y_2 = \lambda_{21} Z_1 + \lambda_{22} Z_2 + \cdots + \lambda_{2p} Z_p,$$
$$\vdots$$
$$Y_p = \lambda_{p1} Z_1 + \lambda_{p2} Z_2 + \cdots + \lambda_{pp} Z_p,$$

where $\lambda_{ij} = a_{ji}$ and

$$\mathbf{y} = \mathbf{\Lambda}_p \mathbf{z}, \tag{7.5}$$

in which $\mathbf{y}$ and $\mathbf{z}$ are $p \times 1$ vectors and the subscripts on $\mathbf{Z}$ and $\mathbf{\Lambda}$ emphasize the numbers of columns they have. All that is involved here is a mathematical transformation of the data.

In factor analysis the model

$$Y_1 = \lambda_{11} Z_1 + \lambda_{12} Z_2 + \cdots + \lambda_{1q} Z_q + \varepsilon_1,$$
$$Y_2 = \lambda_{21} Z_1 + \lambda_{22} Z_2 + \cdots + \lambda_{2q} Z_q + \varepsilon_2,$$
$$\vdots$$
$$Y_p = \lambda_{p1} Z_1 + \lambda_{p2} Z_2 + \cdots + \lambda_{pq} Z_q + \varepsilon_p$$

is assumed and the ε_i are 'error' terms. This model can be written as

$$\mathbf{y} = \mathbf{\Lambda}_q \mathbf{z} + \boldsymbol{\varepsilon}, \tag{7.6}$$

where $\mathbf{z}$ is now a $q \times 1$ vector and $\boldsymbol{\varepsilon}$ is a $p \times 1$ vector of errors.

Despite the superficial similarity of equations (7.5) and (7.6), there are important differences, the most fundamental of which is that PCA simply involves a mathematical transformation from p variables Y_i to p new variables Z_i having specified mathematical properties, whereas factor analysis assumes a model in which the observed Y_i are explained by their relationship to q underlying *latent variables* or *factors*, Z_i.

Another way of looking at the difference is that whereas PCA transforms to new variables that successively account for maximal *variance* in the data, factor analysis postulates a smaller number of latent variables that explain the *covariance* between the observed variables. Assume that the errors are uncorrelated with each other and have variance ψ_i; that the errors are uncorrelated with the factors Z_i; and that the factors are uncorrelated with each other. This implies that the covariance matrix of the observed variables is

$$\mathbf{\Sigma} = \mathbf{\Lambda}_q' \mathbf{\Lambda}_q + \mathbf{\Psi}, \tag{7.7}$$

where $\mathbf{\Psi}$ is a diagonal matrix with diagonal elements ψ_i and that their variances are of the form

$$\sigma_i^2 = \sum_{i=1}^{q} \lambda_{ij}^2 + \psi_i,$$

where the first term on the right-hand side is the communality of variable i, and ψ_i its unique or specific variance.

In applications of PCA the emphasis tends to be on the relationship between cases, whereas with factor analysis the emphasis tends to be on the relationship between the variables and interpretation of the coefficient vectors $(\lambda_{i1} \; \lambda_{i2} \cdots \lambda_{iq})$. Unfortunately these are not uniquely defined. For any suitably conformable orthogonal matrix $\mathbf{T}$ we have

$$\mathbf{y} = \mathbf{\Lambda}_q \mathbf{T}\mathbf{T}'\mathbf{z} + \boldsymbol{\varepsilon} = \mathbf{\Lambda}_q \mathbf{z} + \boldsymbol{\varepsilon},$$

which leads to the same covariance structure (7.7), so that any rotation, $\mathbf{\Lambda}_q \mathbf{T}$, of an estimate of λ_q is an equally valid solution of the model.

Texts such as Mardia *et al.* (1979), Seber (1984) and Jolliffe (2002) give details of the various estimation procedures available. Statisticians tend to prefer the method of maximum likelihood (Everitt and Dunn, 2001: 277) which assumes sampling from a multivariate normal distribution, is scale-invariant, and permits tests of whether or not q factors are adequate. There are few applications of this method in the archaeological literature that I know of. Principal factor analysis can be viewed as PCA applied to an estimate of the reduced covariance matrix $\mathbf{\Sigma} - \mathbf{\Psi}$ (Seber, 1984: 219). It will be equivalent to PCA if $\psi_i = 0$ for all i, and will be similar to PCA if the ψ_i are small.

Rotation of factors (components) was discussed in the previous section. Rotation is more naturally motivated in the context of factor analysis than PCA, since the former starts (or should start) from the assumption that there are indeed underlying and interpretable factors and that solutions of the form $\mathbf{\Lambda}_q \mathbf{T}$ and $\mathbf{\Lambda}_q$ are equally valid. This indeterminacy is also one of the problems with the idea of rotation. There are a large number of criteria that can be used to define the particular choice of $\mathbf{T}$, with no guarantee that they will lead to similar interpretations. This was seen in the example of the previous subsection where different, but interpretable, factors were obtained according to whether $\mathbf{T}$ was defined by varimax rotation, or taken to be the identity matrix, $\mathbf{I}$. A major practical problem with rotation is that the true number of factors, q, is usually not known and the rotated solutions can be sensitive to the choice of q. Statisticians are often skeptical about the value of factor analysis (e.g., Krzanowski, 1988: 503). Compared to the 1970s and 1980s, PCA, possibly with rotation and described as such, is now much more common in the archaeological literature than analyses described as 'factor analysis'.

7.4 Multidimensional scaling

7.4.1 The main ideas

Multidimensional scaling (MDS) is defined in a 'narrow' sense by Cox and Cox (2001: 1) as 'the search for a low-dimensional space, usually Euclidean', in which distances between cases in the space match, as closely as possible, a set of dissimilarities or *proximities* between cases. Different definitions of 'as closely as possible' lead to different scaling techniques.

In classical or metric scaling, let δ_{ij} be a measure of dissimilarity between cases i and j. A configuration is sought, equivalently a set of coordinate axes, such that $\hat{d}_{ij} \approx \delta_{ij}$, where $\hat{d}_{ij}$ is the Euclidean distance between cases in the configuration. In the particular case where $\delta_{ij} = d_{ij}$, the p-dimensional Euclidean distance between cases, an exact solution is possible, and the data may be visualized using plots based on the first two or three axes. This can be shown to be equivalent to PCA and, in the context of MDS and in some archaeological applications (e.g., Tubb *et al.*, 1980), is called *principal coordinate analysis.*

More generally, some model for the distances, $d_{ij} = f(\delta_{ij}) + \varepsilon_{ij}$, may be assumed, such as

$$d_{ij} = \alpha + \beta \delta_{ij} + \varepsilon_{ij}$$

where ε_{ij} is an error term, and some goodness-of-fit criterion such as

$$S^2 = \frac{\sum_{i<j}(d_{ij} - \hat{d}_{ij})^2}{\sum_{i<j} d_{ij}^2} \tag{7.8}$$

minimized, where S is generally known as the *stress*. Non-metric MDS methods are obtained when, rather than assuming some functional form $f(\cdot)$, one simply requires that the fitted values, $\hat{d}_{ij}$, preserve the rank ordering of the δ_{ij}, equation (7.8) being minimized subject to this constraint.

There was something of a vogue for MDS in the early days of quantitative archaeology, with some of the earliest papers to use multivariate analysis employing MDS (Doran and Hodson, 1966; Hodson *et al.*, 1966), and with several papers on MDS included in the seminal collection of Hodson *et al.* (1971). This latter book includes an expository article by Kruskal (1971), whose earlier work did much to lay the foundations for MDS. Full details of the methodology are available in Seber (1984) and Cox and Cox (2001). A common use of MDS in archaeology has been for the purposes of seriation, and this is discussed separately in Chapter 16.

7.4.2   Example: Analysis of ceramic thin-section data

Both PCA and correspondence analysis can be viewed as forms of MDS. More recent archaeological applications of MDS that do not fit within this framework are comparatively rare. To illustrate the methodology, the data of Table 7.2 will be used. This shows data for 25 samples of pottery from a ceramic assemblage in the cistern of the Punic and Roman site of Ses Paises de Cala d'Hort in Eivissa, also known as Can Sora (Cau, 1999). Nineteen categorical variables that reflect different aspects of the petrography of the specimens determined from thin-section analysis were recorded, with entries in the table showing which category for each variable a specimen is categorized into. The second column shows which of five types a specimen has been classified as, plutonic (x), phyllite (p), volcanic (v), muscovite (m) or Pantelleria ware (f), with 'loners' or outliers coded as 'o'. The aim of collecting these data was to investigate whether or not 'objective' statistical methods could recover the 'subjective' typology derived from traditional methods of petrographic classification. Full details, including definitions of the variables and categories used, are given in Cau *et al.* (2002).

There are $p = 19$ categorical variables. If variable k has L_k levels it may be recoded as L_k dummy or indicator variables, giving rise to an $n \times L_k$ matrix of 0s and 1s. Treating each variable in this fashion gives rise to an $n \times L$ data matrix, $\mathbf{G}$, where $L = \sum_{k=1}^{p} L_k$. PCA may be applied directly to $\mathbf{G}$ to investigate the relationship between cases (Gower and Hand, 1996: 67), and is equivalent to

Table 7.2 Categories for 19 variables used to define the petrographic characteristics of 25 samples of pottery from Can Sora (Eivissa). The second column indicates the petrographic classification, details of which are given in the text. Source: Cau *et al.* (2002)

Id	Type	1	2	3	4	5	6	7	8	9	10	11	12	13	14	15	16	17	18	19
CS-2	x	1	1	1	4	11	3	1	1	1	5	3	3	1	3	7	2	1	25	4
CS-3	x	1	1	1	4	11	3	1	1	1	5	3	3	1	3	7	2	6	26	4
CS-4	x	3	1	1	5	9	3	1	1	1	5	3	3	1	3	7	2	1	25	4
CS-5	x	1	2	1	4	11	3	1	18	1	5	3	3	1	3	7	2	6	27	4
CS-6	x	3	2	1	4	11	3	1	1	1	5	3	3	1	3	7	2	1	16	4
CS-7	o	3	1	1	3	9	2	1	4	1	6	3	3	1	3	7	2	1	16	4
CS-8	o	3	2	1	5	9	3	1	1	1	5	3	3	1	3	6	2	6	27	4
CS-9	v	3	1	1	4	11	1	2	1	29	5	2	3	1	1	7	2	6	20	4
CS-10	v	1	1	1	4	11	1	2	1	28	5	2	2	1	1	7	2	1	16	4
CS-11	v	1	1	1	4	11	1	2	1	22	5	2	3	1	2	7	2	1	16	4
CS-13	o	1	1	1	5	11	1	1	1	7	6	3	3	1	2	5	2	5	27	3
CS-14	x	1	1	1	4	11	3	1	1	1	5	3	3	1	3	7	2	1	27	4
CS-15	v	3	1	1	4	11	1	2	1	30	5	2	3	1	2	7	2	4	16	4
CS-16	v	1	1	1	5	11	1	2	1	29	5	2	3	1	2	7	2	4	16	4
CS-17	v	1	1	1	5	11	1	2	1	3	5	2	3	1	1	7	2	4	16	3
CS-18	m	1	1	1	5	1	2	1	5	1	6	3	3	1	1	5	2	4	28	4
CS-19	m	1	1	1	5	1	2	1	5	1	6	3	3	1	2	5	2	4	16	4
CS-20	m	1	1	1	5	1	2	1	5	1	6	3	3	1	1	5	2	4	29	4
CS-21	p	3	1	1	4	11	1	1	3	1	6	1	2	1	1	5	2	4	16	5
CS-22	p	3	2	1	4	11	1	1	3	1	6	1	2	1	1	5	2	4	16	5
CS-23	x	3	2	1	5	9	3	1	1	1	5	3	3	1	2	7	2	4	16	4
CS-24	o	1	1	1	4	9	3	1	1	1	5	3	3	1	2	7	2	4	16	4
CS-25	o	3	1	1	3	1	1	1	5	1	6	2	3	1	1	5	2	4	27	4
CS-26	f	3	1	1	4	9	1	5	1	1	5	3	3	6	1	5	2	4	30	3
CS-27	f	2	1	1	4	9	1	5	1	1	5	3	3	6	1	5	2	4	30	3

classical metric MDS applied to the dissimilarity matrix whose elements are the square roots of the elements of $\mathbf{P} - \mathbf{GG}'/p$, where $\mathbf{P}$ is an $n \times n$ matrix of 1s. This dissimilarity matrix may also be used as the starting point of a non-metric MDS analysis.

Figure 7.5 shows component plots based on a PCA of $\mathbf{G}$ that is equivalent to classical metric MDS. Cases are labeled according to their type, and the first three components account for 51.5% of the variation in the data. Apart from the fact that not all the petrographic 'loners' are clearly isolated, the analysis does a reasonable job of separating the different types, suggesting the possibility that this kind of analysis can reproduce groups defined more 'subjectively'.

In Figure 7.6 the same data are analyzed via non-metric isotonic MDS using the isoMDS function in S-Plus from the MASS library of Venables and Ripley (1999: 335). This leads to essentially the same conclusions as the previous analysis. A three-dimensional solution has been extracted, for which the stress given by equation (7.8) is 11.3%. Using the guidelines summarized in Everitt and Dunn (2001: 108) suggests that this is a fair fit. To get a good fit a five- or six-dimensional solution with a stress of about 5% is needed.

A fuller analysis of these data, and similar but larger and more complex data sets, is given in Cau *et al.* (2002). Among the other methods they explore is multiple correspondence analysis, which is illustrated using the Can Sora data in Section 11.4.3.

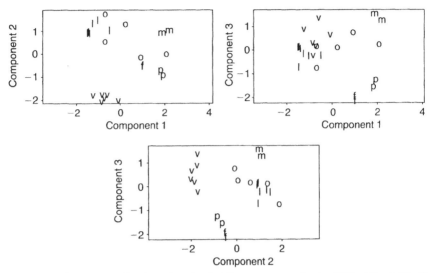

Figure 7.5 Component plots based on the first three components of a PCA of the matrix **G** derived from the Can Sora data of Table 7.2. This is equivalent to a metric MDS. Labeling of cases is given in the text.

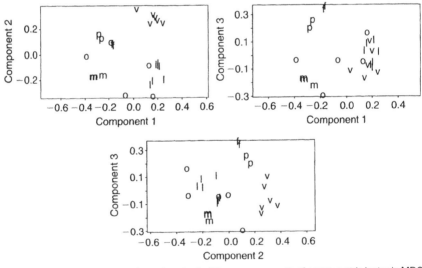

Figure 7.6 Component plots based on the first three components of a non-metric isotonic MDS analysis of the dissimilarity matrix based on the matrix **G** derived from the Can Sora data of Table 7.2. Labeling of cases is given in the text.

7.5 Projection pursuit

In PCA the components are defined to maximize variance, subject to their being uncorrelated. In most archaeological applications results are presented as plots based on the first few components, in the hope of revealing interesting structure in the data. *Projection pursuit* (PP) might be similarly described, except that linear combinations of variables are sought that optimize an index other than variance that attempts, more directly, to measure 'interesting' structure in the data. It has been claimed that PCA is 'something of a blunt instrument' for detecting interesting structure because large variation need not be interestingly structured variation (Jones and Sibson, 1987: 2).

An overview of projection pursuit is provided by Ripley (1996: 296–303). The central idea is to find k linear functions (projections) of the original variables that, when plotted, show interesting structure in the data. Usually $k = 1$ or $k = 2$ functions are sought, though $k = 3$ is possible (Nason, 1995; Glover and Hopke, 1992, 1994). Many approaches begin by equating *uninteresting* structure with the normal distribution and seek projections that are as non-normal as possible. This involves optimizing some index of 'interestingness'. Simonoff (1996: 117) has noted that any reasonable test statistic for normality is a candidate index.

A large number of different projection indices have been proposed (Krzanowski and Marriott, 1994: 94–6). Some of these take the form of a weighted distance between the probability density of the random variable(s) of interest and the normal distribution. Several of these are available in the XGobi package (Swayne *et al.*, 1998), and were investigated by Westwood and Baxter (2000) in an empirical study of the use of PP for finding structure in artifact compositional data. This is the most extensive investigation of PP methodology in the archaeological literature that I know of. Generally either PP did no better than PCA, or suggested additional clustering that was archaeologically uninterpretable. One problem with typical data is the sparsity (low n/p ratio) arising from the relatively small sample size. Cook *et al.* (1993: 244–5) provide examples to show that for sparse, high-dimensional data PP can suggest spurious structure in random data. One of their cautionary examples shows apparent structure in a 100×5 data set, generated randomly from a normal distribution. Many archaeometric data sets have smaller n and larger p, so that particular heed should be paid to their warning that 'exploratory projection pursuit will always find structure, albeit weak, but care must be taken when emphasizing the significance of that structure'. Westwood and Baxter (2000: 88) did not recommend PP for the *routine* examination of data.

One-dimensional PP has proved useful in the study of lead isotope data (Chapter 18). The final graph in Figure 18.1 is based on a linear combination of the ratios that produces the most non-normal projection of the data as measured by the Shapiro–Wilk test of normality. It can thus be viewed as an example of one-dimensional PP, though not originally interpreted in this manner (Baxter, 1999a).

8

Cluster analysis

8.1 Introduction

Cluster analysis is a generic term for a wide range of methods for discovering homogeneous groups or clusters in a set of data. Methods of cluster analysis are among the most widely used statistical techniques in archaeology, and both Shennan (1997) and Baxter (1994a) devote considerable space to them. Books devoted to cluster analysis include Gordon (1999) and Everitt et al. (2001).

Given n cases and p variables, many methods of cluster analysis are defined by specifying (a) a measure of (dis)similarity between pairs of cases and (b) an algorithm for grouping cases on the basis of their (dis)similarity, which typically involves specifying how the similarity between two clusters is to be measured. There are many ways of specifying (a) and (b) and hence many different ways in which a cluster analysis might be carried out.

Hierarchical agglomerative methods of cluster analysis begin with each case treated as a separate cluster, and cases are successively merged until only a single cluster remains. The clustering procedure may be conveniently represented in the form of a *dendrogram*, the appearance of which is often used to decide on the number of clusters or groups in the data. An example is shown in Figure 8.1, based on the standardized chemical compositional data of Table 2.2, where dissimilarity is measured by Euclidean distance, and the average-link method of cluster analysis is used. The dendrogram suggests that there are three main groups in the data and three outliers, cases 4, 35 and 36. The three groups correspond to three regions of origin, and that they are genuinely distinct can be confirmed by a PCA plot such as that on the left in Figure 7.2. This is an 'easy' data set to analyze, and interpretation is usually less straightforward.

Many archaeological applications of cluster analysis, particularly in the archaeometric literature, are very similar to this one, varying mainly in the choice of clustering algorithm. Similar considerations concerning data transformation and standardization to those that apply in using PCA arise (Section 7.2), and another common approach is to use unstandardized logged (to base 10) data (Bieber et al., 1976; Glascock, 1992).

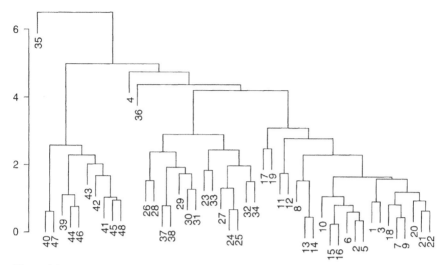

Figure 8.1 The dendrogram arising from an average-link cluster analysis of the standardized data of Table 2.2, using Euclidean distance as the measure of dissimilarity.

The average-link method of cluster analysis is widely used, not for any theoretical reason but simply because it frequently provides what are felt to be interpretable results, and is readily available. A competing approach is Ward's method. Suppose at a given stage of clustering there are $M < n$ clusters, and define the variability of the mth cluster as

$$S_m = \sum_i \sum_j (x_{ij,m} - \bar{x}_{j,m})^2,$$

where the additional subscript, m, indicates membership of cluster m. The total variability of the clustering is $T_M = \sum_{m=1}^{M} S_m$. In the clustering process the two clusters are amalgamated for which $T_{M-1} - T_M$ is minimized, after recalculating the S_m. The statistic T_M is a measure of how good the clustering is at any given stage, including that finally selected. It is not necessarily optimized in a hierarchical cluster analysis.

Relocation methods often take as their starting point a clustering derived from a hierarchical analysis and then attempt to further reduce T_M or some similar criterion, by moving cases between clusters. If changes occur, this results in a partition of the data not readily represented in the form of a dendrogram. More generally, partitioning methods may seek directly to optimize some measure of clustering success. Under the title *k-means analysis* such methods have been widely used to cluster spatial data in archaeology (Chapter 13). Archaeologists have been aware of k-means methodology at least since Hodson's (1970, 1971) advocacy of it, but outside the field of spatial archaeology it has been underused.

In the next sections the major clustering methods that have been used in archaeology, with some examples, are reviewed. A fuller discussion is available in Baxter (1994a: 140–4).

8.2 Hierarchical clustering methods

8.2.1 Hierarchical agglomerative clustering

Continuous data

Although the use of cluster analysis in archaeology is common, there is considerably more skepticism about what it can be expected to achieve than was the case in the 1960s and 1970s. In the field of artifact typology, for example, cluster analysis is now much less used than was the case, though it continues to be used routinely in other areas, such as provenance studies and related analyses of artifact compositional data.

Archaeologists are relatively unadventurous in their use of cluster analysis. There are, for example, few uses of more 'modern' model-based clustering methods. For continuous data the use of a hierarchical agglomerative method, such as average linkage (e.g., Schutkowski *et al.*, 1999), complete linkage (e.g., Mallory-Greenhough *et al.*, 1998) or Ward's method (e.g., Hall *et al.*, 1998) is common. Typically data are standardized and Euclidean distance or its square (often with Ward's method) used as the measure of dissimilarity, though see Chapdelaine *et al.* (2001) for a use of Manhattan distance. Single linkage is rarely used as the method of choice, and when it is used and produces clear clustering it is likely that the other methods will produce similar results. Baxter (1999c) illustrates its possible use for outlier detection.

The methods mentioned above differ in the way the (dis)similarity between two clusters is defined, but can be viewed within the same framework. Let i, j and k indicate three groups; let (ij) indicate the group formed by the fusion of groups i and j; and let d_{ij} be the 'distance' between groups i and j, not necessarily Euclidean distance. In the initial stages of clustering, i and j correspond to single cases and d_{ij} is the 'distance' between cases as defined by the dissimilarity coefficient used. A general formula for the distance between the clusters labeled by k and (ij) is (Lance and Williams, 1967)

$$d_{k(ij)} = \alpha_1 d_{ki} + \alpha_2 d_{kj} + \beta d_{ij} + \gamma |d_{ki} - d_{kj}|, \tag{8.1}$$

with different clustering methods corresponding to different choices of the α_i, β and γ.

If $\alpha_i = 1/2$, $\beta = 0$ and $\gamma = -1/2$ equation (8.1) gives

$$d_{k(ij)} = (d_{ki} + d_{kj} + |d_{ki} - d_{kj}|)/2.$$

If $d_{ki} > d_{kj}$ this evaluates to d_{kj} and if $d_{ki} < d_{kj}$ it evaluates to d_{ki}, so that overall it evaluates to $\min(d_{ki}, d_{kj})$. This defines the single-link or nearest neighbor cluster method in which the distance between the cluster indexed by k, C_k say, and that indexed by (ij) is the shortest distance between a case in C_k and a case in either C_i or C_j.

The complete-link, or furthest neighbor, method is obtained in a similar way with $\gamma = 1/2$, when equation (8.1) evaluates to $\max(d_{ki}, d_{kj})$. A feature of both the single- and complete-link methods is that the distance between two clusters is defined by just two cases, one from each cluster, and takes no account of cluster structure. This is commonly thought to be a disadvantage of the two methods.

In average linkage the distance between two clusters is taken to be the average distance over all possible pairs of cases, one from each cluster. In terms of equation (8.1), $\alpha_i = n_i/(n_i + n_j)$ where n_i is the number of cases in cluster i, and $\beta = \gamma = 0$. In Ward's method, discussed in the previous section, $\alpha_i = (n_k + n_i)/(n_k + n_i + n_j)$, $\beta = -n_k/(n_k + n_i + n_j)$ and $\gamma = 0$.

Other methods of hierarchical agglomerative clustering that are occasionally reported, such as centroid linkage (Buxedai Garrigós, 1999), can also be placed in this framework; however, average linkage, Ward's method and, to a lesser extent, complete linkage, tend to be the methods of choice. A subjective impression is that Ward's method is used less now than it was up to the mid-1990s. This may be because the package CLUSTAN, in which it is the default, is less widely used, or the impression may be incorrect.

Ward's method dendrograms tend to be clearer and 'easier' to interpret, and some users prefer the method for this reason. The drawback is that the method can suggest clear structure, even when the data are random (see Baxter, 1994a: 161, for an example), and the method can be poor at identifying even quite obvious outliers. Average-linkage dendrograms tend to be 'messier', but this may be a more realistic reflection of structure. One approach is to use Ward's method to get an initial idea of the number of clusters in the data and then see whether or not these can be recognized in an average-linkage dendrogram. Once clusters have been tentatively identified, labeled principal component plots, not necessarily just of the first two components, can be used to explore how separate clusters are. This exercise may benefit from removal of the more obvious outliers from an analysis.

Formal methods for identifying the number of clusters in a data set have been little used by archaeologists. Model-based methods (Section 8.4), which provide statistics for assessing the number of clusters, have also been little used. In studies of archaeological provenance it is sometimes possible to include reference material known to comprise a group in an analysis, and terminate clustering at the point when this group is broken up (e.g., Djingova and Kuleff, 1992). It is common to identify the number of clusters by subjectively 'cutting' the dendrogram at a single point, but may be better to cut it at several different joins (Digby and Kempton, 1987: 138), an approach that has been explored in the archaeological literature by Whallon (1990). Aldenderfer (1982) provides a discussion of cluster validation aimed at archaeologists.

Binary data

All the examples cited so far have involved the analysis of continuous data, often of artifact compositions. When data arise in binary (presence/absence) form the same clustering algorithms can be used, but a different coefficient of (dis)similarity is required. Many coefficients are based on Table 8.1. If there are p variables (attributes), for two cases i and j, a is the number of attributes that both cases possess; b is the number of attributes possessed by i but not j; and so on.

Table 8.1 For two cases, i and j, and p attributes, a is the number of attributes that the cases have in common; b is the number that i, but not j, has; etc.

	Present (1)	Absent (0)
Present (1)	a	b
Absent (0)	c	d

Wishart (1987) lists 25 possible coefficients based on this table, most of which have seen little or no use in archaeology.

Perhaps most common is Jaccard's coefficient, defined as

$$S = a/(a + b + c),$$

though the simple matching coefficient

$$S = (a + d)/(a + b + c + d)$$

is also sometimes used. These differ in whether or not mutual absence of a character is to be regarded as indicative of similarity. Some publications that have used the Jaccard coefficient, with various forms of hierarchical clustering, include classifications of rock art (Soggnes, 1987); of Aztec sculptures (Baquedano and Orton, 1990) based on stylistic or iconographic attributes; of sites, based on presence or absence of artifacts (Pyszczyk, 1989); and of insect assemblages (Perry *et al.*, 1985). Applications to the study of burial assemblages are noted in Section 8.2.3.

Mixed data
In order to combine data of mixed type (e.g., continuous, binary, categorical) in an analysis a coefficient of (dis)similarity needs to be defined. One such is Gower's (1971) coefficient. Ottaway (1981) seems to have been one of the first applications, and Rice and Saffer (1982), Phillip and Ottaway (1983), Palumbo (1987) and Howell and Kintigh (1996) are others, but sightings are comparatively rare.

Struyf *et al.* (1996) extend Gower's (1971) coefficient slightly, defining the dissimilarity between cases *i* and *j* as

$$d(i,j) = \frac{\sum_{f=1}^{p} \delta_{ij}^{(f)} d_{ij}^{(f)}}{\sum_{f=1}^{p} \delta_{ij}^{(f)}} \in [0, 1],$$

where $\delta_{ij}^{(f)}$ is the weight of variable f, and $d_{ij}^{(f)}$ is the contribution of variable f to $d(i,j)$, and depends on the variable type. Given a dissimilarity coefficient, various ordination or clustering techniques can be applied to investigate the relationship between cases.

Beardah *et al.* (2002) specialized to the case where variables were binary (mineralogical data) or continuous (chemical compositional data) in an investigation of methods for combining the two kinds of data. Assuming that there are no missing data, for continuous data

$$d_{ij}^{(f)} = |x_{if} - x_{jf}|/r_f$$

where r_f is the range of variable f, so that the contribution of the variable is between 0 (identical) and 1 (most different). Binary variables may be treated symmetrically or asymmetrically, depending on whether or not 0–0 matches are considered to contribute to the similarity between cases. In either case define

$$d_{ij}^{(f)} = \begin{cases} 0, & \text{if } x_{if} = x_{jf}, \\ 1, & \text{otherwise.} \end{cases}$$

The weight $\delta_{ij}^{(f)}$ is 1 unless a variable is asymmetric binary, in which case it is equal to 0.

Beardah *et al.* (2002) presented an example, based on data for 115 specimens of Late Roman cooking ware from the Balearic Islands and the eastern Iberian peninsula that had previously been analyzed in a variety of ways in Cau (1999) and Cau *et al.* (2002). A variety of multivariate analyses, including cluster analysis and ordination methods, revealed patterns in the data that were not isolated by analysis of the mineralogical or chemical data only.

A possible reason for the relative paucity of analyses of mixed-mode data in the archaeological literature, using Gower's coefficient or something similar, is the suspicion that too much weight is given to binary or categorical variables, relative to continuous data. This was investigated in Beardah *et al.* (2002) who used different weightings for the continuous data by, rather crudely, duplicating it $k \geq 2$ times to obtain a series of 'views' of the data for different weights. This highlighted, for $k = 2$, a cluster of cases more clearly than the unweighted analysis, but the general approach needs to be investigated more fully with other data sets.

8.2.2 Hierarchical divisive clustering

In contrast to agglomerative methods, hierarchical divisive methods initially view the sample as a single cluster that is successively subdivided into smaller clusters. Divisive methods have not been much used in archaeology, except in the context of burial studies, discussed in the subsection to follow. Monothetic divisive approaches have been most widely used.

In the monothetic divisive approach a case is characterized by the presence or absence of p types or attributes. At any given stage of clustering a cluster is split into two, on the basis of a single attribute, in order to increase the homogeneity of the clustering as much as possible. Different algorithms result from different definitions of homogeneity. One of the more favored approaches to divisive clustering uses the information statistic. Let n_i be the size of the ith cluster and let n_{ij} be the number of cases possessing attribute j. The information associated with the attribute is

$$n_i \log n_i - n_{ij} \log n_{ij} - (n_i - n_{ij}) \log(n_i - n_{ij}),$$

which is zero if all cases possess the attribute and at its maximum if half possess the attribute. Summing over j gives the total information in a cluster and summing over i gives

$$\sum_i pn_i \log n_i - \sum_i \sum_j [n_{ij} \log n_{ij} - (n_i - n_{ij}) \log(n_i - n_{ij})],$$

the total amount of information in the classification. This is a minimum of 0 when each cluster consists of an identical set of cases, so that at any stage of clustering the split is chosen that leads to the greatest reduction in total information. Other methods, such as association analysis (Whallon, 1972), have sometimes been used but have been found to perform poorly (Baxter, 1994a: 173).

8.2.3   Hierarchical clustering methods in burial studies

Parker-Pearson (1999: 12) observes that

> Cemeteries reveal much more than grave good variation and chronology, and
> may provide evidence about kinship, gender and other indicators of social
> status. Detailed analyses of these issues require the use of statistical methods
> on large samples of sometimes many hundreds of graves. The identification
> of intra-cemetery clusters often require statistical techniques such as clus-
> ter analysis, principal co-ordinate and principal component analysis, and
> significance tests.

Burials have been often been classified on the basis of presence/absence data
relating to the artifacts found in a burial, and possibly body position. Very simply
put, one idea is that the burials of individuals from different strata of society
will be characterized by different types of artifact, so that cluster analysis based
on the presence or absence of types will group the burials into different social
strata.

This idea has motivated the use of divisive clustering methods in burial studies,
possibly because the methodology has been seen by some as congruent with
archaeological theory and expectations. Studies that use this approach include
Peebles (1972), Tainter (1975), O'Shea (1984, 1985) and O'Shea and Zvelebil
(1984). It is noticeable from this list, and the review of cluster analysis in burial
studies in McHugh (1999: 64–73), that much of the relevant literature pre-dates the
1990s. McHugh (1999: 66) observes that an 'awareness of the problems involved
[in using cluster analysis] has reduced its popularity to some extent, in archaeology
at least'.

Earlier critiques, either of specific clustering approaches or cluster analysis in
general include Doran and Hodson (1975), Braun (1981) and Brown (1987). These
are reviewed briefly in Baxter (1994a: 172–3) and at greater length in McHugh
(1999: 66–73). Some of these focus on the relative merits of monothetic divisive
clustering and polythetic agglomerative clustering, in which cluster formation at
any stage is not determined by a single attribute. Studies of the latter kind, employing the Jaccard coefficient with some form of hierarchical clustering, include
Rothschild (1979) and Pearson *et al.* (1989). McHugh (1999: 69) takes the view
that the 'debate over the use of polythetic agglomerative and monothetic divisive
clustering methods is . . . linked very much with theoretical perspectives, rather
than the practical utility of the methods'. His analyses of a range of simulated
data sets, of known structure, lead him to conclude that multivariate statistical
techniques, including cluster analysis, often produce 'quite reasonable results',
but that it is impossible to make 'sweeping statements' about the effectiveness
of any particular method (McHugh, 1999: 140). Parker-Pearson (1999: 88) states
that 'however systematic, rigorous and objective the method of pattern sorting
and cluster recognition, the interpretation of these results is forever ambiguous
and open to question'. McHugh (1999: 144) advocates a pragmatic standpoint,
suggesting that the ease with which cluster analysis can be applied makes it a
sensible tool 'even as a preliminary investigation of the data'. He further suggests
that 'it does not provide a definitive answer, but rather is a pointer to potential
groupings in the data, which can be compared with the results from other methods,
or a manual consideration of the data'.

8.3 Relocation/partitioning methods

8.3.1 Main ideas

In a hierarchical clustering procedure, where an attempt is made to optimize some criterion of clustering success, relocation of cases between clusters in order to improve the value of the criterion seems natural. When Ward's method is used, for example, $T_M = \sum_{m=1}^{M} S_m$ is the measure of variability across all clusters and could be computed and plotted (possibly after transformation) for $T_k, k = 1, 2, \ldots$, in a scree plot to try and identify the true number of clusters. At any given stage of clustering relocation may reduce T_k to $T_k^* < T_k$, and plots based on T_k^* may be used to identify the true number of clusters (though this is often not obvious). This form of methodology often goes under the name of k-means clustering. The description above assumes that a hierarchical clustering is used to get an initial partition of the data, but random or purposively selected starts are equally possible. Because an analysis may result in a local rather than global minimum of T_k^* it is advisable to conduct analyses from different starting positions and desirable that the initial partition be a good one.

Hodson (1970, 1971) suggested that k-means analysis was a potentially useful tool for archaeologists, and in the latter paper reported on a pilot study involving an attempt to classify 488 British handaxes characterized by six variables. The k-means approach has also been used to cluster two- and three-dimensional spatial data, and an example is discussed later in this chapter, where it is viewed as a model-based method.

In provenancing studies using artifact compositional data, under the heading of group evaluative or refinement procedures, relocation methodology was pioneered at the Brookhaven National Laboratory (Bieber *et al.*, 1976), and further developed at the Missouri University Research Reactor (MURR) (Glascock, 1992; Neff, 1992). There is nothing specific about this methodology that requires the use of this particular type of data, and it could equally well be applied in other contexts. Typically, initial groups are defined using cluster analysis based on Euclidean distance, and a preliminary decision made about the number of groups in the data. If a group is sufficiently large, and assuming it is a sample from a multivariate normal distribution, the Mahalanobis distance of each case from the group centroid is calculated, using equation (6.16) or (6.13) according to whether or not a case is a member of the group. Probability calculations based on equation (6.15) or (6.14) can then be used to assess whether a case is a plausible member of the group or not. This is done for each group in turn and, if necessary, cases are reallocated to another group or declared to be outliers. The process can then be repeated until stability is achieved. The case study to follow provides a detailed illustration of this sort of use.

For similar problems, the method devised by Beier and Mommsen (1994) for grouping pottery by chemical composition can be viewed as a form of cluster analysis. Conceptually it is simple. Starting from a small group (possibly a single case), the distances of all other cases from the group centroid are calculated. Those close to it are added to the group and the process repeated. A group is 'grown' in this way until no more cases are added, at which point the process is repeated from a different starting point. The outcome is a classification in which all cases are either assigned to a group or declared to be chemical outliers. The technical details are more complex. Initial group formation is rather *ad hoc*, and may be based on inspection of dendrograms arising from more standard clustering procedures. In the early stages of cluster formation Beier and Mommsen (1994: 289) suggest using weighted Euclidean distance between cases as a measure of closeness, the weights being provided by the error variances of the measurements. The chi-squared distribution is suggested as a reference distribution to determine if cases are 'close', but this does not work well in the initial stages of cluster formation, and the groups may have to be forced to grow to a reasonable size before statistical criteria can be used with confidence to determine group membership. Once a group is sufficiently large a switch is made to using a modified Mahalanobis 'distance' to measure the distance of a case from a cluster. The modification takes the form of equation (6.17), for $k = 1$, where Σ_w is the $p \times p$ diagonal matrix of measurement errors for a case. The complexity of the procedure is such that it seems to have been little used outside of the Bonn laboratory where it was developed (e.g., Mommsen *et al.*, 2002).

8.3.2   A case study

As noted, statistics based on MD calculations have been used for what is called group evaluation or refinement, associated with some methodologies used in provenance studies (Glascock, 1992). A detailed illustration is provided in the investigation by Slane *et al.* (1994) of the chemical homogeneity and similarity of different pottery wares from Tel Anafa, Israel. Two of the wares, represented by 72 and 13 specimens respectively, were classified as Eastern Sigillata A (ESA) and a black-slipped predecessor (BSP). Previous research had raised questions about whether or not the ESA specimens fell into two chemical groups, and whether or not the BSP ware was chemically similar to the ESA ware. Preliminary multivariate analyses, based on logarithmically transformed measurements for 28 elements, suggested that, as a first approximation, the ESA wares could be treated as a single chemical group, so this was treated as a reference group.

The MDs of ESA cases from their group centroid, and associated probabilities, were calculated using both d_i^2 and $d_{(i)}^2$ (equations (6.13) and (6.16)). In 14/72 cases the difference in probabilities calculated from the two methods was greater than 0.6, and in 33/72 cases it was greater than 0.3 (Table 3 of Slane *et al.*, 1994). Using $d_{(i)}^2$, eight cases returned probabilities of less than 0.05, six of which were sufficiently small to suggest that the associated cases were not members of the ESA group. No unusual cases were suggested using d_i^2. These rather large differences are attributable to the sparsity of the data set, the n/p ratio of 2.6 being somewhat lower than Harbottle's (1976) recommendation of a minimum of 3, and preferably 5. In comparing cases in the BSP group to the ESA reference group, 4/13 probabilities based on d_i^2 were less than 0.02, three of which were effectively zero.

Repeating this analysis with BSP as the reference group was not directly possible, as there were fewer cases than variables, and the solution adopted was to work with variables defined by the first eight principal components of the data. Subsequent calculations suggested that only one BSP case was not plausibly a member of its group, and that all the ESA specimens could plausibly be members of the BSP group. Some caveats have to be entered here. Apart from the sparsity of the data, a similar analyis undertaken for the ESA group showed large differences in the probabilities based on $d_{(i)}^2$ when using all the variables, and when using the principal components. This suggests that calculations based on the principal components are not wholly to be trusted.

Slane *et al.* (1994: 58) ultimately concluded that ESA was a subset of BSP and could be treated as a single chemical group. It is, however, equally possible to conclude, on the basis of the analyses outlined earlier, that at least 10 specimens (six originally classified as ESA) cannot comfortably be regarded as members of a core ESA group. This points to one of the problems in using MD for group refinement in this sort of analysis, which is that, ideally, group sizes need to be large in relation to the number of variables used, in order to have confidence in the assignments. Confidence in any final judgment, in the present example, needs to be tempered by knowledge of the sparsity of the data, the small BSP sample size, and discrepancies between calculations based on the principal components and the original variables.

8.4 Model-based methods

8.4.1 Classification maximum likelihood

The group refinement procedures used at MURR (Glascock, 1992) and grouping procedures of Beier and Mommsen (1994) can be viewed as positing a statistical model from which the data are sampled. This is that there are G groups represented within the data and that, for cases within the gth of these,

$$\mathbf{y}_i \sim MVN(\boldsymbol{\mu}_g, \boldsymbol{\Sigma}_g),$$

where $MVN(\boldsymbol{\mu}_g, \boldsymbol{\Sigma}_g)$ is the multivariate normal distribution with mean $\boldsymbol{\mu}_g$ and covariance matrix $\boldsymbol{\Sigma}_g$. If n_g is the sample size for the gth group, so that $\sum_{g=1}^{G} n_g = n$, the full power of the methods is available when $n_g \gg p$ and stable MD-based probability calculations are possible. In archaeometric applications modern instrumentation has led to an increase over time in the values of p typically used, without there necessarily being a commensurate increase in the sample sizes, n, available. Thus, even the weaker requirement that $n_g > p$ may often not be satisfied.

Let $\boldsymbol{\gamma} = (\gamma_1\ \gamma_2 \cdots \gamma_n)'$ be a labeling vector for which $\gamma_i = g$ if a case is sampled from the gth population. The methods discussed above can be viewed as providing estimates of the labels, γ_i, and this is more directly the focus of attention in the classification maximum likelihood approach to clustering developed by Banfield and Raftery (1993). In this approach the log-likelihood to be maximized, expressed as a function of $\boldsymbol{\gamma}$, is

$$l(\boldsymbol{\gamma}) = \text{constant} - \frac{1}{2} \sum_{g=1}^{G} [\text{tr}(\mathbf{W}_g \boldsymbol{\Sigma}_g^{-1}) + n_g \log |\boldsymbol{\Sigma}_g|],$$

where

$$\mathbf{W}_g = \sum_{i \in E_g} (\mathbf{y}_i - \bar{\mathbf{y}}_g)(\mathbf{y}_i - \bar{\mathbf{y}}_g)'$$

and $E_g = \{i : \gamma_i = g\}$.

Usually some constraint needs to be placed on the form of Σ_g in order to obtain useful estimates, and different constraints lead to different clustering criteria. Thus, assuming $\Sigma_g = \sigma^2 \mathbf{I}$ for all g maximizes the log-likelihood by choosing $\mathbf{y}$ to minimize tr($\mathbf{W}$), where

$$\mathbf{W} = \sum_{g=1}^{G} \mathbf{W}_g.$$

This is the criterion that one is attempting to minimize when using Ward's method. It is based on the assumption that covariance matrices within the populations from which the sample is derived are equal and spherical, and explains the often observed phenomenon that clusters derived from a Ward's method analysis tend to be of similar size and spherical. One way of viewing this is that a common method of cluster analysis, often used in archaeology in an exploratory fashion, is a model-based method and based on modeling assumptions that may be inappropriate. The sphericity assumption implies that variables are uncorrelated within the populations that are sampled, and this is transparently not the case in many applications. Gordon (1999: 65–8) has observed that the imposition of spherical structure occurs with other supposedly 'model-free' methods, and explicit modeling of the covariance structure may be needed to avoid this if it is a concern (Krzanowski and Marriott, 1995: 89).

If it is assumed that the covariance matrices are equal but ellipsoidal, $\Sigma_g = \Sigma$ for all g, then Friedman and Rubin's (1967) clustering criterion that minimizes $|\mathbf{W}|$ is obtained. This will tend to produce clusters that are ellipsoidal with the same size, shape and orientation. The development in Banfield and Raftery (1993) permits some of these aspects of the clustering configuration to be varied. The covariance matrix may be parameterized in terms of its eigenvalue decomposition (see equation (6.6))

$$\Sigma_g = \mathbf{V}_g \Lambda_g \mathbf{V}_g'$$

where orientation is controlled by $\mathbf{V}_g$ and size and shape by Λ_g. These last two qualities may be separately controlled by writing $\Lambda_g = \lambda_g \mathbf{A}$, where λ_g is the largest eigenvalue of Σ_g and $\mathbf{A}$ is a diagonal matrix whose diagonal elements are in descending order with the largest equal to 1. With this formulation λ_g controls size and $\mathbf{A}$ controls shape. In the method called S^* in Banfield and Raftery (1993) ellipsoidal clusters with differing orientation and size but the same shape are obtained by fixing $\mathbf{A}$ but allowing other terms to vary. Increasingly restrictive forms of clustering are obtained by constraining the size and orientation parameters. An implementation of some of Banfield and Raftery's methods is available in S-Plus (Venables and Ripley, 1999) and analyses can be carried out using functions within the S-Plus library MCLUST.

8.4.2   Mixture maximum likelihood

The classification maximum likelihood approach to clustering assumes that the sample is selected from a mixture of multivariate normal distributions and focuses on estimation of the labels $\mathbf{y}$. An alternative way of proceeding starts with the mixture model

$$f(\mathbf{y}; \boldsymbol{\pi}, \boldsymbol{\Theta}) = \sum_{i=1}^{G} \pi_i g(\mathbf{y}; \boldsymbol{\mu}_i, \boldsymbol{\Sigma}_i), \tag{8.2}$$

where the $\pi_i > 0$ and $\sum_{i=1}^{G} \pi_i = 1$ are the mixing proportions, with $\pi = (\pi_1 \; \pi_2 \cdots \pi_G)$; $g(\mathbf{y}; \boldsymbol{\mu}_i, \boldsymbol{\Sigma}_i)$ is $MVN(\boldsymbol{\mu}_i, \boldsymbol{\Sigma}_i)$; and $\boldsymbol{\Theta}$ contains the elements of the $\boldsymbol{\mu}_i$ and $\boldsymbol{\Sigma}_i$. From this a system of equations for the maximum likelihood estimates of $\hat{\pi}_i$, $\hat{\boldsymbol{\mu}}_i$ and $\hat{\boldsymbol{\Sigma}}_i$ can be solved using an expectation–maximization (EM) algorithm (e.g., Krzanowski and Marriott, 1995: 84–5; Everitt *et al.*, 2001: 118–20).

Whereas the focus in the classification maximum likelihood approach is on estimation of the cluster labels, in the mixture model just described the focus is on estimation of the parameters of the model from which a probabilistic assignment of a case to a cluster can be obtained via

$$\hat{p}(i \mid \mathbf{y}_j) = \frac{\hat{\pi}_i g(\mathbf{y}_j; \hat{\boldsymbol{\mu}}_i, \hat{\boldsymbol{\Sigma}}_i)}{\sum_{k=1}^{G} \hat{\pi}_k g(\mathbf{y}_j; \hat{\boldsymbol{\mu}}_k, \hat{\boldsymbol{\Sigma}}_k)}.$$

The class for which $\hat{p}(i \mid \mathbf{y}_j)$ is a maximum can be taken to be that to which a case is assigned if a unique assignment is required.

As with the classification maximum likelihood approach, it may often be necessary to impose some form of constraint on the $\boldsymbol{\Sigma}_i$ to get useful results. A good starting position is also desirable in implementing the EM algorithm. The MCLUST library noted above allows estimation of the mixture model via EM under constraints on the covariance matrices similar to those for classification maximum likelihood described in Banfield and Raftery (1993). Fraley and Raftery (2000) advocate an approach in which classification maximum likelihood is used to provide a classification that initiates the EM algorithm in the mixture model, and this is illustrated in the example on spatial k-means clustering to follow.

A potential attraction of model-based clustering is that tests are available for the number of clusters, which Fraley and Raftery (2000) discuss in the context of model selection. They recommend use of the Bayes information criterion (BIC) which approximates $p(D \mid M_k)$, the probability of the data, D, given model M_k, by

$$2 \log p(D \mid M_k) \approx 2 \log p(D \mid \boldsymbol{\Theta}_k, M_k) - \nu_k \log(n) = BIC_k,$$

where $\boldsymbol{\Theta}_k$ is the parameter vector for the model of which ν_k are independent. Fraley and Raftery argue that this has given good results in a range of applications, and that for two models $k = 1$ and $k = 2$, if $BIC_1 - BIC_2 > 10$ this provides very strong evidence in favor of the first model.

Other model-based clustering methods that, in published applications at least, are dependent on the multivariate normality assumption include Bayesian approaches (Buck and Litton, 1991; Buck, 1993; Litton and Buck, 1996; Dellaportas, 1998). These add additional structure, in the form of prior assumptions, to the mixture model and are discussed in Section 14.3.2.

8.4.3 Archaeological applications

Other than special cases, such as Ward's method, the methods discussed here have had limited application to archaeological data. Undoubtedly software availability and the more formidable mathematics involved, compared to exploratory methods, has contributed to this. Other factors may also have militated against the wider use of model-based methods. For high-dimensional and sparse data sets, possibly consisting of several clusters, some with few cases, exploratory as opposed to model-based cluster analysis may be more informative. In a study of ceramic compositions involving 130 cases and 24 variables, Papageorgiou *et al.* (2001) found that the S^* method of Banfield and Raftery (1993) performed comparably to average-link cluster analysis and Ward's method, but only after removing 22 outliers identified using univariate methods and PCA.

The S^* method was also applied to the compositional data of Table 2.2 and failed to recover the rather obvious structure using standardized data, though it did so using log-transformed data. This suggests another potential weakness of model-based methods, which is that they may be sensitive to the modeling assumptions used, and it is not always obvious, without prior knowledge of the clusters, what transformation, if any, will best approximate multivariate normality. Application of the mixture model estimated via EM was unsuccessful, and could not deal with the fact that there were several quite small ($n_g < p$) clusters in the data.

Hall and Minyaev (2002) used the MCLUST software to investigate clustering in an 80×9 pottery compositional data set. They undertook PCA of log-transformed data and fitted a variety of multivariate normal models to the first five components, after deleting 13 outliers detected from inspection of the components. Their preferred model, as indicated by the BIC, favored a solution with three spherical clusters of varying volume. It is interesting that the second best solution was that provided by Ward's method.

It is difficult to draw any firm conclusions from these two studies about the merits of model-based clustering for artifact compositional data. It helps to screen the data for obvious outliers, and remove these before attempting the model fitting. The high dimensionality and relatively sparse nature of the data sets typically available mean that they are not obviously well suited to model-based analyses. That one of the simpler models, equivalent to Ward's method, performed comparably well in both the studies cited suggests that the evidence for the superiority of the more complex models over the clustering methods typically used in applications is weak or non-existent. More experience is needed with similar applications.

Model-based clustering is often applied successfully to low-dimensional data, and k-means spatial clustering is an archaeological application where model-based methods have obvious potential. Kintigh (1990) describes k-means spatial clustering, or 'pure locational clustering' in his preferred terminology, as

> a rigorous, numerical procedure that was intended to approximate the results of an intuitive division of a point distribution into a set of clusters. That is, its principal objective is *not* to provide an index for, or test of, spatial clustering, but to identify spatial clusters with their component points.

Kintigh and Ammerman (1982) introduced k-means spatial clustering as a heuristic approach that was concordant with archaeological aims and data. Further developments are reported in Simek (1984), Simek *et al.* (1985), Ammerman *et al.* (1987) and Koetje (1987), the last of these extending the method to data recorded in three dimensions. Reviews and comparisons with other techniques for intrasite spatial analysis are provided in Kintigh (1990) and Gregg *et al.* (1991). The method has been widely used (e.g., Blankholm, 1991; Savage, 1997b; Vaquero, 1999). It involves an attempt to optimize the criterion that governs a Ward's method cluster analysis and may thus be viewed as an application of classification maximum likelihood cluster analysis, assuming that clusters are circular and of equal size. This tendency to form circular clusters has been recognized as 'the most serious problem with pure locational clustering' (Kintigh, 1990: 190), and model-based

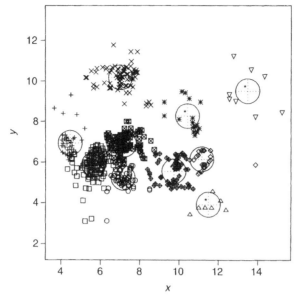

Figure 8.2   A model-based cluster analysis of the Mask Site data showing the 10-cluster result when clusters are constrained to be circular and of equal size.

clustering that relaxes the assumptions of equal size and circularity seems an obvious alternative to explore.

For illustration, the Mask Site data of Table 2.6, discussed in Section 2.2.6, will be used. For comparison with Kintigh (1990) all the data are used. The two-stage procedure suggested by Fraley and Raftery (2000) has been applied. In the first stage classification maximum likelihood is used. This is refined at a second stage using a mixture-model estimated with an EM algorithm. At either stage it is possible to impose constraints on the size, shape and orientation of the clusters.

The first analysis, resulting in Figure 8.2, emulates that reported in Kintigh's (1990) Figure 17, except that the 10-cluster solution suggested by the BIC rather than his eight-cluster solution is reported. The main difference from Kintigh's results is that his cluster 2 is split into the three clusters at the bottom right of Figure 8.2. This makes visual sense and is interpretable. The small cluster indicated with open triangles separates out 6/8 tools subsumed in Kintigh's larger cluster, and that above indicated with open diamonds consists of a group of bone splinters to the right of a known hearth. Other clusterings are almost as good as the 10-cluster solution; that with 18 clusters has a BIC value that differs by less than 1. At the 18-cluster level two distinct groups of wood scrap are separated out well in comparison to the 10-cluster solution.

Figure 8.3 shows an 11-cluster solution where the first-stage clusters are constrained to be circular and of equal size, and at the second stage elliptical clusters of the same size and shape but with different orientation are used. There are a number of differences in detail from the clustering shown in Figure 8.2, mainly occurring in the right-center of the figure. In Figure 8.3 the cluster identified by open diamonds consists almost entirely of bone scrap associated with a hearth, while that immediately beneath it mixes bone scrap, large bones and projectiles;

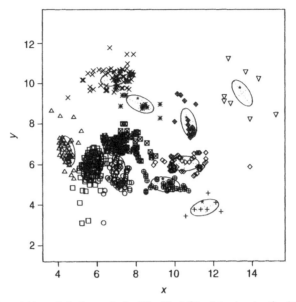

Figure 8.3 A model-based cluster analysis of the Mask Site data showing the 11-cluster result when clusters are constrained to be of the same shape and size.

this is a different partition than that shown in Figure 8.2 which divides the first of these clusters, amalgamating part of the result with the second. The cluster in Figure 8.3 identified by filled diamonds isolates, slightly more successfully than the equivalent cluster in Figure 8.2, a concentration of wood scrap. Overall the two analyses are not that dissimilar.

Other analyses are reported in Baxter (2002). An analysis using unconstrained clusters at both stages produced a seven-cluster solution, six of them small and dense, and one very large one covering most of the site. This was interpretable statistically, since it consisted of points in non-dense regions of the plot that could be viewed as 'outliers' not belonging to clearly defined dense clusters. Archaeologically, however, the cluster does not make sense, hence the constraints imposed to get Figure 8.3.

It is clear that model-based clustering does allow one to overcome the major problem associated with spatial k-means clustering as commonly applied in archaeology. It is equally evident that, in selecting a solution, criteria other than purely statistical ones may be needed for the results to make archaeological sense.

9

Discrimination and classification

9.1 Introduction

Discriminant analysis is another widely used multivariate method in archaeology. Graham's (1970) paper on discrimination between Paleolithic hand-axe groups is an early application. McKern and Munro (1959) is earlier, and used physical anthropological data to classify archaeological sites by phase but, as with other multivariate methods, use of discriminant analysis only 'took off' from the early 1970s.

The methods discussed in this chapter take, as their starting point, an $n \times p$ data matrix $\mathbf{X}$ divided into G known groups of size n_i for $i = 1, \ldots, G$, with $\sum_{i=1}^{G} n_i = n$. Questions that may be asked include 'are the G groups distinct with respect to the p variables that have been measured?', 'which variables best discriminate between groups?', and 'into which, if any, of the G groups should a new case be allocated on the basis of its p measurements?'. A characteristic archaeological problem would be to take a sample of pottery, classified by site of origin for example, and ask if the groups so defined can also be distinguished in terms of their chemical composition (e.g., Adan-Bayewitz et al., 1999). If this can be done successfully it may be possible to infer the site of origin for cases not in the sample, on the basis of their chemical composition. Such problems of discrimination or, more generally, classification and pattern recognition, have attracted a considerable statistical literature (e.g., McLachlan, 1992; Ripley, 1996; Hand, 1997). They are to be distinguished from problems of clustering, also sometimes called classification, where the prime aim of the analysis is to identify groups in the data.

Archaeological practice has been largely untouched by recent developments in the statistical literature. Fisher's linear discriminant analysis (LDA) is overwhelmingly the method of choice, with occasional forays into quadratic discriminant analysis (QDA). Logistic discriminant analysis has found a niche role in the predictive modeling of site locations but is not otherwise much used. Other methods, such as classification tree methodology and neural networks, have had little impact as yet.

9.2   Linear and quadratic discriminant analysis

9.2.1   Normal-theory LDA and QDA

Consider the problem of allocating a case **x** to one of G populations. Let π_g be the prior probability of membership of class g, with $f(\mathbf{x} \mid g)$ the distribution within class g assumed to be multivariate normal:

$$f(\mathbf{x} \mid g) = (2\pi)^{-p/2} |\mathbf{\Sigma}_g|^{-1/2} \exp -[(\mathbf{x} - \boldsymbol{\mu}_g)' \mathbf{\Sigma}_g^{-1} (\mathbf{x} - \boldsymbol{\mu}_g)]/2. \qquad (9.1)$$

Let $\bar{\mathbf{x}}_g$ and $\hat{\mathbf{\Sigma}}_g = (n-1)\mathbf{S}_g/n$ be the maximum likelihood estimates of $\boldsymbol{\mu}_g$ and $\mathbf{\Sigma}_g$, where $\mathbf{S}_g$ is defined as in equation (6.1) for class g. The posterior distribution of class g given **x** is

$$f(g \mid \mathbf{x}) \propto \pi_g f(\mathbf{x} \mid g), \qquad (9.2)$$

and the *Bayes rule* allocates **x** to the class for which this is greatest. On taking logarithms, this allocation rule is equivalent to minimizing

$$Q_i = d_i^2 + \log |\hat{\mathbf{\Sigma}}_g| - 2\log \pi_g \qquad (9.3)$$

where d_i^2 is Mahalanobis distance, defined as in equation (6.13) with respect to class g and maximum likelihood estimates of the parameters. Ripley (1996: 37) calls this the plug-in estimate of the best quadratic rule.

Equation (9.3) simplifies in various special cases.

1. If there are just two groups then the allocation rule assigns a case to class 1 if $Q_2 - Q_1 > 0$, or

$$d_2^2 - d_1^2 - \log \frac{|\hat{\mathbf{\Sigma}}_2|}{|\hat{\mathbf{\Sigma}}_1|} > 2\log \frac{\pi_2}{\pi_1}. \qquad (9.4)$$

Such a *quadratic discriminant function* does not necessarily work well in practice. It is sensitive to non-normality (Lachenbruch, 1975: 20), and requires all $n_i > p$ simply for estimation of the $\hat{\mathbf{\Sigma}}_i$ to be possible. Ideally $n_i \gg p$ is needed for stable estimation of the covariance matrices. This is an important consideration in applications.

2. Assuming that $\mathbf{\Sigma}_g = \mathbf{\Sigma}$, for all g, gives rise to a linear discriminant function that assigns a case to class 1 if

$$d_2^2 - d_1^2 > 2\log \frac{\pi_2}{\pi_1}$$

in the two-group case. That this is linear in **x** can be seen by writing the rule as

$$2\mathbf{x}' \hat{\mathbf{\Sigma}}^{-1} (\bar{\mathbf{x}}_1 - \bar{\mathbf{x}}_2) + \bar{\mathbf{x}}_2' \hat{\mathbf{\Sigma}}^{-1} \bar{\mathbf{x}}_2 - \bar{\mathbf{x}}_1' \hat{\mathbf{\Sigma}}^{-1} \bar{\mathbf{x}}_1 > 2\log \frac{\pi_2}{\pi_1}. \qquad (9.5)$$

This can be rewritten in the form

$$[\mathbf{x} - (\bar{\mathbf{x}}_1 + \bar{\mathbf{x}}_2)/2]' \hat{\mathbf{\Sigma}}^{-1} (\bar{\mathbf{x}}_1 - \bar{\mathbf{x}}_2) > 2\log \frac{\pi_2}{\pi_1}$$

which is given in, for example, Mardia *et al.* (1979: 309) and Everitt and Dunn (2001: 252–3).

3. When the probabilities of class membership are assumed equal, then terms involving the π_i in equation (9.4) and those derived from it become equal to zero. The two-group allocation rule discussed above is then the same as that used in Fisher's LDA and amounts to assigning a case to the class with the nearest centroid, as measured by Mahalanobis distance.

9.2.2 Fisher's LDA

The discussion in the previous subsection assumed a normal distribution for the data, within groups. It is quite common to read in the archaeological literature that discriminant analysis assumes normality. This is not necessarily so. The original approach developed by Fisher (1936) for two groups and generalized by Rao (1948) makes no such assumption, and is distribution-free, requiring only the assumption that class covariance matrices are equal.

Where confusion may have arisen is because, in a two-group analysis, Fisher's approach gives rise to the same discriminant function as a normal-theory approach assuming equal prior class probabilities. Thus, Fisher's approach is optimal if the data are normally distributed, and valid, though not necessarily optimal, if the data are not normal. If LDA is used primarily to demonstrate group separation graphically, then there is no need to invoke normality.

In Fisher's approach, extended to G groups and assuming equal population covariance matrices, define the within-group sample covariance matrix as the generalization of equation (6.11) to be

$$S_w = \sum_{i=1}^{G} (n_i - 1)S_i/(n - G),$$

where $n = \sum_{i=1}^{G} n_i$. Let the weighted between-groups covariance matrix be defined as

$$S_b = \sum_{i=1}^{G} n_i(\bar{x}_i - \bar{x})(\bar{x}_i - \bar{x})'/(G - 1). \tag{9.6}$$

Discriminant functions, or canonical variates, successively maximize the ratio

$$a'S_b a/a'S_w a,$$

and can be derived as the eigenvectors of

$$S_w^{-1}S_b.$$

Equivalent results are obtained if, as in some treatments, S_w and S_b are replaced by $W = (n - G)S_w$ and $B = (G - 1)S_b$. In other treatments an unweighted version of S_b is used (equivalent to replacing n_i with 1 in equation (9.6)). If $p > G$ there are $G - 1$ such functions, giving rise to a system of linear equations such as those of equation (6.2) with $q = (G - 1)$. For $G > 2$ it is common to present results in the form of a bivariate plot based on the scores for the first two functions, as in Figure 9.1.

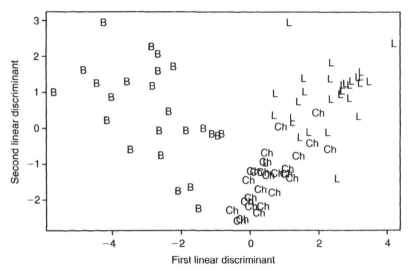

Figure 9.1 Discriminant analysis plot based on samples of steatite from three quarries. See Section 9.2.7 for quarry identifiers and fuller discussion.

The Mahalanobis distance of the ith case from the centroid of the jth group can be written as

$$d^2_{i(j)} = (\mathbf{x}_i - \bar{\mathbf{x}}_j)'\mathbf{S}_w^{-1}(\mathbf{x}_i - \bar{\mathbf{x}}_j) = -2[\bar{\mathbf{x}}_j'\mathbf{S}_w^{-1}\mathbf{x}_i - c_j] + c$$

for constants c_j, dependent on the group, and c. The term in square brackets, which is a linear function of $\mathbf{x}_i$, is sometimes called Fisher's linear discriminant function for the jth group. Allocating a case to the class whose centroid is closest in terms of Mahalanobis distance is equivalent to allocating it to the class for which Fisher's linear discriminant function is largest.

For the two-group case the difference between the two linear discriminant functions has the form

$$a_0 + a_1X_1 + a_2X_2 + \cdots + a_pX_p = a_0 + Q, \qquad (9.7)$$

which is an unstandardized linear discriminant function. A standardized (canonical) linear discriminant function, in which the constant term in equation (9.7) becomes zero, is obtained if scaled variables of the form

$$Y_i = (X_i - \bar{X}_i)/s_{wi}$$

are used, where s^2_{wi} is the ith diagonal element of $\mathbf{S}_w$.

9.2.3 Assessing the success of a classification

Commonly, the success or otherwise of a discriminant analysis is assessed informally by reference to a plot showing group separation on the first few discriminant functions (for $G > 2$). More formally, the success of a classification rule can be measured by estimating the *error rate*, which is the proportion of cases misclassified by the rule. The *actual* or *true* error rate can be thought of as 'the error rate which would be obtained if the classifier were applied to an infinite test set from the same distribution as the data set used in its construction' (Hand, 1997: 120). If e is the true error rate the problem in practice is to obtain a reasonably unbiased estimate of it.

A common estimate of e is the *resubstitution* or apparent error rate, e_R, which is the error rate observed when the classification rule is applied to the data set that was used to construct it. For small to moderate-sized data sets this will typically lead to a biased and optimistic estimate of the error rate, since the cases classified have influenced the rule used to classify them.

Another way of assessing the error rate is to divide the data into a training set, used to derive the classification rule, and an independently determined test set, used to 'validate' the rule. The success can be measured by the percentage of cases correctly classified in the test set. Ripley (1996: 67) notes that this is sometimes called the *hold-out* method. What appears to be the hold-out method has been used in a number of archaeological publications concerned with the predictive modeling of site locations (Parker, 1985: 192; Warren, 1990a: 109; Kvamme, 1990b: 279; Warren and Asch, 2000: 20) where it has sometimes been described as 'cross-validation'. Ripley (1996: 348) describes this as an abuse of the term, since cross-validation does not mean the use of a test set.

In *cross-validation* the data set is randomly divided into K subsets. Each subset is withheld from the analysis in turn and the cases within it classified on the basis of the rule derived from the remainder of the data. The error rate is obtained by averaging across the K analyses. This is K-fold cross-validation. In the 'leave-one-out' (LOO) approach each case in turn is withheld from the analysis, and then allocated on the basis of the rule derived from the rest of the data. The LOO approach is an extreme, and balanced, version of cross-validation, in which the subsets are of equal size, consisting of single cases (Ripley, 1996: 70). Cross-validation produces a less biased estimate of the true error rate than the resubstitution method. The LOO approach is increasingly reported in archaeological applications, though I am unaware of applications of the more general K-fold approach. Often the difference between the resubstitution and LOO methods will be small for large data sets. For smaller data sets quite large differences of up to 20% have been reported in the literature (Knutsson *et al.*, 1988; Baxter, 1994c).

The *jackknife* estimate of the error rate is also based on the successive withholding of cases from the analysis, but has a different rationale from the LOO method. Let e_{-iR} be the resubstitution error rate when the ith case is withheld from the analysis, with mean $\bar{e}_{R\cdot}$. The jackknife estimator of the true error rate, with bias of order n^{-1}, is then

$$e_J = e_R + (n-1)(e_R - \bar{e}_{R\cdot}).$$

An estimator with bias of order n^{-2} is

$$e_{J\cdot} = e_R + (n-1)(e_{R(\cdot)} - \bar{e}_{R\cdot})$$

where $e_{R(\cdot)}$ is the mean of $e_{(-i)R}$, the proportion of the *total* sample misclassified by a rule derived from the sample withholding the case i (Hand, 1997: 125). Hand shows that the relation of this to the estimated LOO error rate, e_L, is

$$e_{J\cdot} = e_L + (e_{R(\cdot)} - \bar{e}_{R\cdot})$$

where the bracketed term might often be expected to be small. Jackknifing is discussed in a more general context in Section 12.3.

In software packages error rates are typically calculated assuming that all cases belong to one of the classes represented in the sample, and this will not always be true. For continuous data, inspection of Mahalanobis distances from group centroids may suggest cases that belong to none of the groups. If one is prepared to invoke normality it is possible to convert the Mahalanobis distance estimates to probabilities to get absolute rather than relative estimates on which to assess likely group membership. Illustrations of this kind of use are given in Section 6.4.3 and Section 9.2.7 below, where examples are given of cases that are 'correctly' classified by the allocation rule but, on the basis of Mahalanobis distance, cannot really be considered to belong to any group.

Archaeologists have been aware of the benefits of cross-validation for some time (Benfer and Benfer, 1981). Baxter (1994a: 202) observed that it seemed not to have been used much up to the early 1990s, possibly because of lack of availability in software packages used by archaeologists, such as SPSS (e.g., Shott, 1997a: 92). Since then the situation has changed, and use of cross-validation, in the form of LOO methods, has become more noticeable (e.g., Truncer *et al.*, 1998; Chen *et al.*, 1999; Hall and Minyaev, 2002).

9.2.4   Two-group discriminant analysis and regression

It is well known that two-group discriminant analysis can be formulated in terms of regression analysis. This goes back to Fisher (1936) and is discussed in Flury and Riedwyl (1988), Ripley (1996) and Hand (1997). In the regression approach the dependent variable can be coded as 0 for one group and 1 for the other and a regression run on the X_i (other codings are permissible). This gives rise to a fitted regression of the form

$$\hat{\beta}_0 + \hat{\beta}_1 X_1 + \hat{\beta}_2 X_2 + \cdots + \hat{\beta}_p X_p = \hat{\beta}_0 + V$$

in which V is proportional to Q in equation (9.7). That is, up to a constant term, a regression analysis on dummy variables is proportional to the linear discriminant function.

The advantage of this formulation is that software available for regression analysis can be used for discriminant analysis, and this can provide greater flexibility. Baxter (1994c) exploits this by using variable selection procedures in MINITAB to show that different methods of variable selection can give rise to disjoint subsets of variables being selected. Another application is Lukesh and Howe (1978), who use regression analysis to discriminate between two types of Italian Bronze Age cups on the basis of dimensional measurements.

The ANOVA associated with testing for the significance of the regression is equivalent to testing whether there is significant separation between the groups, and the Mahalanobis distance between group centroids, d_{12}^2, can be expressed in terms of the coefficient of determination, R^2, as

$$d_{12}^2 = \frac{R^2}{1 - R^2} \frac{(n_1 + n_2)(n_1 + n_2 - 2)}{(n_1 n_2)}.$$

In archaeological applications the analogy between regression and discriminant analysis has been more often exploited in the context of logistic discrimination (Section 9.3).

9.2.5 Variable selection

Variable selection is quite common in reported applications, often based on automatic forward stepwise selection procedures available in the software package used by the researcher(s). The SPSS package, for example, offers five such procedures, described in Klecka (1980). Baxter (1994c) makes the general points that different selection procedures may select different, and different numbers of, variables, and that it is often possible to find subsets of variables of size similar to or smaller than that selected that perform as well or better in terms of the success of classification. This can happen particularly when there are high correlations among some of the variables in an analysis (Baxter, 1994a: 206–9). It is a mistake to assume, as is implied in some publications, that variable selection procedures lead to a selection of the most effective discriminating variables. The number of variables selected is also sensitive to the criterion used to determine whether further variables should be entered/deleted.

Baxter (1994c) discussed an example based on glass compositional data, where a cluster analysis divided the 87 specimens of Romano-British glass into two compositional groups. Ten variables were used and, taking the two groups as given, all five variable selection procedures in SPSS identified three of these (Fe, Cr, Sr) as the 'best' discriminating variables. Using a regression approach, this subset was also identified using forward and stepwise selection procedures in MINITAB. Using backward elimination, forced to continue until only three variables remained, resulted in a disjoint subset of variables, (Mg, Ca, Ti), selected as being 'best'. As judged by R^2, this latter subset was the best-fitting three-variable subset available. The former subset was not among the best five, differing by an R^2 of 1.1% from the best-fitting model.

Although the use of discriminant analysis to 'validate' the results of a cluster analysis is quite common, simply noting that the success rate is 'high' is not particularly informative, since this is to be expected given that the original cluster analysis is designed to produce separate groups. In the example just described the first subset successfully classified 86/87 and 85/87 cases for resubstitution and cross-validation respectively, while the second subset resulted in the successful classification of 86 cases using both approaches.

More usefully, a discriminant analysis can be used to identify cases that do not really belong to the clusters to which they are assigned. Pollard and Hatcher (1986) did this in their compositional study of Oriental Greenwares where, in one application, 53 specimens were grouped into three clusters; five cases with high values of Mahalanobis distance from their group centroids were removed as outliers; and then variable selection was used to identify five (of eight) variables as the best discriminators.

9.2.6 Quadratic discriminant analysis

The quadratic discriminant rule for the general case was given in equation (9.3) and for the two-group case in equation (9.4). Because of the requirement that $n_i \gg p$ for reliable estimation of the within-group covariance matrices to be possible,

QDA has not been widely used in archaeology, since this condition is frequently not satisfied.

Leese (1988) discussed the possibility of using QDA in the statistical treatment of stable isotope data, where p is small. In marble provenancing studies, for example, source quarries may often be characterized by the isotope ratios of just two variables, carbon and oxygen. Leese (1992) has also discussed the possibility of using QDA in the analysis of lead isotope ratio data, when $p = 3$ (see Chapter 18). Although the low dimensionality of these kinds of data means that, for reasonable sample sizes, QDA is feasible, the data are also amenable to graphical analysis and display, which may suffice for many purposes. Current thinking, in the analysis of lead isotope data at least, seems to be that discriminant analysis, quadratic or otherwise, is an unnecessarily complex tool for answering the questions that such data are collected to ask (Chapter 18).

Attanasio *et al.* (2000) contrast LDA and QDA in their study of marble provenance in which, in addition to the isotope ratios of oxygen and carbon, nine other variables derived from petrographic and electron spin resonance measurements were used. Discrimination between marble from three regions, with $n_i = 37, 46$ and 29, was attempted. Results using the full set of variables and a subset of six of these were reported, using both resubstitution and jackknifing to estimate the error rate. The most useful results were judged to be those using QDA with the six-variable subset, where the jackknifed error rate of 19% compared with 27% for LDA, and 21% for QDA using all variables (where concerns about overfitting because of the smallish n_i/p ratios were expressed).

Hall and Minyaev (2002) used LOO cross-validated QDA to contrast the success of two different clusterings of a 67×9 data set of pottery compositions, divided into three clusters. The error rates were 6% and 16%, and for the former preferred clustering the group sizes were $n_i = 22, 22$ and 23.

9.2.7   Examples

Steatite quarry sources
For illustration, a subset of the data used in Truncer *et al.* (1998) will be used to compare LDA, QDA and other methods discussed later in this chapter. The resubstitution, LOO and 10-fold cross-validation methods of assessing success will also be compared. Truncer *et al.* (1998) were interested in identifying the geological provenance of prehistoric steatite (soapstone) artifacts on the basis of chemical composition, a problem that had previously proved difficult. Their data consisted of measurements of 17 elements, determined using instrumental neutron activation analysis, on specimens from eight quarries. For the purposes of this section data from six of the quarries are used. These are Boyce Farm (B), Chula (Ch), Clifton (Cl), Lawrenceville (L), Orr (O) and Susquehanna (S).

In the first analyses only six variables will be used (Co, Cr, Fe, Mn, Sc, V). These are transition metals that were found to be useful discriminators and for which there are no missing data (Truncer *et al.*, 1998: 38). Logged data (to base 10) are used in all the analyses to follow.

Table 9.1 includes summaries of the results from applying LDA and QDA to these data, assuming equal prior probabilities of group membership. QDA with resubstitution suggests a 90% success rate; however, this drops sharply to 73% when LOO cross-validation is used. Using LOO cross-validation, QDA and LDA perform almost equally well. Results for 10-fold cross-validation are slightly

Table 9.1   Numbers of cases successfully classified for each of six quarries using different methods. Results for linear discriminant analysis (LDA), quadratic discriminant analysis (QDA), logistic discrimination (LD) and classification tree (CT) methods are reported. The resubstitution (RS), leave-one-out (LOO) and 10-fold cross-validation (CV10) approaches to assessing the success of a classification are compared, numbers and percentages referring to cases successfully classified

Method	B	Ch	Cl	L	O	S	All	%
LDA RS	16	24	19	25	17	20	121	76
LDA LOO	13	24	19	24	15	19	114	72
LDA CV10	12	23	18	23	15	19	110	69
QDA RS	21	26	24	28	20	24	143	90
QDA LOO	16	23	20	24	16	17	116	73
QDA CV10	17	22	19	23	16	17	114	72
LD RS	17	21	20	29	18	20	125	79
LD LOO	16	21	18	25	16	17	113	71
LD CV10	15	21	17	25	17	18	113	71
CT	15	21	16	30	17	21	120	75
n	24	26	26	31	25	27	159	

Table 9.2   Summary of the success of LDA for all possible pairwise analyses of quarries. Figures in the upper triangle show the percentage success of classification, as measured by resubstitution. Figures in the lower triangle show the percentage success as measured by LOO cross-validation

Quarry	B	Ch	Cl	L	O	S
B	–	98	92	100	86	90
Ch	98	–	94	90	98	96
Cl	90	90	–	95	92	87
L	96	84	93	–	96	93
O	82	94	88	96	–	94
S	88	89	85	90	90	–

worse than for the LOO approach. Figures for the number of cases successfully classified for each quarry suggest that Chula is the quarry best separated from the others.

A plot based on the first two discriminant functions, attempting to display quarry separation, is something of a mess as there is a lot of overlap between quarries. Table 9.2 summarizes the results from LDA applied to all possible pairs of quarries. It can be seen, using the cross-validated success rates, that the success rate is better than 80% in every analysis and is nearly perfect for Boyce Farm and Chula, for example. If a graphical demonstration of the success of classification is needed, analyses such as these can be used to identify subsets of the groups that can be expected to show reasonable separation. Figure 9.1 shows a discriminant analysis plot for the quarries Boyce Farm, Chula and Lawrenceville, which suggests that Boyce Farm can be successfully separated from the other two, which show some overlap.

A more sensitive analysis can be based on estimates of Mahalanobis distance. For example, for Boyce Farm using LOO calculations, one case has a Mahalanobis distance of 135.1 from the centroid, the next largest being 22.8. These give probabilities of 0.00 and 0.04 if multivariate normality is assumed. The extreme case is within the main body of cases labeled 'B', so the plot does not draw attention to it. Since the extreme case will affect the MD calculations for all other cases, it is worth seeing what happens when it removed entirely from the analysis. This results in the identification of another case with a probability of 0.01. If one chooses to remove this, no further cases are highlighted. The discriminant analysis plot provides some indication that group separation is possible, but it cannot be inferred from it that all cases within a group comfortably belong to it. Similar analyses for Chula and Lawrenceville suggest at least 2 and 5 cases respectively that are chemically distinct from the bulk of cases in the groups.

Sand-temper compositions
Heidke and Miksa (2000) undertook a study of the provenance of sand-tempered pottery in the Tonto Basin, central Arizona. In summary, 206 sand samples were collected and characterized by a point-counting method leading to an expression of the composition of a sample in terms of the percentages of 27 'parameters' (grain and lithic fragment types). With the aid of correspondence analysis, discussed in Section 11.4.4, sand-temper resource procurement zones or *petrofacies* were defined. Discriminant analysis was used to assess how well separated the petrofacies were, and the discriminant functions derived were used to provenance 154 pottery sherds on the basis of their sand-temper composition as determined by point-counting.

The analysis has several interesting features. One is that the data were log-ratio transformed, because of their fully compositional nature (Section 7.2.2). Zero values were replaced by 0.5. This use of discriminant analysis in conjunction with log-ratio data is uncommon in archaeology, Baxter's (1988) rather different study of the morphology of post-medieval wine bottles being another example. A 'nested' approach to discriminant analysis, described as 'somewhat novel' (Heidke and Miksa, 2000: 285), was then used. Initially the assignment of sands to either a mineral-rich group or rock-fragment-rich group of petrofacies was tested. This resulted in a success rate of 96.6%. Separate discriminant analyses were then undertaken within these two broad groups, using more specific petrofacies, eight in each case. For these analyses the success rate was about 84% in each case, resubstitution rather than cross-validation being used because of the unavailability of the latter. In the archaeological application the provenance (to petrofacies) of 154 sherds was inferred on the basis of means other than point-counting, and compared with the provenances assigned on the basis of point counts, using the discriminant functions determined from analysis of the sands. The level of agreement was 90.2%.

9.3 Logistic discrimination

Logistic discrimination can be motivated in several ways. For the two-group case, suppose one wishes to model $f(1 \mid \mathbf{x})$ and $f(2 \mid \mathbf{x}) = 1 - f(1 \mid \mathbf{x})$ directly. A linear model of the form $f(1 \mid \mathbf{x}_i) = \mathbf{x}'_i \boldsymbol{\beta}$ is possible, but unappealing in that predictions outside the range $[0, 1]$ are possible (Hand, 1997: 42). To avoid this a possible model is based on the logistic transform

$$\pi_i = f(1 \mid \mathbf{x}_i) = \frac{\exp(\mathbf{x}_i'\boldsymbol{\beta})}{1 + \exp(\mathbf{x}_i'\boldsymbol{\beta})}.$$

This is the logistic regression model of equation (5.13), which can be linearized, using the logit transformation

$$\log\left(\frac{\pi_i}{1 + \pi_i}\right) = \mathbf{x}_i'\boldsymbol{\beta}.$$

Another way of motivating logistic discriminant analysis is to note that, from equation (9.2), and assuming two groups with equal population covariance matrices and the multivariate normal density of equation (9.1), the logit transformation gives rise to a linear discriminant function of the form (9.5) (Hand, 1997: 43). When the normality and equal covariance matrix assumptions hold, logistic regression and LDA might be expected to give similar results. When the assumptions do not hold, rather different classification rules may be obtained (Ripley, 1996: 44).

In applications of predictive modeling in archaeology, where the aim is to discriminate between archaeological sites and 'non-sites' on the basis of environmental and other variables, logistic discrimination has been regarded as a more robust procedure than LDA. Parker (1985: 201), for example, suggests that 'archaeological data for predicting site locations will probably not have multivariate normal distributions with equal covariance matrices'. Indeed, the variables used for prediction may often be categorical or binary, and while LDA may work perfectly well with such data the assumption of normality will clearly be invalid. An example was given in Section 5.4.2, and a general discussion is provided in Section 13.3.

The extension of two-group logistic discrimination to several groups is discussed in Ripley (1996: 109). For G groups, and assuming normality and equal covariance matrices, a relationship of the form

$$\log f(g \mid \mathbf{x}) = \log f(1 \mid \mathbf{x}) + \alpha_g + \mathbf{x}'\boldsymbol{\beta}_g$$

can be derived, where group 1 is arbitrarily chosen as the 'base' group for comparison. The $f(g \mid \mathbf{x})$ can be modeled as

$$f(g \mid \mathbf{x}) = \frac{\exp(\alpha_g + \mathbf{x}'\boldsymbol{\beta}_g)}{\sum_j \exp(\alpha_j + \mathbf{x}'\boldsymbol{\beta}_j)} \tag{9.8}$$

and estimated by maximum likelihood.

Ripley (1996) and Venables and Ripley (1999) use, as a running example, a data set in which 214 fragments of glass, divided into six groups, have their chemical composition measured with respect to their refractive index and the percentage weight of eight oxides. The refractive index apart, this data set is identical in structure to the steatite compositional data described in Section 9.2.7 (and many similar compositional data sets that have been subjected to discriminant analysis). Venables and Ripley (1999: 360–5) subject their data set to a variety of classification methods, including logistic discrimination, and their routines have been used in S-Plus to analyze the steatite data.

Table 9.1 summarizes the results of using logistic discriminant analysis to analyze the steatite data of Truncer *et al.* (1998). It can be seen that, using LOO cross-validation, logistic discrimination performs comparably with LDA and QDA. Judged by 10-fold cross-validation, logistic discrimination and QDA are similar, and outperform LDA.

9.4 Classification trees

Assume that a classification exists for a set of data. Classification trees may be used as a type of variable selection procedure (Venables and Ripley, 1999: 303) to identify which variables best discriminate between the groups in the classification. Discussions of the methodology are given in Breiman *et al.* (1984), Clark and Pregibon (1992), Ripley (1996) and Venables and Ripley (1999). Classification trees have been used very little in archaeology. Kuttruff (1993) used the CART methodology of Breiman *et al.* (1984) but gave few technical details. The following account and example is based on Baxter and Jackson (2001).

Figure 9.2 shows a classification tree for the steatite source data used earlier in the chapter. Six of 17 elements for which there are no missing values have been used (Co, Cr, Fe, Mn, Sc, V). The idea behind tree construction is very simple. At each stage a binary split on one variable is chosen that most improves the purity of the tree as a whole. The tree is grown until all terminal nodes are pure or until some preset minimum node size is reached. Algorithmic details differ, for example, in how the purity of a tree is defined. The rpart library of Therneau and Atkinson (1997) with the S-Plus package (Venables and Ripley, 1999) was used here.

If the sample of size n is partitioned into G classes with n_i cases in class i, the *root node* is defined by G probabilities $(p_1, p_2, \ldots, p_G)$ where $p_i = n_i/n$. A node

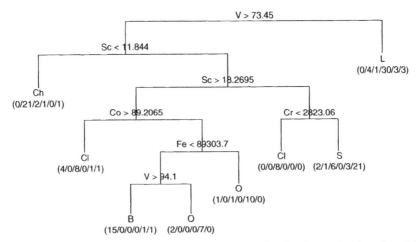

Figure 9.2 A classification tree for steatite compositional data for six quarries, based on data from Truncer *et al.* (1998). Six variables, for which full information is available, have been used and the tree has been pruned.

is labeled according to the most dominant class. The misclassification rate at a node is the number of cases not in the dominant class. For the root node call this $R(T_0)$. The root node is split to form two nodes, for each of which probabilities and misclassification rates can be defined after labeling the node. Thus, in the example, the first split is made on $V > 73.45$, cases for which this is true going to the left. Splitting continues until terminal nodes are reached. These will either contain cases from a single group or have reached some minimum allowable size. For the analysis here no attempt has been made to split a node with fewer than 20 cases and a terminal node must contain at least 7 cases. There is one pure terminal node, associated with Cl to the right.

The *impurity* of a node can be defined in different ways; in the example the *Gini index*

$$1 - \sum_i p_i^2$$

was used. For a pure node, with all cases in a single class, this takes the value zero, and the smaller the value of the index the purer the node. At the root node there are n cases and any continuous variable, X say, can be ordered as $(x_1, x_2, \ldots, x_n)$. For $i = 2, \ldots, n$, splits may be contemplated such that cases with values less than x_i are separated out from cases with values greater than or equal to x_i. New nodes are defined by the two subsets of cases thus defined, and the impurity of these, and hence their average impurity, measured. All possible splits for all variables are examined and the split that most reduces the average impurity is chosen. Each new node is treated in a similar way and the process continues until terminal nodes are reached.

This approach may 'overfit' the data and give an over-optimistic assessment of the success of the classification. The tree in Figure 9.2 originally had 11 terminal nodes. *Cost–complexity pruning* is used to reduce the size of the tree. Let $R(T)$ be the number of misclassifications in a tree. Let $|T|$ be the 'size' of a tree measured by the number of terminal nodes. A cost–complexity measure of the form

$$R_{cp}(T) = R(T) + cp|T|R(T_0)$$

is defined, where cp is a complexity parameter. For fixed cp, as leaves are pruned, $|T|$ will decrease and $R(T)$ will increase. The degree of pruning is chosen to minimize $R_{cp}(T)$. As cp increases the size of the pruned tree will tend to decrease.

To choose cp, cross-validation is used. The data are split into 10 groups of roughly equal size. Nine of the groups are used to grow a tree and it is tested out on the remaining group to obtain a measure of the error involved. This can be done in 10 ways and the results averaged to get an estimate of the error and its standard deviation. Results may be summarized graphically, as in Figure 9.3 based on the unpruned tree which, after pruning, led to Figure 9.2. Plotted points show the mean number of errors across the 10 analyses divided by $R(T_0)$ (i.e., the vertical axis is scaled to lie between 0 and 1), with the bars showing the estimated standard error of the mean. The horizontal reference line is located at the mean error for the largest tree plus one standard error. The tree selected is the smallest one whose mean error lies below and within one standard error of the line. In this case this rule results in the unpruned tree of size 11 being pruned to one of size 8. Therneau and Atkinson (1997: 13) observe that this method has proved good at screening out 'pure noise' variables. The classification success rate of 78%

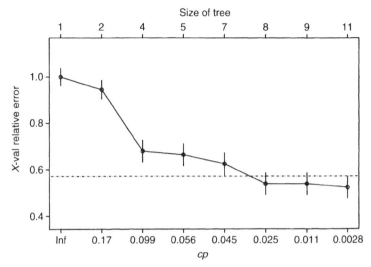

Figure 9.3 A plot of error rate against tree size based on 10-fold cross-validation, used to determine the degree of pruning to apply to the tree that, once pruned, gives rise to Figure 9.2.

compares well with that obtained using other methods of discriminant analysis (Table 9.1).

Classification trees can be viewed as an alternative to stepwise discriminant analysis, but have several potential advantages. One is the transparency of the method. Others are the fact that the method can handle missing data (illustrated in Baxter and Jackson, 2001) or mixed data, and that it is not necessary to worry about data transformation because of the monotonic relationship between raw and log-transformed data.

10

Missing data and outliers

10.1 Missing and censored data

10.1.1 Introduction

Missing data are common in archaeology. Objects such as bone, stone, ceramic and other artifacts are frequently found in a broken condition, and the properties one might wish to measure are not necessarily available for measurement. In chemical compositional studies data may be 'missing' because values are unrecorded or below the level of detection. Data 'missing' for this latter reason are commonly referred to as censored in the geochemical literature, though this usage is less common in archaeometry. Here, a distinction will be made between missing and censored data.

I am unaware of any systematic treatment of the problem in the archaeological literature. There may be good reasons for this, since any solution may be problem-dependent, specific to the questions being asked, the structure of the data set to hand, and the statistical techniques used. The simpler methods for handling missing data may be unsatisfactory, whereas more complex methods may require strong assumptions about the mechanisms that generate missing data (e.g., Krzanowski, 1988: 316–18; Gordon, 1999: 26–7). Krzanowski and Marriott (1995: 68–9) distinguish between data that are missing at random, data that are missing for structural reasons, and data that are unrecorded, not necessarily randomly. Schafer (1997: 10–11) distinguishes between data missing completely at random, where the missing data are a simple random sample of all values, and data missing at random, which requires that the probability of the data being missing does not depend on the missing data themselves but may depend on observed data.

10.1.2 Missing data

One way of dealing with missing data, often the default in software packages, is case deletion, in which any case for which any information is missing is omitted from analysis. Unless either very few cases are affected in this way, or measurements are missing completely at random, this will not usually be a satisfactory solution, because of biases that may arise in estimates of quantities determined from the data.

If missing values are concentrated in a small number of variables there may be little option other than to omit the offending variables. Brothwell and Krzanowski (1974: 250), in their study of biological differences between early British populations, base analyses on 11 of 48 craniometric measurements 'selected because they were less riddled with missing values than most in the series available for study'. If an omitted variable is strongly correlated with another, for which fairly complete information is available, little may be lost by the omission. Similarly, if an omitted variable is unimportant for the task to hand – for example, in a discriminant analysis where it is not a useful discriminator – omission will also not matter. It will often be difficult to make a prior judgment about importance.

Some of the more satisfactory approaches to dealing with missing data involve estimating the missing values, a process known as *imputation*. A simple way to do this is to replace missing values with variable means. If there are groups in the data and discovery of these groups is the purpose of an analysis this may not be sensible, since within-group means may differ markedly from the overall means. Even when group structure is assumed, and missing data are replaced by within-group means, this will underestimate variability in groups with many missing values.

Where group discovery is the object of the exercise, imputation may be something to be avoided. Kraznowski (1988: 29) observes that imputing missing values when calculating a (dis)similarity coefficient is to be discouraged as it can exaggerate the similarity between cases, and that it is better to work only with those variables that are present for both cases and to rescale the measure. This is common archaeometric practice (e.g., Bieber *et al.*, 1976) where, rather than using (squared) Euclidean distance, defined in equation (6.9), as a measure of dissimilarity between two cases, the measure

$$d_{ij}^2 = \sum_{k=1}^{m} (y_{ik} - y_{jk})^2 / m$$

is used, where summation is over the $m \leq p$ variables for which measurements are available for the two cases.

Where the structure of the data is such that mean imputation is sensible, typically where one is prepared to assume that the data form a homogeneous group, better methods are available. One possibility is to estimate missing data using multiple regression methods. To exploit this it may be necessary to 'fill in' the data missing for the independent variables used to predict missing values in the variable treated as dependent. Consideration of how to fill in these values leads naturally to the idea of iterated regression and an EM algorithm, and this is discussed more fully in Section 10.1.4.

Several forms of multivariate analysis depend on the calculation of means and covariance or correlation matrices, among them PCA and techniques using Mahalanobis distance. It is possible to calculate these quantities using those data for which complete information is available so that, for example, the correlation between two variables is estimated using those cases for which values for both variables are available. This can lead to strange results, since the matrices involved may no longer be positive definite, and phenomena such as negative Mahalanobis distances may arise (Brothwell and Krzanowski, 1974; Sayre, 1975). As a general solution this is not satisfactory.

In Bayesian analyses (Chapter 14), if the proportion of missing values is not large, missing values can be treated as parameters to be estimated (Baxter and Buck, 2000: 737). There are few examples of this kind of analysis in the archaeological literature, though an extended example is provided by Dellaportas (1998).

10.1.3 Censored data

Censored data occur commonly in archaeometric studies, based on artifact compositional data, where an element is present, but below the limit of detection. It is desirable to include such data in statistical analysis since it can be informative. Baxter *et al.* (1995) studied the compositional variability of colorless Roman vessel glass for four different vessel types that were also chronologically distinct, found in excavations at Colchester, UK. One of the larger and later groups, of cylindrical cups, was noteworthy in that nearly all the measurements for lead (Pb) were recorded as zero. That is, they were below the limit of detection. This (as later unpublished work confirms) seems to be a characteristic of this particular type of vessel. There are various ways in which such data might be handled in statistical analysis, including the following:

1. Treat censored data as having a value of zero. This can be satisfactory if the cases involved are associated with genuinely distinct groups and if analysis using untransformed data is judged to be reasonable. If an analysis using logarithmically, or log-ratio, transformed data is needed (see Section 7.2.2) then there is a problem, since the zero must be substituted with a non-zero value and, on a log scale, the substitution can have an undue effect on the outcome of the analysis.
2. Substitution is the strategy most commonly used, but practice varies. If *LOD* is the limit of detection, censored values are replaced by αLOD where $0 < \alpha < 1$. For secondary analyses of published data, when *LOD* is not known, *LOD* might be replaced by the minimum available measurement for a variable. Hall (2001: 63) replaces censored data with 0.5 *LOD*, a common practice, not just in archaeometric study (Thompson *et al.*, 2000: 103). In other areas of study other replacement values, such as 0.75 *LOD*, are apparently common (Sanford *et al.*, 1993). Sanford *et al.* (1993) evaluate objective substitution methods for censored data, using maximum likelihood and assuming a lognormal distribution for the data. For a sufficiently small number of censored observations (less than 10%, say) they suggest that as an empirical procedure using 0.55 *LOD* is reasonable.
3. Sanford *et al.* (1993) use distributional, as opposed to substitution, methods as a way of estimating censored values. Typically, measurements for a variable are assumed to follow a log-normal distribution so that, with censored data, what is observed is a sample from a truncated log-normal distribution. Maximum likelihood methods are used to estimate values for the censored data. There is an extensive literature on this, but its relevance to archaeometric practice is questionable. Often the underlying assumption, and reason for an analysis, is that there are groups in the data that one wishes to detect, and the assumption of log-normality in these circumstances is not reasonable.

4. Where substitution is used and there is a concern about the effect of the choice
of α on an analysis, it is easy enough to experiment with different values of α
to monitor the effect. This is a concern in analyses based on log or log-ratio
transformed data, where low values on the original scale (whether observed
or arising from substitution) become outliers on the transformed scale that
dominate subsequent multivariate analyses (Baxter, 1995). An example is given
in Baxter (1989) where PCAs of standardized and log-ratio transformed glass
compositional data were compared. Two variables had values recorded as zero.
Replacing these with substitute values, to permit the log-ratio analysis, it was
found that for sufficiently small α, the log-ratio analysis suggested a small
group not evident in the standardized analysis. This small group consisted of
the cases for which the values on one variable were zero.

10.1.4 An EM algorithm

A common approach to dealing with missing data in the statistical literature is
estimation using an EM algorithm. The seminal paper on the EM algorithm is
that of Dempster *et al.* (1977). Sayre (1975), in an influential but unpublished
paper, anticipated one application of the EM algorithm (though not called that)
for estimating missing archaeometric data that is sometimes used (e.g., Truncer
et al., 1998). I have not seen the technical details given in Sayre's (1975) report
reproduced in the archaeological literature. They are given below, in more general
form than the original.

Using the notation of Section 6.2 and writing $y_i = x_i - \bar{x}$, the Mahalanobis
distance of the ith case from the centroid of the group defined by the data matrix
X is

$$d_i^2 = y_i' S^{-1} y_i,$$

where S is the estimated variance–covariance matrix for the data. Partition y_i' as
$[y_{1i} \mid y_{2i}]'$ and partition S conformably as

$$\begin{bmatrix} S_{11} & S_{12} \\ S_{21} & S_{22} \end{bmatrix},$$

and its inverse, S^{-1} as

$$\begin{bmatrix} S^{11} & S^{12} \\ S^{21} & S^{22} \end{bmatrix}.$$

The Mahalanobis distance can then be written as

$$d_i^2 = y_{1i}' S^{11} y_{1i} + 2y_{2i}' S^{21} y_{1i} + y_{2i}' S^{22} y_{2i} \qquad (10.1)$$

using the fact that $y_{2i}' S^{21} y_{1i} = y_{1i}' S^{12} y_{2i}$. If y_{1i} is equated with observed data
and y_{2i} is equated with missing data, Sayre's (1975) idea was to estimate $\hat{y}_{2i}$ to
minimize the Mahalanobis distance of y_i from the group centroid, *assuming* the
availability of estimates of S and $\bar{x}$. Returning to this point shortly, differentiating
equation (10.1) with respect to y_{2i} and equating the result to 0 leads to

$$\hat{y}_{2i} = -(S^{22})^{-1} S^{21} y_{1i}, \qquad (10.2)$$

which is the generalization of the results given in Sayre (1975).

In practice $\hat{\mathbf{y}}_{2i}$ is estimated iteratively as follows:

1. Estimate all missing values, $\hat{\mathbf{y}}_{2i}^{(0)}$ with the means for the appropriate variables, using the data for which these are observed, to obtain a complete data matrix.
2. Estimate starting values $\tilde{\tilde{\mathbf{x}}}^{(0)}$ and $\tilde{\mathbf{S}}^{(0)}$ from the complete data matrix.
3. Using equation (10.2), reestimate all the missing values to get new estimates, $\hat{\mathbf{y}}_{2i}^{(1)}$, and use these to determine new $\tilde{\tilde{\mathbf{x}}}^{(1)}$ and $\tilde{\mathbf{S}}^{(1)}$.
4. Repeat the cycle until the $\tilde{\tilde{\mathbf{x}}}^{(k)}$, $\tilde{\mathbf{S}}^{(k)}$ and $\hat{\mathbf{y}}_{2i}^{(k)}$ converge after k iterations.

This is a particular application of an EM algorithm for estimating missing values in multivariate normal data (Beale and Little, 1975; Dempster *et al.*, 1977). Using results on the inverse of partitioned matrices (e.g., Graybill, 1983: 184) gives

$$\mathbf{S}^{21} = -\mathbf{S}^{22}\mathbf{S}_{21}\mathbf{S}_{11}^{-1};$$

substituting into equation (10.2) and rearranging gives

$$\hat{\mathbf{x}}_{2i} = \bar{\mathbf{x}}_2 + \mathbf{S}_{21}\mathbf{S}_{11}^{-1}(\mathbf{x}_{1i} - \bar{\mathbf{x}}_1),$$

which is the expression given in Seber (1984: 515) for determining missing values using the EM algorithm.

Scott and Hillson (1988) is an early archaeological application of an EM algorithm. They were interested in using discriminant analysis to distinguish between skulls from two Egyptian cemeteries, using 13 cranial measurements, and used EM to estimate missing data, which affected 30% of cases of known sex.

More recent approaches to estimating missing data have been developed in the statistical literature, such as multiple imputation (e.g., Schafer, 1997). I am not aware of any applications of this methodology in archaeology. For typical multivariate problems, where grouping in the data is explored using methods such as PCA and cluster analysis, both specifying a suitable model for the data and comparing results across analyses is non-trivial.

10.2 Outliers

Outliers frequently occur in multivariate archaeological data, and can easily dominate an analysis. In the plots often produced in conjunction with a PCA analysis, for example, extreme outliers can dominate and determine the scale of the plots, making any structure in the rest of the data difficult to discern. It frequently makes sense to identify and remove outliers from an analysis, in order to get a better idea of the main features of the data. This is not to say that outliers should be ignored, as they need to be explained and may be of interest in their own right.

When outlier removal is justified, *ad hoc* methods, based on univariate data inspection, PCA or cluster analysis, will sometimes be all that is needed to identify them. More generally, the problem of detecting multiple multivariate outliers is difficult. The following discussion looks at some of the methods available, and their limitations for many archaeological problems.

Several approaches to multiple outlier detection that have been proposed in the statistical literature exploit Mahalanobis distance, possibly estimated 'robustly' (e.g., Rousseeuw and van Zomeren, 1990; Atkinson and Mulira, 1993; Atkinson, 1994). The simplest approach is perhaps that of Atkinson and Mulira (1993) who exploit Mahalanobis distance in its usual form, but calculated for outlier-free subsets of the data identified by a forward search procedure. Their algorithm is as follows:

1. Select $p + 1$ observations and calculate $\bar{x}$ and S using these.
2. Calculate the d_i^2, increment the sample size by some small integer k, and select a new sample to consist of those cases with the smallest values of d_i^2.
3. Use the new sample to recalculate $\bar{x}$ and S and repeat stage 2, selecting $p + 1 + sk$ cases at stage s, until the data set is exhausted.
4. Identify outliers and display results for a suitable choice of s, so that the distances are calculated using a hopefully outlier-free subset of $m = p + 1 + sk$ cases.

As a cut-off point to define an outlier Atkinson and Mulira (1993: 29) suggest the maximum expected value from a sample of n chi-squared random variables on p degrees of freedom, so a case is declared as an outlier if

$$d_i^2 > \chi_p^2[(n - 0.5)/n].$$ (10.3)

When m is small this can result in too many cases being declared as outliers, and Atkinson and Mulira (1993: 31–2) describe a 'normalization' procedure designed to avoid this. Let $d_i^2(m)$ be the (squared) distances obtained using a subset of size m. Define

$$T(m) = \sum_{i=1}^{n} d_i^2(m),$$

so that $T(n) = p(n - 1)$ and generally $T(m) > T(n)$ for $m < n$. Now simulate an $n \times p$ data set from a standard multivariate normal distribution, calculate the $T(m)$ for these data, repeat N times, and calculate the average values of $T(m)$, say $\bar{T}(m)$, across the N simulations. Normalized squared distances may then be calculated as

$$\tilde{d}_i^2(m) = p(n - 1)d_i^2(m)/\bar{T}(m)$$

and, for any subset of size m, cases declared to be outliers if

$$d_i^2 > \frac{\bar{T}(m)}{p(n - 1)}\chi_p^2[(n - 0.5)/n],$$ (10.4)

which will, in general, result in fewer outliers.

Various options exist for displaying results, including index plots of the d_i for suitably chosen m, and chi-squared probability plots for the d_i^2 for chosen m. The latter two options are illustrated in Figure 10.1. The data used are a 108×7 data matrix of pottery chemical compositions, obtained by high-precision X-ray fluorescence analysis (Adan-Bayewitz et al., 1999). Ratios of seven elements to Fe are used. In this particular data set it was believed that the majority of pottery was from the same production site, and the expectation was that the data were

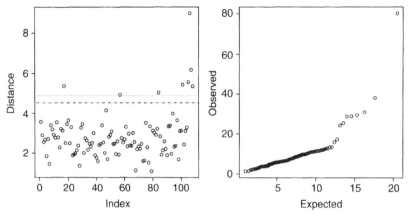

Figure 10.1 The left-hand figure is an index plot of the Mahalanobis distances, d_i, of a sample of 108 cases from the centroid of a group defined by $m = 98$ of the cases. The right-hand figure is a chi-squared probability plot of the squared distances.

dominated by a single group. Detecting and removing outliers from the data was considered to be important, in order to ensure the integrity of the group which was to be used for comparison with other pottery groups, including groups apparently made within a few kilometers of the above-mentioned production site.

The index plot of the distances, d_i, on the left of the figure is based on $m = 98$, a subset of about 90% of the data used in the Mahalanobis distance calculations. The lower and upper reference lines are the square roots of the criteria given in equations (10.3) and (10.4). For either criterion, and both plots, eight outliers are indicated.

Results for this example are fairly clear-cut, with different choices of m leading to similar conclusions. The starting set of $p + 1$ cases is chosen randomly so that, in principle, different runs may lead to different subsets of outliers being identified. This did not appear to happen here, but can do so for sparser data sets with smaller n/p ratios. For a similar but smaller data set than the one used for Figure 10.1, in 100 analyses two patterns of outliers accounted for 98% of the results, one occurring about 70% of the time, with considerable overlap between the sets of outliers identified. It is usually worth checking whether this is happening or not. With very sparse data, results may be meaningless, unless the number of variables can be reduced. Another practicality is that the $p + 1$ cases used to initiate the analysis may give rise to a singular covariance matrix, $\mathbf{S}$, and their number may need to be increased.

Much of the statistical literature on multiple outlier detection in multivariate data assumes a model in which there is a single, normally distributed, dominant group in the data. This is inappropriate for many archaeological problems where analysis is often undertaken in the expectation of finding several groups in the data. In these circumstances informal methods of outlier detection, including cluster analysis and PCA, may be as or more effective at finding outliers, though they are not foolproof (Baxter, 1999c). Analysis of the data in Table 2.2 is instructive. The three outliers highlighted in the table are clearly picked out using the average- and single-link methods of cluster analysis and less clearly by the complete-link method, using standardized data with Euclidean distance as the

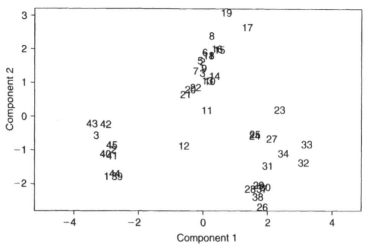

Figure 10.2   Plot based on a PCA of the standardized data of Table 2.2, omitting cases with highlighted outliers.

dissimilarity measure. Scanning the table or using univariate plots also highlights these cases. Omitting them and running a PCA on standardized data results in the plot of Figure 10.2, where case 12 is a multivariate outlier. It belongs to the central group (group 1 in Table 2.2) and relative to this group has the smallest value for eight out of nine variables. Baxter (1999c: 332–4) reports on a number of analyses of these data. Average-, complete- and single-link cluster analyses all identify three groups but not the outlier. Once groups have been established more formal methods can be applied within groups to see if they include outliers. For example, using Mahalanobis distances and 'leave-one-out' calculations (equation (6.16)) suggests that, within group 1, cases 1 and 12 are outliers, a conclusion also reached using the sequential test of Caroni and Prescott (1992).

A practical strategy, where detecting outliers is a concern and where their removal from analysis is justified, is to identify the more obvious ones from simple graphical analyses that may include PCA and cluster analysis. If, after their removal, clear groups are revealed these can be subjected to analyses, such as that of Atkinson and Mulira (1993), that are underpinned by the assumption of a single main group in the data. This is effectively what is done with the group evaluation procedures practiced by some researchers (Glascock, 1992; Section 6.4.3). Caveats to be entered here are that, ideally, groups should be distinct from each other, and should be large enough relative to the number of variables for Mahalanobis distance calculations to be sensible. This may often not be the case.

Another practical strategy is to identify outliers iteratively. For example, univariate methods may identify the very obvious outliers as cases that are extreme on several variables. A PCA might then reveal further outliers, or small groups, on plots of the first few components. Removing these and repeating analyses may then reveal yet further outlying cases that could also be removed. An example of this approach is described in Papageorgiou *et al.* (2001), where a 130 × 24 ceramic compositional data set had 22 outliers identified and removed in the manner just described before being subjected to various forms of cluster analysis. Statistically this can be justified by noting that cases removed from analysis because they are remote from other cases in principal component space will be remote in the p-dimensional space defined by the chemical variables used. As it happened this data set had been analyzed and classified petrographically, and the cases identified as chemical outliers had nearly all been classified as petrographic outliers as well.

11

Analysis of tabular data

11.1 Introduction

Cross-tabulated data, frequency table data and contingency table data are all terms used to describe tables of counts based on cross-classified categorical variables. If two variables have I and J categories, n_{ij} is the number of cases falling into category i of the first variable and j of the second. The term *level* will be used for the categories defining a variable, and n_{ij} is the count in the (i,j)th *cell*. The $I \times J$ table of these counts will be denoted by $\mathbf{N}$; $n_{i\cdot}$ is the sum for row i; $n_{\cdot j}$ is the sum for column j; and $n_{\cdot\cdot} = N$ is the overall total or sample size. The notation extends to higher-order tables. The subject of this chapter is the statistical analysis of such tables.

Spaulding (1953) was one of the earliest archaeological examples of such analysis. The importance of this paper is discussed in Section 1.2.2. The illustrative examples in Spaulding (1953) were based on small two- and three-way contingency tables in which the categorical variables used related to qualities of ceramic vessels, such as tempering (two levels), surface decoration (two levels) and shoulder decoration (three levels) giving, in this example, a $2 \times 2 \times 3$ contingency table. In Spaulding's notation $d = O - E$ was the difference between an observed value in a cell, O, and its expectation, E, estimated under the null hypothesis that the categorical variables were independent. The expectation may be written in the form $E = N\hat{p}$, where N is the sample size and $\hat{p}$ the estimated proportion for a cell under the independence hypothesis. The estimated standard error of E is then $\hat{\sigma} = \sqrt{N\hat{p}(1 - \hat{p})}$, and $(d/\hat{\sigma})^2$ was used as a test statistic, referred to the chi-squared distribution with 1 degree of freedom, to determine if an observed cell value departed significantly from expectations under the independence hypothesis.

The archaeological argument was that a vessel type was defined by 'a consistent assemblage of attributes whose combined properties give a characteristic pattern', and that such assemblages of attributes could be identified from those combinations of categories defining cells in the contingency table departing significantly from expectations under the independence hypothesis (Spaulding, 1953: 305). This will be illustrated numerically later.

Since 1953 the use of chi-squared in the analysis of two-way tables in archaeology has become routine (Shennan, 1997: 104–15). The next section reviews some of the underlying ideas in order to establish notation. There are difficulties in using Spaulding's method when dealing with more than a very few categorical variables. The statistical technology for dealing with multiway tables, in the form of log-linear models, was only developed at a later date (e.g., Bishop *et al.*, 1975), and Spaulding (1976), along with Clark (1974, 1976) and Read (1974), was an early advocate of such models in archaeology. This is dealt with in Section 11.3. Despite advocates in the 1980s (e.g., Lewis, 1986; Cowgill, 1989b: 87), log-linear modeling has not been widely adopted in archaeology. Correspondence analysis has had a greater impact and is discussed in Sections 11.4 and 11.5.

11.2 Chi-squared analysis of contingency tables

11.2.1 Two-way tables

Under the null hypothesis that the two categorical variables that define a two-way table are independent in the population, the expected value in the (i,j)th cell is

$$E_{ij} = n_{i\cdot}n_{\cdot j}/N, \tag{11.1}$$

and the null hypothesis is tested by referring

$$X^2 = \sum_{ij} \frac{(n_{ij} - E_{ij})^2}{E_{ij}} \tag{11.2}$$

to the χ^2 distribution with $(I-1)(J-1)$ degrees of freedom.

The justification for using χ^2 as a reference distribution is an asymptotic one, and may break down if too many of the E_{ij} are too small. What is to be understood by 'too many' and 'too small' has been a matter for debate, though the commonly quoted rule requiring all $E_{ij} > 5$ is usually too stringent (Everitt, 1992: 25). A more forgiving rule is to require fewer than 20% of the cells to have $E_{ij} < 5$ and all to have $E_{ij} > 1$. For 2×2 tables with small expected frequencies Fisher's exact test (Everitt, 1992: 14–19; Cox and Snell, 1989: 48–9) is available, and sometimes surfaces in archaeological applications (e.g., Cowgill, 1989a; Schutowski *et al.*, 1999).

If the null hypothesis of independence is rejected it is important to identify which cells of the table contribute most to the lack of independence, and for this purpose standardized residuals, defined by

$$e_{ij} = (n_{ij} - E_{ij})/\sqrt{E_{ij}}, \tag{11.3}$$

are sometimes used.

11.2.2 Three-way tables

An alternative formula for the expected values in equation (11.1) is

$$E_{ij} = N\hat{p}_{i\cdot}\hat{p}_{\cdot j} \tag{11.4}$$

where $\hat{p}_{i\cdot} = n_{i\cdot}/N$ and $\hat{p}_{\cdot j} = n_{\cdot j}/N$. For a three-way table, with K categories for the third variable, expected values under the null hypothesis of independence generalize from equation (11.4) to

$$E_{ijk} = N\hat{p}_{i\cdot\cdot}.\hat{p}_{\cdot j\cdot}.\hat{p}_{\cdot\cdot k} \tag{11.5}$$

and $\hat{p}_{i\cdot\cdot} = n_{i\cdot\cdot}/N$, $\hat{p}_{\cdot j\cdot} = n_{\cdot j\cdot}/N$ and $\hat{p}_{\cdot\cdot k} = n_{\cdot\cdot k}/N$, where the notation generalizes in an obvious way from that of the previous subsection. The chi-squared statistic (11.2) to test for independence generalizes as

$$X^2 = \sum_{ijk} \frac{(n_{ijk} - E_{ijk})^2}{E_{ijk}} \tag{11.6}$$

and is tested with $IJK - I - J - K - 2$ degrees of freedom. The general rule for determining the degrees of freedom is (number of cells -1) $-$ (number of probabilities estimated for the hypothesis being tested). For the independence hypothesis this last quantity is $(I - 1) + (J - 1) + (K - 1)$ since, for the first variable for example, once the $(I - 1)$ terms $\hat{p}_{1\cdot\cdot}, \ldots, \hat{p}_{I-1\cdot\cdot}$ are estimated, $\hat{p}_{I\cdot\cdot}$ is fixed by the requirement that the probabilities sum to 1. The definition of the standardized residuals (11.3) generalizes in an obvious way.

To illustrate methods discussed here the artificial data set used by Spaulding (1953), given in Table 11.1, will be examined. Table 11.2 shows the observed and expected values calculated from equation (11.5), the standardized residuals, and the deviance residuals to be discussed later. The final column shows the signed square root of the statistic used by Spaulding (1953) to identify significant departures from the expected values obtained under the independence hypothesis. These are similar to the standardized residuals, so Spaulding's analysis will be interpreted in terms of the latter.

The value of X^2 is 178.4 which, on 7 degrees of freedom, is highly significant at the levels normally used. Spaulding (1953) identified three combinations with particularly high residuals, GStP, SSmR and SSmC. Lumping the last two together, along with SSmP, allows a 'shell-tempered, smooth-surfaced' type to be identified, while a second 'grit-tempered, stamped-surface, plain-shouldered' type was also identified. An alternative way to analyze the data, not available in 1953, would be through the use of log-linear models. This is the subject of the next section.

Table 11.1   Data from Spaulding (1953) showing the relationship between temper type, surface and shoulder decoration for a hypothetical assemblage of pots. For the type of shoulder decoration 'Rect.' is rectangular and 'Curv.' is curvilinear

Shoulder	Surface			
	Grit temper		Shell temper	
	Smooth	Stamped	Smooth	Stamped
Plain	14	51	41	5
Rect.	0	3	38	6
Curv.	2	0	26	0

Table 11.2 Results from a chi-squared and log-linear analysis of the data in Table 11.1 on fitting the independence model. Following the observed and expected values and standardized residual are the deviance residuals, d_{ijk}, and the signed and square-rooted values of the statistic used by Spaulding (1953). The code for the types identifies two kinds of temper (Grit or Shell), two types of surface (Stamped or Smooth) and three types of shoulder decoration (Plain, Rectangular or Curvilinear)

Type	n_{ijk}	E_{ijk}	e_{ijk}	d_{ijk}	r_{ijk}
GSmP	14	27.2	−2.5	−2.8	−2.7
GSmR	0	11.5	−3.4	−4.8	−3.5
GSmC	2	6.9	−1.9	−2.2	−1.9
GStP	51	14.6	9.5	7.4	9.9
GStR	3	6.2	−1.3	−1.4	−1.3
GStC	0	3.7	−1.9	−2.7	−1.9
SSmP	41	45.0	−0.6	−0.6	−0.7
SSmR	38	19.1	4.3	3.8	4.6
SSmC	26	11.4	4.3	3.7	4.5
SStP	5	24.2	−3.9	−4.8	−4.2
SStR	6	10.2	−1.3	−1.4	−1.4
SStC	0	6.1	−2.5	−3.5	−2.5

11.3 Log-linear models

11.3.1 Three-way tables

Expositions of log-linear models written for an archaeological audience include Lewis (1986) and Shennan (1997: 201–13). Statistical texts that deal with the subject include Bishop *et al.* (1975) and Dobson (1990). The following discussion is couched in terms of models for a three-way table.

One possible model for counts, n_{ijk}, is that they are sampled from independent Poisson distributions with means μ_{ijk}. That is,

$$n_{ijk} \sim P(\mu_{ijk})$$

from which

$$E_{ijk} = E(n_{ijk}) = \mu_{ijk}.$$

If μ_{ijk} is modeled as a function of parameters, such that its logarithm is linear in the parameters, then a particular form of generalized linear model (see Section 5.4) known as a *log-linear* model is obtained. More realistically, models based on the multinomial distribution may be more appropriate for frequency data; however, use of the Poisson assumption leads to the same results (e.g., Venables and Ripley, 1999: 226).

An ANOVA-type formulation is usual (see Section 5.3.5). Thus one possible model, with no interactions, on the logarithmic scale, is

$$\log E_{ijk} = \mu + \alpha_i + \beta_j + \gamma_k. \tag{11.7}$$

This is equivalent to postulating a multiplicative model of the form

$$E_{ijk} = e^{\mu + \alpha_i + \beta_j + \gamma_k} = e^{\mu} e^{\alpha_i} e^{\beta_j} e^{\gamma_k}$$

on the frequency scale. Equating α_i, β_j and γ_k with the ith, jth and kth levels of the three categorical variables gives rise to parameters $(\mu, \alpha_1, \ldots, \alpha_I, \beta_1, \ldots, \beta_J, \gamma_1, \ldots, \gamma_K)$ that must be estimated. As with ANOVA models in regression the model is overparameterized so that constraints need to be imposed on each set. The particular constraint used may be software-dependent. One common choice is

$$\alpha_1 = \beta_1 = \gamma_1 = 0,$$

and another is

$$\sum_i \alpha_i = \sum_j \beta_j = \sum_k \gamma_k = 0. \tag{11.8}$$

Model (11.7) corresponds to a model of independence between the categorical variables used, and the fitted values are identical to those obtained in the chi-squared test of independence. The chi-squared statistic given by equation (11.6) can be used to test whether the model is reasonable but, more usually in the context of log-linear models, the *deviance* or log-likelihood ratio statistic

$$G^2 = 2 \sum_{ijk} [n_{ijk} \log(n_{ijk}/E_{ijk}) - (n_{ijk} - E_{ijk})] \tag{11.9}$$

is used. If a constant term is included in the model $\sum(n_{ijk} - E_{ijk}) = 0$, so G^2 simplifies. The deviance is asymptotically distributed as χ^2 with the same degrees of freedom as the X^2 statistic. The square-rooted components of the contribution of the (i, j, k)th cell to equation (11.9),

$$\sqrt{2[n_{ijk} \log(n_{ijk}/E_{ijk}) - (n_{ijk} - E_{ijk})]}, \tag{11.10}$$

define the *deviance residuals*, where they have the same sign as $(n_{ijk} - E_{ijk})$ and where $n_{ijk} \log(n_{ijk}/E_{ijk}) = 0$ if $n_{ijk} = 0$. These residuals may be used in much the same way as the standardized residuals from a chi-squared analysis. They are given, for comparison, in Table 11.2, for which $G^2 = 165.3$ and $X^2 = 178.4$, both leading to strong rejection of the independence hypothesis.

The independence hypothesis is often of little interest in itself. What is more useful is to explore the structure of a table and the relationships between the variables, and it is to this that log-linear models, rather than chi-squared, lend themselves.

In contrast to the independence model (11.7), a *saturated* model has the form

$$\log E_{ijk} = \mu + \alpha_i + \beta_j + \gamma_k + (\alpha\beta)_{ij} + (\alpha\gamma)_{ik} + (\beta\gamma)_{jk} + (\alpha\beta\gamma)_{ijk}$$

and fits the data exactly. The *null* model is of the form $\log E_{ijk} = \mu$.

If the independence model does not fit the data well, some model intermediate between this and the saturated model is sought that does provide a good fit. The α_i, β_j and γ_k are the *main effects*. Terms of the form $(\alpha\beta)_{ij}$, $(\alpha\gamma)_{ik}$ and $(\beta\gamma)_{jk}$ are *first-order interactions* and $(\alpha\beta\gamma)_{ijk}$ is a *second-order interaction*. A general rule is that if a higher-order term is in a model all the lower-order terms that can be formed from it must also be in the model. Thus, if $(\alpha\beta\gamma)_{ijk}$ is in the model then $(\alpha\beta)_{ij}$, $(\alpha\gamma)_{ik}$ and $(\beta\gamma)_{jk}$ must be included; and if $(\alpha\beta)_{ij}$ is in the model then α_i and β_j must be included. With this in mind, a simplified notation for representing

Table 11.3 Data from Leese and Needham (1986) showing the relationship between decorative combinations on the side and face of Early Bronze Age axes from southern Britain, and their phase

Face	Side					
	Early phase			Late phase		
	Undec.	C1-4	C7-10	Undec.	C1-4	C7-10
Undecorated	29	34	1	16	36	11
Furrowed (A1-2)	8	15	2	8	9	19
Furrowed (A3)	1	0	1	0	2	6
Punched (B1-2)	3	16	3	2	2	3
Punched (B3)	1	0	1	0	1	4

models has been developed. Associating [1] with the α parameters, [2] with the β parameters and [3] with the γ parameters, the independence model can be written as [1][2][3]. The saturated model is written as [123], with lower-order terms of the form [1], [2], [3], [12], [13] and [23] present by implication. The model

$$\log E_{ijk} = \mu + \alpha_i + \beta_j + \gamma_k + (\alpha\beta)_{ij} + (\beta\gamma)_{jk} \qquad (11.11)$$

would be written as [12][23], with all main effects present by implication.

Given two models, A and B, model B is *nested* within A if the terms that define it are a subset of those contained in model A. Thus, the model [12][3], given by

$$\log E_{ijk} = \mu + \alpha_i + \beta_j + \gamma_k + (\alpha\beta)_{ij},$$

is nested within model (11.11), whereas the model [13][2] given by

$$\log E_{ijk} = \mu + \alpha_i + \beta_j + \gamma_k + (\alpha\gamma)_{ij}$$

is not, since it contains terms $(\alpha\gamma)_{ij}$ not in the first model.

This is relevant to the way models are sometimes selected. Let G_A^2 and G_B^2 be the deviances for two models, with degrees of freedom df_A and df_B, respectively. If model B is nested within model A then $G_B^2 - G_A^2$ should be distributed approximately as χ^2 with $df_B - df_A$ degrees of freedom, if model A can be simplified to model B without making the fit significantly worse. A similar statistic was discussed in the context of logistic regression in Section 5.4.2.

To illustrate some of these ideas, data given in Leese and Needham (1986) relating to the decoration of Early Bronze Age axes from southern Britain, reproduced in Table 11.3, are used. The axes are from two phases and are categorized by the type of decoration on the face and on the side of the axe. For the face the most complex decorations are types A3 and B3, while for the side, type C7-10 is most complex.

Table 11.4 shows the outcome of fitting all possible models to these data. The difference in deviance between the null model and independence model, [1][2][3], is $324.7 - 96.6 = 228.1$ and, with $28 - 21 = 7$ d.f., is clearly highly significant; however, the deviance for the independence model of 96.6 with 21 d.f. indicates that the fit is poor. For model [12][13][23] the deviance is 11.5 with 7 d.f. and has

Table 11.4   The deviance, G^2, and degrees of freedom (d.f.),
for different log-linear models fitted to the data of Tables 11.3.
The first set of results uses all the data, the second set omits
an outlier corresponding to the B1-2/B1-4/Late cell

Model	G^2	d.f.	G^2	d.f.
Null	324.7	29	318.4	28
[1][2][3]	96.6	22	89.6	21
[12][3]	56.4	14	42.6	13
[13][2]	81.4	18	80.1	17
[1][23]	64.5	20	59.4	19
[12][13]	41.1	10	36.8	9
[12][23]	24.3	12	13.9	11
[13][23]	49.3	16	48.7	15
[12][13][23]	12.3	8	11.5	7
[123]	0.0	0	0.0	0

a p-value of 0.12, suggesting a reasonable fit to the data. This is not surprising since it differs from the saturated model only by the omission of the second-order interaction term, and little simplification is achieved. The best of the models that omit one of the first-order interactions is [12][23], and this differs in deviance from the model [123] by 10 with 4 d.f., the p-value for which is 0.04. The deviance is 24.3 with 11 d.f. and a p-value of 0.01, which Leese and Needham (1986: 6) described as 'almost low enough to be acceptable'.

There is a conflict here between choosing a model that describes the data well but may be too complex to be informative, and choosing a simple model that is more interpretable but does not fit the data as well as one would like. Leese and Needham (1986: 5) resolved this by undertaking a residual analysis in which the standardized residuals were used to identify three cells with 'rather high' residuals. The second largest of these was omitted on the grounds that its removal effected the greatest reduction in the deviance. This was the cell B1-2/C1-4/Late. It is evident on inspection of Table 11.3 that this is far lower than would be expected by comparison with the corresponding cell for the early phase. Leese and Needham (1986: 6) could not find a 'convincing reason for this apparent anomaly', but omitted the cell in the interests of obtaining a good overall fit to the rest of the data. It is easier to identify the cell as anomalous using the deviance residuals (11.10), for which the absolute value for cell B1-2/C1-4/Late is 2.49, compared to the next largest of 1.83. The cell has a standardized residual of 2.08, and there is one other cell with a larger value of 2.12.

Interpreting the results in more detail can be tricky. Leese and Needham (1986) did so by presenting the estimated interaction terms for the model, as in Table 11.5. They used a parameterization similar to equation (11.8) for the interactions so that, for example, estimates sum to 0 across the rows of the table, and down columns for the separate interactions. Focusing on the larger positive values in the table, and remembering that cell B1-2/C1-4/Late has been omitted from the fit, it can be seen that complex side designs (C7-10) are relatively more common in the later phase. The complex side designs also tend to be associated with the complex face designs (A3, B3) more than would be expected by chance, while the moderately complex side and face designs (C7-10, B1-2) tend to be found together, as do the simplest, undecorated designs.

Table 11.5 Table from Leese and Needham (1986) showing the estimated interactions for their preferred model for the data of Table 11.3

	Side		
	Undec.	C1-4	C7-10
Side/phase interaction			
Early phase	0.4	0.2	−0.6
Late phase	−0.4	−0.2	0.6
Side/face interaction			
Undecorated	0.7	0.5	−1.2
Furrowed (A1-2)	0.2	−0.1	−0.1
Furrowed (A3)	−0.5	−0.5	0.9
Punched (B1-2)	−0.3	0.9	−0.6
Punched (B3)	−0.1	−0.9	0.9

Table 11.6 Table from Maschner and Stein (1995) showing the qualities associated with 93 archaeological site locations

Island size	Beach quality	Sheltered		Exposed	
		Southern exposure	Northern exposure	Southern exposure	Northern exposure
Large	Sand/gravel	23	8	8	0
	Rocks/boulders	0	2	0	1
Medium	Sand/gravel	14	6	2	1
	Rocks/boulders	1	1	0	0
Small	Sand/gravel	9	3	6	1
	Rocks/boulders	4	2	1	0

11.3.2 Four-way tables – an example

As an example of an analysis of a four-way table, data from Maschner and Stein (1995) given in Table 11.6 are used. These data were collected as part of an investigation into the decision-making processes that influenced site location in Tebenkof Bay, Kuiu Island in southeast Alaska. The 93 archaeological sites involved differ with respect to the size of the island on which they are located, beach quality, the cardinal exposure of the site, and whether or not the site is sheltered or seasonally exposed to storms. The variables will be abbreviated as *size*, *beach*, *exposure* and *shelter*. Logistic regression analysis was used to establish that these variables were useful for differentiating between archaeological site and non-site locations. Following this, log-linear modeling was used to investigate the structural relationship between the four variables.

Certain features of the data are obvious on careful examination of the data, for example that sand/gravel beaches, southern exposure and sheltered locations are favoured. What is less obvious is whether or not there are any interactions (or associations) among the variables. Maschner and Stein (1995: 69) stated that 'a model of complete independence is highly unlikely based on the nature of the ethnographic data'. Their preferred model, 'using a stepwise selection procedure in SPSS' to select a 'best' model, was [123][4]. This contains what is described

Table 11.7   The deviance, G^2, and degrees of freedom for different log-linear models fitted to the data of Tables 11.6

Model	G^2	d.f.
Null	141.3	23
[1][2][3][4]	25.5	18
[12][13][23][4]	15.0	13
[12][13][14][23][24][34]	10.2	9
[123][4]	10.1	11
[123][124][134][234]	0.0	2

as a 'significant three-factor interaction' between the variables *size, beach* and *exposure.*

The choice of this preferred model is debatable, and illustrates some of the problems involved in choosing parsimonious models for high-dimensional tables of data. To facilitate discussion of this, the fits from some selected models are shown in Table 11.7. The results for models [1][2][3][4] and [123][4] agree with those of Maschner and Stein (1995), who also report the X^2-values. The model [1][2][3][4] has $G^2 = 25.5$ with 18 d.f., for which the *p*-value is 0.11. This is not significant, even at the 10% level, and one might reasonably conclude that, despite the expectations of Maschner and Stein (1995), an independence model does fit the data adequately. That the somewhat less parsimonious model [123][4] is preferred is a function of the unspecified selection procedure used. It could be argued that the difference in G^2 from the independence model of 15.4 with 7 d.f. has a *p*-value of 0.03 and thus is significant; however, one could equally well argue that dropping the three-factor interaction from [123][4] to get the model [12][13][23][4] increases G^2 by 4.9, which has a *p*-value of 0.09 on 2 degrees of freedom. This would suggest that in the model [123][4] the three-factor interaction is not particularly significant.

What this discussion shows, at the very least, is that the model selected can be highly dependent on the selection procedure used, and that reliance on automatic selection procedures associated with particular software packages may be unwise. If the deviance residuals for the independence model, [1][2][3][4], are examined, only that of -2.49 for the large/rocks boulders/southern/sheltered cell stands out at all, with an observed value of 0 and fitted value of 3.1. If this is omitted, as in the analysis of the previous subsection, G^2 for [1][2][3][4] is 17.9 on 17 d.f. and for [123][4] is 10.1 on 10 d.f. The difference in deviance, of 7.8 on 7 d.f., is not significant and would lead one to prefer the simpler independence model.

11.4   Correspondence analysis

11.4.1   Introduction

In this section some applications of correspondence analysis (CA) in archaeology are discussed. A more mathematical account of the method is sketched in the next section.

Orton (1999: 32) has identified CA as the most notable statistical technique introduced into archaeology in the 1980s. Baxter (1994a: 133–9) gives an account of the rather curious way CA has diffused through the archaeological literature, and provides an extensive bibliography of applications up to the early 1990s. A summary version of this development is provided in the discussion of the paper by Bølviken *et al.* (1982) in Section 1.2.2.

At its simplest the attraction of CA is not hard to understand. It is an exploratory data-analytic technique, essentially principal component analysis for tables of counts, which enables one to obtain a graphical view of the structure of the table. Typically a CA results in the graphical representation of the relationship between the rows of a table and between the columns of a table, the joint interpretation enabling one to identify the association between rows and columns. Its use will be introduced through two examples.

Example 1: Seriation
One common use of CA is for seriation. Typically the rows and columns of the table correspond to assemblages and types, and the aim is to use CA to suggest an ordering of the assemblages that, it is often hoped, has a chronological interpretation (Chapter 16). In this example we use the burial data from Table 2.5.

The left-hand plot in Figure 11.1 is for the column data with assemblages labeled *a–p* having the same ordering as the columns of the data matrix. An analysis is usually considered successful if it recovers a relative ordering of the data that, using non-statistical evidence, can be equated with a relative chronological sequence. 'Horseshoe'-shaped plots, in which the ordering can be read around the horseshoe, are often considered by archaeologists to be evidence for a good seriation. In the present case one can read around the horseshoe from *c* to *m*, with *e* departing from the main trend and the precise sequencing of small clusters of assemblages such as *n*, *o* and *p* being difficult to determine. The *inertias* of the first two components, which correspond to the proportion of variance explained in a PCA, are 28% and 17%. These are not especially high, but a plot such as that shown would normally be regarded as evidence of a satisfactory analysis since there is clear archaeological evidence that the right-hand extremity of the

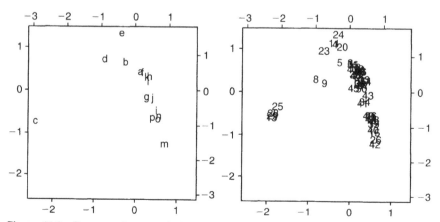

Figure 11.1 Correspondence analysis of the burial data from Table 2.5, the plot to the left being that for columns and that to the right for rows of the data matrix. Labeling, a–p and 1–52, corresponds to the row and column ordering.

horseshoe corresponds to early burials and the left-hand to late burials. McClellan (1979) was particularly interested in the sequencing of tombs 532, 542, 552 and 562 (*g* to *j* on the plot). His proposed ordering, from earliest to latest of 552 (*i*), 562 (*j*), 532 (*g*) and 542 (*h*) is consistent with the evidence of Figure 11.1.

The right-hand plot produces a seriation of pottery types. Assemblages to the left of the left-hand figure (principally *c*) are characterized by a relatively high proportion of those pottery types to the left of the right-hand figure. The labels 13, 25, 28 and 29 correspond to types 79, 111, 126 and 127 that are readily seen in Table 2.5 to form a high proportion of assemblage *c* in comparison to other assemblages.

Example 2: Romano-British glass assemblages

As a second example, illustrating a different kind of use of CA to investigate patterning in data other than sequencing, data from Table 1 of Cool and Baxter (1999: 80) are used. The data relate to 18 Romano-British vessel-glass assemblages in which the entries were amounts of vessel glass for each of six vessel categories. The assemblages were classified as first to third century AD and fourth century AD, labeled as E (early) and L (late) in Figure 11.2.

The CA of Figure 11.2 shows a distinction between early and late sites, with the latter characterized by a relatively higher proportion of cups and lower proportion of other vessel types. A style of presentation involving overlaying of the two graphs has been used. Although this analysis distinguishes between early and late sites it is not a seriation, since no attempt is made to sequence the sites within period, nor is such a sequence suggested. In Cool and Baxter (1999), on the basis of this analysis, early sites were separated from late ones and subjected to more detailed scrutiny.

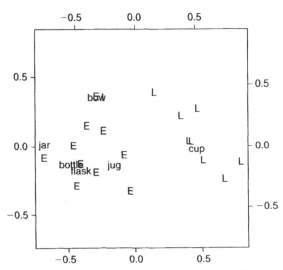

Figure 11.2 Correspondence analysis of vessel-glass assemblage data taken from Table 1 of Cool and Baxter (1999). Labels for assemblages indicate whether the assemblage was early (E) or late (L).

The successive subdivision of the data, by separating out obvious subgroups and/or outliers, to reveal further structure was likened by Cool and Baxter (1999) to the peeling of an onion and is characteristic of the way CA is often used in an exploratory manner. Such an approach was used to good effect in Moreno-García *et al.* (1996) and Orton (1996) who exploited CA for comparing animal bone assemblages. Their analysis was of three-way data characterized by site, species and bone element, which could be collapsed into two-way tables across any one of these dimensions. For example, the CA in Figure 23.2 of Orton (1996), for the species by site data for Roman contexts, showed a very clear difference between sites with a high proportion of horse bones and other sites. Omitting horses and the associated sites from the analysis then revealed a contrast between sites associated with sheep/goat and sites associated with cattle, not immediately evident in the original analysis.

11.4.2 Detrended correspondence analysis

When used successfully for seriation, CA typically gives rise to the horseshoe pattern noted earlier and the relative ordering of assemblages is read off from this. Correspondence analysis has been widely used in ecology and, in the contexts that arise there (see Gauch, 1982: 152–60), the horseshoe effect has sometimes been considered undesirable both because it reflects a systematic relationship between the first two axes (well illustrated, for ideal data, in Madsen, 1988b), and because the scale is compressed at either end of the horseshoe. This means that distances between points in the plot do not have a consistent interpretation in terms of differences between assemblage compositions. This has led to the development of methods of detrended correspondence analysis to try and deal with these perceived problems (e.g., Hill and Gauch, 1980).

Algorithms have been developed to accomplish the detrending and rescaling. For example, the first axis is divided into segments and, within segments, values on the second axis are adjusted to remove any trend. Commentators on some of this methodology have worried that it seems 'rather arbitrary' (Digby and Kempton, 1987: 97) and may introduce spurious detail into the results (Greenacre, 1984: 232). Where a relative ordering of the data is all that is required there seems little need to worry about the horseshoe effect or detrending. This is the view of many archaeological users, who welcome the effect as a sign that a seriation has 'worked'.

Lockyear (2000a) is one of the few archaeological papers to have studied the use of detrended CA in any detail. His assemblages corresponded to 241 Roman republican coin hoards with types corresponding to the year of issue of coins in the hoard. Correspondence analysis produced a good seriation of the data, predictable given that the existence of a time gradient was already known, and detrended CA was used to see if it revealed features masked by the strong time gradient in the original CA.

As applied to the full data set, detrended CA had little to offer over the original CA, because of the size and complexity of the data set. Both methods were also applied to a smaller set of 24 hoards, from Italy and Romania, which had previously been grouped into three clusters by a cluster analysis (Lockyear, 1996). The detrended CA highlighted, more so than the CA, an isolated pair of Romanian hoards and, within one of the original clusters, appeared to separate out the Romanian from Italian hoards, which the CA did not. Close inspection of the data by other means led Lockyear to believe that the phenomena highlighted by detrended CA were genuine, though he concluded that he 'would not recommend the use of [detrended CA] without prior analysis by ordinary CA, and careful reference back to the original data is also a necessity' (Lockyear, 2000a: 17).

11.4.3   Multiple correspondence analysis – example

Correspondence analysis can be applied to two-way tables of incidence or abundance data. In the former case the data consist of 1s and 0s, indicating whether or not a particular type is present in a particular context. For higher-order tables multiple correspondence analysis (MCA) is ordinary CA applied to an indicator matrix of 1s and 0s derived from the higher-order table. The mathematics of this is described more fully in Section 11.5.4.

If the kth of p categorical variables has L_k levels, then L_k dummy or indicator variables may be derived from these. For a particular case and categorical variable, a code of 1 is used for the category into which the case falls, and 0 for the remaining categories. If L is the total number of categories across all p variables and n is the number of cases, this coding gives rise to an $n \times L$ incidence or indicator matrix in which the row sum for each case is p.

Two papers that use MCA, and where data are given, are Holm-Olsen (1985) and Engelstad (1988). The data from the latter study, of the typology of pit-houses in Arctic Norway, are also analyzed in Baxter (1994a). Wilson's (1998) analysis of Pacific rock-art anthropomorphs and geometrics is another example of MCA. For example, one characteristic of an anthropomorph is the 'head', which may be classified as 'round', 'square' or 'triangular', so that 1 is recorded for the appropriate option, and 0 otherwise. The numerous correspondence analyses in Wilson (1998) are used to investigate similarities between rock-art motifs found in different regions.

To illustrate, the data of Table 7.2, previously analyzed in Section 7.4.2 using various forms of multidimensional scaling, will be used. The outcome of MCA applied to these data is shown in Figure 11.3, and may be compared with the outputs from the analyses by classical and non-metric MDS in Figures 7.5 and 7.6. For comparison, and because it is the relationship between cases that is mainly of interest only the row plot is shown.

The separation between the different petrographic groups is reasonably clear, and arguably slightly better than that achieved in the previous analyses, though there is not much in it. This analysis is given, in slightly modified form, in Cau *et al.* (2002). For the larger and more complex data set studied in that paper it was found more informative to use CA and classical MDS together than either method in isolation, with other forms of MDS adding little extra to the analysis.

11.4.4   Some applications

Data transformation
Lockyear (2000b) was a response to a challenge laid down in Reece's (1995) analysis of coin finds from Roman sites in Britain, for 140 assemblages characterized by the numbers of coins from 21 different periods. Using 'informal, simple and intuitive' methods (Lockyear, 2000b: 399), Reece produced results that he stated could not readily be obtained by more formal statistical methods, which were sensitive to sample size. Using a combination of cluster analysis and CA,

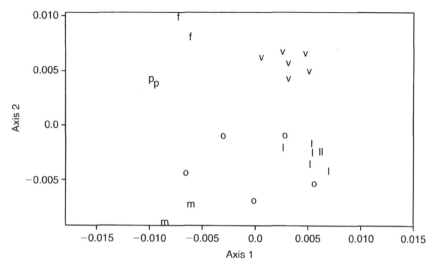

Figure 11.3 A multiple correspondence analysis of the petrographic data of Table 7.2. This may be compared with Figures 7.5 and 7.6, where the labeling is given.

Lockyear was able to produce an analysis of the data that Reece, in the postscript to Lockyear's paper, acknowledged to 'mark a major step forward in the numerical analysis of coin hoards'.

Lockyear's CA had a number of interesting features. The initial CA of the 140 assemblages was unsuccessful, in the sense that the column plot did not show the expected sequencing of the time periods successfully. A more satisfactory analysis was obtained when four influential outliers were removed, and the data subjected to a square-root transformation. Figure 8 in Lockyear (2000b), after this treatment, showed the expected horseshoe pattern for the periods quite clearly and in the expected sequence. Given this clear evidence of structure from the column plot, the subsequent interpretation of the row plot was described as 'pleasing'. In Lockyear's (2000b: 418) words:

> we have a dirty and mixed data set which despite all its problems can still show broad-scale patterning between and within sites ... the explicit expansion of the use of coinage from things called 'towns', to 'villas' and then to 'rural sites' is here made explicit.

The two technical aspects of the analysis of interest are the effect of outlier removal, and the use of data transformation. The value of an iterative approach, successively 'peeling' the data to reveal more subtle structure, was discussed earlier, and can involve the omission of outliers. Orton (1997) provides another example of the application of CA to similar coin data that benefits from the removal of a single extreme point. The main effect of the square-root transformation in Lockyear's analysis was to downweight the effect of the more extreme cases, producing a more readable plot, and to reduce the effect of different sample sizes on the analysis. Other uses of data transformation prior to CA can be found in the completely different context of the analysis of artifact compositional data (e.g., Underhill and Peisach, 1985; Bollong *et al.*, 1997). Transformations used include the square root of the elemental compositions, division by element means, rescaling so that the element sum is 100%, and use of ratios to some element. Studies comparing the use of CA for artifact compositional analysis with other multivariate methods include Baxter *et al.* (1990), Baxter (1991) and Bollong *et al.* (1997), some conclusions from the first of these being given in Section 7.2.2.

Some applications of correspondence analysis in the North American literature
In reviewing the use of CA in archaeology up to about 1992, I could locate very few applications published in the North American journal literature or by North American scholars (Baxter, 1994a: 135). Cowgill (2001) has suggested that CA 'was virtually unheard of in the US until publications about it began to appear in English' around the late 1980s. In a paper published in *American Antiquity* on the use of CA for seriation, Duff (1996: 90) wrote that CA was 'not well established in Americanist literature'.

The situation has changed since then. The following selection of papers that have subsequently appeared serves, as much as anything, to illustrate the uses to which CA has been put. Duff's (1996) application was to problems of micro-seriation intended to show relative intersite or intrasite relationships, over short timespans. He focused on the problem of seriating assemblages using ceramic types, as opposed to attributes, types being defined as constellations of attributes. His results were compared with earlier detailed seriations of LeBlanc (1975) using multidimensional scaling of attribute data, and it was concluded that his results were comparable with those of LeBlanc.

Shott (1997b) studied Great Lakes Paleoindian assemblages characterized by counts of eight tool types. Analysis of the row plot for one set of 10 assemblages suggested some grouping of assemblages, though this grouping appeared to depend on which tools were used in the analysis. Less grouping was evident in analyses of a second set of 22 assemblages. An interesting feature of the analyses was that omission of several assemblages with small tool counts had little effect on the patterns observed.

Clouse's (1999) application was to historical archaeological data derived from excavations at Fort Worth, Minnesota. The fort was initiated in AD 1820 and continued in active existence for 125 years. Assemblages were associated with 'context units' divided into three classes, 'defense' (e.g., towers, wall lines), 'support' (e.g., shops, offices) and 'habitation' (e.g., barracks, latrines). These were cross-tabulated with functional groups that were aggregates of counts of artifact classes, examples being 'leisure' (games, toys, teaware, musical instruments), 'construction tools' and 'writing utensils'. Correspondence analyses were undertaken separately for each class of context unit, and used to suggest clusters of units that, on the basis of the artifact evidence, had a similar function. For example, five of seven contexts in the 'defense' category appeared similar, with two outliers – both towers – having rather different profiles, dominated by a small number of functional groups associated with later use rather than their original function. One tower, for instance, was initially constructed as what is described as a 'last stand' defensive bastion, but was later used as an ordnance storehouse and was dominated by artifacts classified as 'armaments'. As is often the case with CA, the patterns revealed are frequently evident on careful inspection of the tabular data on which they are based, but the graphical display of CA makes them much clearer. In several of Clouse's analyses, plots were dominated by a small number of contexts/artifact classes. This use of CA for integrating the analysis of structural and artifact evidence in order to investigate function is foreshadowed in the British literature in Cool and Baxter (1995) in their analysis of small finds assemblages from excavations in Roman York.

A novel use of CA is provided in Robertson's (1999) spatial and multivariate analysis of pottery assemblages from Teotihuacan, Mexico. The background to this is described more fully in Section 14.3.1. To anticipate, Robertson used Bayesian methods to estimate the mean proportion of 12 ceramic types calculated within 4441 assemblages. Essentially such estimates produce a smoother map of the distribution of types than is achieved using observed proportions, which may be distorted by small sample sizes. Application of CA to the *estimates* enabled Robertson to group together ceramic categories having a similar functional interpretation in a more satisfactory way than analysis of the observed proportions did.

11.5 The mathematics of correspondence analysis

11.5.1 Chi-squared distance

In the following discussion the initial focus will be on obtaining a graphical representation of the rows of N. In many archaeological applications the rows would correspond to assemblages and columns to types. Many applications of CA are concerned with the joint representation of the rows and columns of N. Analogous formulas are available for the columns or, most simply, the following results could be applied to the transpose of N.

The chi-squared statistic of equation (11.2) can, on substituting for the E_{ij} defined in the denominator of equation (11.1), be rearranged as

$$I = X^2/N = \sum_{ij} (n_{ij} - E_{ij})^2/n_i.n_{\cdot j} \tag{11.12}$$

a quantity known as the *total inertia* of the data. Under the null hypothesis of no association between the categorical variables defining the rows and columns of the table, it is expected that I will be 'small' since the n_{ij} and E_{ij} should be similar. The more 'heterogeneous' the data, when there is association, the larger I is expected to be. The right-hand side of equation (11.12) can also be written as

$$\sum_i \frac{n_i.}{N} \sum_j \left(\frac{n_{ij}}{n_i.} - \frac{n_{\cdot j}}{N}\right)^2 / \frac{n_{\cdot j}}{N} = \sum_i p_i. \sum_j (p_{ij} - p_{\cdot j})^2/p_{\cdot j} \tag{11.13}$$

where $p_{ij} = n_{ij}/n_i.$, $p_i. = n_i./N$ and $p_{\cdot j} = n_{\cdot j}/N$.

The vector $(p_{i1}\, p_{i2} \cdots p_{iJ})$ is the row profile of row i. The vector $(p_{\cdot 1}\, p_{\cdot 2} \cdots p_{\cdot J})$ is the average row profile. The second term in equation (11.13), $\sum_j (p_{ij} - p_{\cdot j})^2/p_{\cdot j}$, is defined as the *chi-squared distance* between row i and the average row profile. The chi-squared distance between the profiles for rows i and k is defined as

$$d_{ik}^2 = \sum_j \frac{1}{p_{\cdot j}} (p_{ij} - p_{kj})^2. \tag{11.14}$$

It can be seen from equations (11.13) and (11.14) that chi-squared distance is a weighted Euclidean distance based on row proportions. The distance between two rows of $\mathbf{N}$ will be zero if they have the same proportions. It can also be seen from the weighting factors $1/p_{.j}$ that rare categories of the column variable have a greater influence on the distance than common ones.

11.5.2   Correspondence analysis as PCA

Correspondence analysis can be presented in a variety of ways (Gower and Hand, 1996: 175–90). Viewed as a form of PCA for counted data, the interpoint distances in the usual PCA plot should approximate the chi-squared distances defined by equation (11.14). This can be achieved by basing a PCA on

$$y_{ij} = \frac{n_{ij} - n_{i.}n_{.j}/N}{\sqrt{n_{i.}n_{.j}}} \tag{11.15}$$

and rescaling the component scores for row i by dividing by $n_{i.}^{1/2}$.

The inertia plays the same role in CA as total variance does in PCA, and the proportion of inertia explained by a component in CA is analogous to the proportion of variance explained by a component in PCA. Other ways of decomposing the inertia are possible and sometimes used. The square of equation (11.15) is the contribution to inertia of an individual cell. Summing across a row gives the contribution of the row to the inertia. Viewed as a PCA of (11.15), giving rise to components of the form

$$Z_i = a_{i1}Y_1 + a_{i2}Y_2 + \cdots + a_{iJ}Y_J,$$

the a_{ij}^2 can be interpreted as the contribution of column j to the inertia of component i, and measure the importance of the column in defining the component.

A practical point of importance is that, in the same way that the size of X^2 can be unduly influenced by rows or columns of $\mathbf{N}$ with small totals, a CA can similarly be unduly influenced. Underhill and Peisach (1985) define the mass of a row as $\sum_j n_{ij}/\sum_{ij} n_{ij}$ and counsel against using rows for which the mass is 'small'. Similar comments apply to columns. There are no hard-and-fast rules for defining what is meant by 'small', and different practitioners tend to operate different rules of thumb. My own practice is to undertake analyses with and without problematic rows and columns, and see if the the interpretation of the main patterns in the analyses is the same. One should not overinterpret features of plots primarily associated with rows or columns having a small mass, though these may suggest hypotheses worthy of future examination.

11.5.3   Matrix formulations

For a thorough account of the matrix algebraic approach to CA, Greenacre (1984) or Gower and Hand (1996) can be consulted. The prime purpose here is to summarize some results used in Section 12.2.2 on bootstrapping a CA.

Define $\mathbf{R} = diag(n_{i.})$ to be an $I \times I$ diagonal matrix, with ith diagonal $n_{i.}$, and $\mathbf{C} = diag(n_{.j})$, a $J \times J$ diagonal matrix. Centering the matrix $\mathbf{N}$ as

$$\mathbf{X} = \mathbf{N} - \frac{\mathbf{R}\mathbf{1}_I \mathbf{1}_J' \mathbf{C}}{N}$$

where $\mathbf{1}_I$ is an $(I \times 1)$ vector of 1s and $\mathbf{1}_J$ is a $(J \times 1)$ vector of 1s, undertaking a PCA on $\mathbf{Y} = \mathbf{R}^{-1/2}\mathbf{X}\mathbf{C}^{-1/2}$, and basing plots on the columns of $\mathbf{R}^{-1/2}\mathbf{Z}$ where $\mathbf{Z}$ holds the component scores, is equivalent to the PCA analysis described above.

Equivalently, in terms of the singular value decomposition (equation (6.5)) of $\mathbf{X}$ defined as above, the plotting coordinates for the rows are obtained from $\mathbf{A} = \mathbf{R}^{-1/2}\mathbf{UD}$, while those for columns are obtained from $\mathbf{B} = \mathbf{C}^{-1/2}\mathbf{VD}$. A little algebra (Gower and Hand, 1996: 181) shows that $\mathbf{A}$ and $\mathbf{B}$ are related by the following transition formulas

$$\mathbf{A} = \mathbf{R}^{-1}\mathbf{XBD}^{-1}, \tag{11.16}$$

$$\mathbf{B} = \mathbf{C}^{-1}\mathbf{X}'\mathbf{AD}^{-1}, \tag{11.17}$$

which arise in the method of reciprocal averaging, and can be viewed as an iterative procedure for calculating the singular value decomposition.

11.5.4 Multiple correspondence analysis – theory

Suppose observations are available for n cases, classified by p categorical variables, and variable k has L_k levels. Then L_k dummy variables corresponding to the levels may be defined giving rise to an $n \times L_k$ matrix of dummy variables. If this is done for each variable in turn an $n \times L$ matrix is obtained, where $L = \sum_k^p L_k$. Multiple correspondence analysis is ordinary CA applied to this matrix. It can be thought of as principal component analysis applied to a derived matrix and is described in this way in Gower and Hand (1996) whose treatment is followed below.

Let $\mathbf{G}$ be the $n \times L$ matrix described above; let l_i be the count of 1s in the ith column; and let $\mathbf{L}$ be the $n \times n$ diagonal matrix with diagonal elements l_i. Multiple CA is equivalent to PCA applied to the derived data matrix $\mathbf{GL}^{-1/2}$. In the derived data matrix the squared Euclidean distance between two cases is

$$d_{ij}^2 = (\mathbf{g}_i - \mathbf{g}_j)\mathbf{L}^{-1}(\mathbf{g}_i - \mathbf{g}_j)',$$

where $\mathbf{g}_i$ is the ith row of $\mathbf{G}$. The contribution to d_{ij}^2 of variable k is 0 if the categories of i and j are the same and $l_i^{-1} + l_j^{-1}$ otherwise (Gower and Hand, 1996). This shows that rare categories have higher weight than common categories in the distance calculation.

Euclidean distance for $\mathbf{GL}^{-1/2}$ can be viewed as chi-squared distance applied to $\mathbf{G}$. The main justification for using chi-squared distance arises in the context of simple CA and the appropriateness of applying it to a matrix of dummy variables, with the differential weighting of rare categories, has been questioned. Noting this, Gower and Hand (1996) consider other forms of distance that might be used. One that they favor is the extended matching coefficient that, for two cases, is the number of matches among the p variables. One reason for favoring this is that it gives equal weight to all variables. The matrix giving the matching coefficient between all pairs may be written as $\mathbf{GG}'$, an $n \times n$ similarity matrix. PCA may be applied to the matrix $\mathbf{G}$ to investigate the relationship between cases (Gower and Hand, 1996: 67), and this and the MCA can be regarded as special cases of a PCA analysis of $\mathbf{GL}^{-a}$ for $a = 0$ or 0.5. PCA is equivalent to classical metric multidimensional scaling. Non-metric multidimensional scaling techniques can also be used to investigate structure in the data. Where these require a dissimilarity matrix as input this is equivalent to basing analysis on $\mathbf{P} - \mathbf{GG'}/p$, where $\mathbf{P}$ is an $n \times n$ matrix of 1s. Examples of such analyses were given in Section 7.4.2.

12

Computer-intensive methods

Monte Carlo analysis has not been used extensively in archaeology . . . .
Although it is intuitively appealing and allows users to address otherwise
intractable statistical problems, Monte Carlo testing is not even mentioned
by many texts in archaeological statistics. (Fisher *et al.*, 1997: 584–5)

12.1 Introduction

One dictionary of statistics (Everitt, 1998: 216) defines Monte Carlo methods as:

> Methods for finding solutions to mathematical and statistical problems
> by *simulation*. Used when the analytic solution of the problem is either
> intractable or time consuming.

(The emphasis is in the original.) The same source (Everitt, 1998: 307) defines
simulation as:

> The artificial generation of random processes (usually by means of *pseudo-
> random* numbers and/or computers) to imitate the behaviour of particular
> statistical models.

The terms 'Monte Carlo method' and 'simulation' seem to be used interchangeably
in the archaeological literature and, if this equation is allowed, their use in archaeology is more widespread that the opening quotation from Fisher *et al.* (1997)
suggests. Some applications of Monte Carlo methods in archaeology, dating to
the 1970s and early 1980s are discussed in Sections 20.4 and 19.2.3.

Manly (1997: 68) observes that with a Monte Carlo test 'the significance of an
observed test statistic is assessed by comparing it with a sample of test statistics
obtained by generating random samples using some assumed model', and adopts
the view that bootstrap and jackknife methods and randomization tests, to be
discussed below, are particular applications of the Monte Carlo method. Rather
than worrying about definitional niceties, all these approaches can be subsumed
under the heading of *computer-intensive* methods.

Computer-intensive studies, often described as simulation, have been used in archaeology for some time (Renfrew and Cooke, 1974; Doran and Hodson, 1975; Hodder and Orton, 1976; Hodder, 1978; Freeman, 1987; Bell *et al.*, 1990; Aldenderfer, 1991). Many applications involve the (mathematical) modeling of archaeological processes rather than statistical analysis, and only the latter is of concern here. Some of the ideas of bootstrapping, jackknifing and other randomization tests are discussed in the sections to follow, largely through examples. The final section provides a very brief introduction to some of the simpler aspects of Markov chain Monte Carlo methods, important for some of the Bayesian applications discussed in Chapters 14 and 15 and elsewhere. The reader is referred to texts such as Efron and Tibshirani (1993), Davison and Hinkley (1997) and Manly (1997) for more technical details.

The increased use of computer-intensive statistical methods to address archaeological problems is likely to be one of the most important developments in archaeological statistics in the early 21st century. Other computer-intensive methods that have found applications in archaeology are discussed elsewhere in the book, and include the use of cross-validation to assess the success of discriminant and classification tree analyses (Chapter 9) and to determine the degree of smoothing in a trend-surface analysis using the lowess smoother (Section 13.5.2); non-parametric smoothing (Chapter 3) and regression (Section 5.6); the detection of multivariate outliers (Section 10.2); projection pursuit methods (Section 7.5); and model-based clustering (Section 8.4).

12.2   The bootstrap

12.2.1   Basic ideas

The *bootstrap* is defined by Efron and Tibshirani (1993: 5) as 'a data-based simulation method for statistical inference'. First considered systematically by Efron (1979), it has been the subject of intensive research in more recent years (Efron and Tibshirani, 1993; Davison and Hinkley, 1997). In its simplest manifestation, given a population parameter of interest, θ, and an estimate, $\hat{\theta}$, from a sample of size n, a bootstrap sample is generated by sampling from the original sample *with replacement*. A bootstrap estimate, $\hat{\theta}^*$, is generated from this sample. This is repeated a large number of times, say B, and inferences about the properties of the estimate $\hat{\theta}$ are based on the distribution generated by the B values of $\hat{\theta}^*$. This is a computer-intensive approach to statistical inference that frees the analyst from the sometimes restrictive assumptions of classical inferential methods, and can allow problems to be addressed that do not admit of an analytic solution.

12.2.2   Examples of bootstrapping

Bootstrap confidence intervals
For illustration, some data from Pakkanen (1998), given in Table 12.1, on the heights of 60 column drums from the temple of Athena Alea at Tegea in the Peloponnese will be used. Pakkanen was concerned to establish confidence limits for the mean drum height, with a view to estimating limits for column heights, composed of six drums. The temple was built with 36 columns, giving $N = 216$ drums in all, and he treated the surviving $n = 60$ as a sample from this finite

Table 12.1 Heights (m) of 60 column drums from the temple of Athena Alea at Tegea (Pakkanen, 1998)

1.464	1.515	1.444	1.472	1.662	1.413	1.561	1.470	1.349	1.484
1.658	1.399	1.481	1.479	1.331	1.580	1.411	1.438	1.522	1.477
1.472	1.398	1.480	1.473	1.415	1.498	1.511	1.448	1.474	1.478
1.514	1.382	1.493	1.474	1.469	1.479	1.447	1.347	1.473	1.472
1.465	1.469	1.476	1.631	1.643	1.321	1.368	1.484	1.356	1.457
1.668	1.320	1.478	1.708	1.466	1.474	1.500	1.510	1.500	1.482

Table 12.2 Lower (L) and upper (U) bounds for 95% bootstrap confidence intervals for the mean drum height of the data in Table 12.1, using four different methods

Method	L	U
Basic	1.455	1.496
Studentized	1.456	1.499
Percentile	1.457	1.498
BCa	1.457	1.499

population. The mean and standard deviation of heights for the surviving drums are $\bar{x} = 1.4764$ m and $s = 0.082\,75$ m.

Assuming sampling from a normally distributed population, temporarily ignoring the fact that the population is finite, and applying standard theory based on the t-statistic, a 95% confidence interval for the mean height is (1.455, 1.498). The position of a drum within a column is known, and those in the lowest positions, A and B, show much smaller variation in their height than those in positions C–F. Because of the differing variation according to position, the normality assumption is unlikely to be even approximately true. Although the sample size is such that the central limit theorem probably ensures that the calculated intervals are reasonable, Pakkanen checked this using bootstrapping methodology. Table 12.2 shows the results from four different methods for obtaining bootstrap confidence intervals, using the terminology of Davison and Hinkley (1997) and obtained using functions in their S-Plus library, boot. These results, which do not take account of finite population sampling, can be seen to be in good agreement with each other and with the parametric interval (1.455, 1.498).

The results in Table 12.2 were based on 4999 bootstrap samples. Texts typically suggest at least 1000 samples for this kind of confidence interval estimation. The left-hand panel of Figure 12.1 shows a kernel density estimate for the 4999 means estimated from the bootstrap samples. Also indicated are the points $\bar{x}^*_{1-\alpha/2} = 1.457$ and $\bar{x}^*_{\alpha/2} = 1.498$ which, for $\alpha = 0.05$, encompass the central 95% of the bootstrapped distribution. The interval so defined is, in fact, the confidence interval obtained by the bootstrap percentile method, called the first percentile method in Manly (1997). In the above notation $\bar{x}^*_\gamma$ is used to indicate the empirically derived percentile of the bootstrapped distribution exceeded by $100\gamma\%$ of the distribution. For a general parameter, θ, the percentile method gives the $100(1 - \alpha)\%$ interval

$$\hat{\theta}^*_{1-\alpha/2} < \theta < \hat{\theta}^*_{\alpha/2}. \tag{12.1}$$

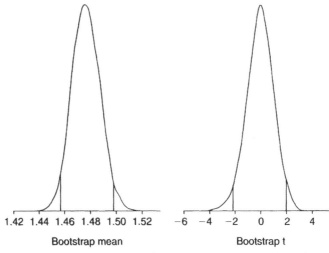

1.42 1.44 1.46 1.48 1.50 1.52   −6 −4 −2   0   2   4

Bootstrap mean   Bootstrap t

Figure 12.1   The left-hand panel shows a KDE for 4999 bootstrap samples of the mean using the data of Table 12.1, together with the 95% confidence limits obtained using the percentile method. The right-hand panel shows a similar figure for the bootstrap-*t* distribution of equation (12.4).

Other forms of percentile interval can be defined. The basic percentile interval of Davison and Hinkley (1997), called the second percentile method in Manly (1997), has the form

$$2\hat{\theta} - \hat{\theta}^*_{\alpha/2} < \theta < 2\hat{\theta} - \hat{\theta}^*_{1-\alpha/2} \qquad (12.2)$$

The texts referenced may be consulted for the logic behind this.

An underlying assumption behind the bootstrap percentile method is that there exists a monotonic increasing function such that transformed values $f(\hat{\theta})$ are normally distributed with mean $f(\theta)$ and standard deviation 1. Manly (1997: 39) discusses how this leads to the interval (12.1) and also provides an illustration where the assumption breaks down. To try and deal with such situations more complex forms of interval have been devised, one of which is the *accelerated bias-corrected* (BCa) interval of Table 12.2. Details are quite complicated, and reference may be made, once again, to the texts cited. The basic assumption (Manly, 1997: 49) is that a transformation exists such that $f(\hat{\theta})$ has a normal distribution with mean $f(\theta) - z_0[1 + af(\theta)]$ and standard deviation $1 + af(\theta)$, where z_0 is a bias-correction term and a allows the standard deviation of $f(\hat{\theta})$ to vary linearly with $f(\theta)$.

The procedure actually used by Pakkanen (1998) was based on the studentized interval of Table 12.2, also known as the bootstrap-*t* interval (Manly, 1997). For a parameter, θ, and an estimate, $\hat{\theta}$, of it define a Student-*t*-like quantity of the form

$$t = \frac{\hat{\theta} - \theta}{\hat{v}^{1/2}}, \qquad (12.3)$$

where $\hat{v}^{1/2}$ is an estimated standard error of $\hat{\theta}$. Under certain assumptions this can then be used to derive a $100(1 - \alpha)\%$ confidence interval of the form

$$\hat{\theta} - t_{\alpha/2}\hat{v}^{1/2} < \theta < \hat{\theta} + t_{1-\alpha/2}\hat{v}^{1/2}.$$

The studentized, or bootstrap-t, interval is obtained by defining a bootstrap-t statistic of the form

$$t^* = \frac{\hat{\theta}^* - \hat{\theta}}{\hat{v}^{*1/2}} \qquad (12.4)$$

estimating $t_{\alpha/2}$ and $t_{1-\alpha/2}$ as $t^*_{\alpha/2}$ and $t^*_{1-\alpha/2}$ from the bootstrap distribution of t^*, and defining the interval as

$$\hat{\theta} + \hat{v}^{1/2}t^*_{1-\alpha/2} < \theta < \hat{\theta} + \hat{v}^{1/2}t^*_{\alpha/2}.$$

For the results in Table 12.2, $\hat{\theta} = \bar{x}$, $\hat{v}^{1/2} = s/\sqrt{n}$, the estimated standard error of $\bar{x}$, $t^*_{1-\alpha/2} = -2.163$ and $t^*_{\alpha/2} = 1.983$. The percentage points of t^* were estimated from 4999 bootstrap samples, and the distribution of these samples with these points is shown in the right-hand panel of Figure 12.1. They contrast with figures given by Pakkanen (1998: 54) of -2.186 and 1.846. Additional simulations showed some variation in the values obtained, which changed the results given in Table 12.2 by up to 0.002, a substantively unimportant difference.

So far we have ignored the fact that the 60 drums are a sample from a finite population of 216. Calculating a classical confidence interval with a finite population correction gives a 95% interval from 1.458 to 1.495 m. To allow for finite population sampling in the bootstrap approach the correction factor may be incorporated into the estimate of $\hat{v}^{1/2}$ in equation (12.4). For our results this gives a bootstrapped studentized interval of (1.458, 1.496) compared with Pakkanen's (1998: 54) result of (1.460, 1.496). Unsurprisingly, agreement with the classical interval is good, and the intervals are shorter than those in Table 12.2 because account has been taken of the finite population size.

It was mentioned that the studentized interval is valid given certain assumptions. One of these is that the distribution of equation (12.3) is (approximately) independent of θ. Efron and Tibshirani (1993: 161–2) suggest that it is particularly suitable for location statistics such as the mean, but less trustworthy for more general problems, for which they prefer some of the percentile methods. Davison and Hinkley (1997: 211) observe that the studentized bootstrap and adjusted percentile methods are inherently more accurate than the basic bootstrap and unadjusted percentile methods.

The bootstrap method is not a panacea. All the texts referenced include examples where bootstrapping does not work well. This can occur when small samples are taken from non-normal distributions and interest centers on parameters other than the mean. For some kinds of data, transformation before applying bootstrapping produces better results.

Bootstrapping and correspondence analysis
The next, more complex, illustration of bootstrapping is based on Ringrose (1992), who was in turn inspired by Greenacre (1984: 214–18). It was designed to investigate the stability of displays produced in a correspondence analysis. From

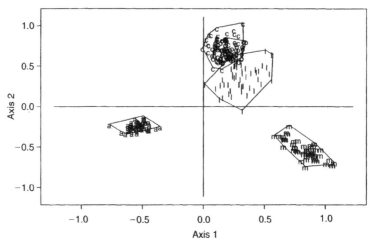

Figure 12.2   A bootstrapped correspondence analysis, based on 50 replicates, for the data of Table 2.4. Different letters label the five different assemblages.

Section 11.5.3 and equations (11.16) and (11.17) the transition formulas in correspondence analysis are $A = R^{-1}XBD^{-1}$ and $B = C^{-1}X'AD^{-1}$, where the first two columns of A and B hold the coordinates for the row and column plots. Here $X = N - R11'C/n_{..}$, N is the matrix of counts, and R and C are diagonal matrices of the row and column sums.

Ringrose's (1992) approach was to obtain a bootstrapped matrix N^*, based on a sample of size $n_{..}$ in which the probability of a case being in the (i,j)th cell was $n_{ij}/n_{..}$. This can be used to derive bootstrap replicates R^* and C^*. If A, B and D are available from the original analysis then bootstrapped coordinates are obtained by plugging the bootstrapped matrices R^* and C^* into the transition formulas.

This is illustrated in Figure 12.2 for the columns of Table 2.4, using 50 bootstrap replicates. It is clear that three of the five assemblages seem genuinely distinct from each other and the remaining two. These last two show considerable overlap and do not differ significantly from each other.

In Figure 12.2 the results from the bootstrapping are displayed as a cloud of points for each assemblage, enclosed by their convex hull. For clarity of presentation, and with more columns of data, or more overlap, the convex hulls only might be presented. A practical limitation of the method is that with very many rows or columns the plot as a whole would be difficult to read. Ringrose (1992) also noted that if $n_{ij} = 0$ for a cell then the bootstrap method used would result in this cell remaining empty for all replicates. He discussed possible ways of avoiding this, but concluded that these introduced problems of their own and did not implement them.

Bootstrapping and biplots
Hopkins (1999) used a *parametric* bootstrap to investigate aspects of variable selection for biplots. Using, as an illustration, a 30×13 data set of ceramic pot measurements published by Impey and Pollard (1985), an initial principal component analysis of standardized data established that there appeared to be four main groups of variables. One variable was selected as representative of each of

these four groups. Assuming a multivariate normal distribution for these variables, with mean vector and covariance matrix the same as that of the original data, 200 samples of size 15 were generated and a biplot analysis generated for each. Superimposing the variable plots from each analysis produced a fan-like pattern for each variable, the spread of the fans allowing an assessment of the stability of the representation of a variable. In the example presented the lack of overlap of the different fans suggested that they were measuring different aspects of the data. Experiments with other sets of variables showed overlapping fans, suggesting that some variables were redundant, and it was concluded that the method could be used to justify measuring fewer variables in future studies.

Bootstrapping and ternary diagrams
Ternary diagrams or triangular graphs are widely used in different subdisciplines of archaeology, and an example was given in Figure 7.1. Given a sample of size n, classified into three categories and expressed as percentages, the sample can be plotted as a point in a two-dimensional ternary diagram because the compositional constraint (that the sum is 100%) reduces the dimensionality of the data from three to two.

Steele and Weaver (2002) describe a bootstrapping method for use with ternary diagrams that allows an assessment of whether or not two or more samples are genuinely distinct. The idea is simple. A large number of 'fictional' samples are generated by sampling with replacement from the original sample of size n. These can be plotted on the ternary diagram, for example by using a 95% contour derived from a kernel density estimate based on the bootstrapped samples, and inspected to see the extent of overlap for contours associated with different samples. The specific application in Steele and Weaver (2002) is to representing age structures in zooarchaeological samples where specimens are classified as 'juvenile', 'prime' or 'old', but the methodology they describe is general.

12.3 The jackknife

The jackknife pre-dates the bootstrap and also provides an approach to testing hypotheses and obtaining confidence intervals when standard theory may not be applicable. The use of the jackknife in discriminant analysis was discussed in Chapter 9. Let θ be a statistic of interest, estimated by $\hat{\theta}$, and let $\hat{\theta}_{-i}$ be the estimate obtained on omitting the ith value. There are n values of $\hat{\theta}_{-i}$ which may be used to define *pseudo-values*

$$\hat{\theta}_i^* = n\hat{\theta} - (n-1)\hat{\theta}_{-i} \qquad (12.5)$$

whose mean, $\bar{\hat{\theta}}^*$, is the jackknife estimate of θ. An approximate $100(1 - \alpha)\%$ confidence interval is

$$(\bar{\hat{\theta}}^* - t_{\alpha/2}s/\sqrt{n}, \bar{\hat{\theta}}^* + t_{\alpha/2}s/\sqrt{n}),$$

where s is the estimated standard error of the pseudo-values. Equation (12.5) and the way it is used is suggested by the result

$$x_i = n\bar{x} - (n-1)\bar{x}_{-i}$$

which holds for the sample mean (Manly, 1997: 24).

The jackknife has occasionally been used in archaeology, and an application by Kaufman (1998) in the study of assemblage diversity is discussed in Section 20.5. Kvamme *et al.* (1996) used the jackknife in their study of standardization in ceramic assemblages. For a sample of 725 pots they were concerned to establish confidence limits for the standard deviation of a variety of dimensions, including the circumference, and on the basis of graphical analysis of the data were unwilling to resort to normal-based theory. Using the circumference as an example, they showed that a jackknifed estimate of a 95% confidence interval, using either raw or logarithmically transformed data, was about 2.5 times wider than an interval estimated using normal theory. For the data of Table 12.1 the 95% confidence interval for the standard deviation based on normal theory is (0.070, 0.101) which is narrower than the jackknife interval of (0.064, 0.102) that, in turn, is similar to the percentile and BCa bootstrap intervals.

In terms of the computing power now available, the jackknife cannot really be regarded as a computer-intensive method. Although the results from the jackknife and bootstrap are similar in the example given above, this is not generally the case for more complex statistics. Efron and Tibshirani (1993: 245) observe that the confidence interval derived via the pseudo-values may not work well, preferring intervals based on the bootstrap, and that the jackknife is generally less efficient than, and an approximation to, the bootstrap.

12.4   Other applications of randomization

12.4.1   Buildings at Danebury

Fletcher and Lock (1984a, 1990) and Bell *et al.* (1990) addressed the problem of identifying square structures within a dense distribution of post-holes revealed in excavations at the Iron Age hillfort of Danebury in Hampshire, England. Having identified a number of possible structures, using a purpose-designed computer program, the problem was to determine if these could have arisen by chance, simply because of the density of post-holes, or whether they were sufficiently numerous to suggest the existence of real structures.

To investigate this, simulation was used in which locations of post-holes were randomly generated, and the number of apparent structures in the simulated data set compared with those from the original data. One problem was to ensure that the simulated data were similar to the real data in terms of their variable density across the site. In the 1984 paper the area studied was subdivided into three parts, to take account of lack of contiguity, and obvious differences in post-hole density. Within each sub-area locations were randomly simulated. In the 1990 paper a method suggested by Litton and Restorick (1983) was used, the idea being to perturb post-hole coordinates (x, y) to new coordinates $(x + \delta x, y + \delta y)$ where δx and δy were uniformly distributed over an interval $(-r, r)$. This was repeated for different values of r. The idea here was that, if the number of apparent structures was purely a chance phenomenon then this number would remain relatively stable as r increased, eventually showing a gradual decline when r was sufficiently large to remove the spatial clustering in the data. If the number of apparent clusters is greater than can be expected by chance, however, then there should be a sharp decline in the values found at relatively small values of r. Bell *et al.* (1990) presented results for the 1976–7 and 1980 excavation seasons, and concluded that there was strong evidence for the existence of buildings in the former case, but not the latter.

12.4.2 Bronze Age cairns on Mull

Geographical information systems have had a considerable impact in archaeology from the early 1990s on. This is discussed more fully in Chapter 13. A number of authors have lamented the lack of statistical rigor in many applications, and Fisher *et al.* (1997: 582) explicitly intended their spatial analysis of visible areas from the Bronze Age cairns of Mull to 'contribute to the quantitative analysis and development of testable hypotheses about archaeological sites in the landscape'.

Their analysis focused on the *viewshed*, which is the area visible from a location, such as an archaeological site, that can be determined within a geographical information system. Of particular interest were the viewsheds from 14 definite or possible Bronze Age cairns from northern Mull, an island off the west coast of Scotland, and the hypothesis that the visibility of the surrounding area may have had some causal influence on the siting of the cairns. For each cairn, quantities such as the area within the viewshed, area of land or sea within the viewshed, and ratio of land to sea area can be calculated, and summarized using statistics such as the mean, maximum and minimum. The question then arises as to whether or not these results differ 'significantly' from what is to be expected by chance.

Data were available in the form of a digital elevation model (DEM) for north Mull, for gridded data with a 50 m resolution. Cairns were associated with the grid intersection to which they were nearest, and for any variable – such as area of viewshed – 14 observations, $\mathbf{x} = (x_1 \, x_2 \cdots x_{14})$, were generated from which some statistic of interest, $\hat{\theta}$, could be defined. Next, a random sample of size 14 was selected from the relevant population of grid intersections and similar calculations undertaken to obtain a statistic $\hat{\theta}_i^*$, repeated for $i = 1, \ldots, 19$. Except for the rather smaller number of simulations undertaken, the idea is similar to that used in bootstrapping, in that the significance of $\hat{\theta}$ is judged in relation to its position relative to the body of the $\hat{\theta}_i^*$.

Defining what is the 'relevant' population requires thought in these types of application. Three approaches investigated in Fisher *et al.* (1997) were to randomly sample all grid intersections within the DEM for north Mull; to select a stratified random sample based on the distance of cairns from the sea; and to sample from locations in close proximity to the cairns. Without going into details, these analyses enabled Fisher *et al.* (1997) to establish, among other things, that the area visible from the cairns was greater than to be expected for randomly selected locations, and that the area of sea visible, and sea to land area ratio, were also greater than expected for random locations.

The analysis reported here is identical in spirit to the use of Monte Carlo methods for statistical inference in *raster* GIS contexts, where spatial data are digitally encoded in gridded format, advocated by Kvamme (1996, 1999: 170).

12.4.3 Artifacts in graves

Manly (1996) considered a number of computer-intensive methods, including randomization tests, for the statistical analysis of artifacts in graves, where artifacts were recorded as present or absent. His data consisted of a 154 × 17, burial by artifact type, occurrence matrix, from the Khok Phanom Di cemetery in central Thailand dating to about 2000–1500 BC. Fifty-nine of the burials included no artifacts. Information was also available on the estimated age and sex of the deceased.

One question of interest was whether pairs of artifacts tended to occur together, or not be found together, more than might be expected by chance. Defining O_{ij} as the observed number of co-occurrences of types i and j, and E_{ij} as the expected number under the hypothesis that the distribution of types among graves was random, a statistic of the form

$$S = \sum_{i=1}^{p} v_i/p,$$

where

$$v_i = \sum_{i=1}^{p} (O_{ij} - E_{ij})^2$$

and p is the number of artifact types, was defined to test the hypothesis of randomness. 'Large' values of S would indicate non-randomness. To assess whether an observed value of S was significantly large or not the elements of the occurrence matrix were randomly changed, subject to the constraint that row and column totals remain unchanged, and S recalculated. Repeating this a large number of times generates a distribution for S against which the significance of the observed value may be judged. Given a significant result, the contributions of the individual artifacts to non-randomness may be assessed in a similar way through examination of the distribution of the v_i. To allow for multiple testing of p artifacts, a value of v_i is only declared significantly large at the $100\alpha\%$ level if the probability of a value as large as that observed is α/p or less (Manly, 1996: 476). In Manly's application S was found to be highly significant, and this could be attributed to four artifact types having a non-random distribution.

To investigate the relationship between burial contents and age and sex, Mantel randomization tests were used. The general idea of a Mantel test, in the context of the paper, is that given two $n \times n$ symmetric matrices of 'distances', **A** and **B**, the strength of relationship between the two matrices may be assessed by regressing the a_{ij} on the b_{ij} (e.g., Manly, 1997: 174–5). A randomized value from the distribution of the regression coefficient can be generated by randomly permuting the rows of one of the matrices, and reordering the columns similarly. The significance of the observed regression coefficient may be assessed with reference to the distribution generated from a large number of randomizations. The idea extends to multiple regression where, for example, a third matrix of distances, **C**, is available. Another application of a Mantel test is described in Section 13.5.3.

In Manly's application the distance between two burials, a_{ij}, was defined as the Euclidean distance between rows which, as the data matrix is an occurrence one, is just the square root of the number of artifacts occurring in one burial but not the other. For analyses involving age, b_{ij} is equal to the absolute difference in estimated age between two burials, while for sex, $c_{ij} = 0$ if two burials are of the same type and 1 otherwise. Separate analyses were undertaken for male burials, female burials and child burials related to age; male, female and child burials related to age and sex; and all burials, including infants. Only the last of these produced significant results, and this appeared to be because the 46 infant burials were mostly devoid of artifacts and, as a group, contained far fewer than the other burials (Manly, 1996: 482).

12.5 Markov chain Monte Carlo

Markov chain Monte Carlo (MCMC) methods have mostly been used in statistics in the context of Bayesian inference, and this is certainly the case in archaeological applications. It is convenient to present a brief account in this chapter, but some readers may prefer to read this after the discussion of Bayesian methods in archaeology, in Chapters 14 and 15.

The prime use of MCMC methods is in Bayesian computation, involving high-dimensional integrals, when analytical or numerical integration are not convenient options. The general idea is to iteratively generate, via simulation, samples from a p-dimensional distribution that can then be used to make summary inferences about the parameters of the distribution. Thus, let $\theta = (\theta_1 \ \theta_2 \cdots \theta_p)$ be a p-dimensional parameter vector, and let $f(\theta)$ be a distribution, such as a posterior density, based on θ. The full conditional distribution of θ_i, given $\theta_{-i} = (\theta_1 \ \theta_2 \ \theta_{i-1} \ \theta_{i+1} \cdots \theta_p)$, is $f(\theta_i \mid \theta_{-i})$. Let θ^j be the jth value of θ, estimated in an iterative approach, with θ^0 a set of arbitrary starting values. In MCMC methodology, for a large number of runs, r, a subset of the simulated values θ^j can be treated as a sample from the distribution $f(\theta)$ and used to draw inferences about that distribution.

There are many different MCMC methods. That most usually encountered in the archaeological literature (as opposed to the statistical literature where archaeological data are sometimes used to illustrate novel MCMC methods) is the *Gibbs sampler* (Buck *et al.*, 1992, 1996) which is now described. Given the arbitrary starting values θ^0, draw random samples from the full conditional distributions as follows to get a new value θ^1 (Smith and Roberts, 1993: 5):

$$\theta_1^1 \quad \text{from} \quad f(\theta_1 \mid \theta_{-1}^0);$$
$$\theta_2^1 \quad \text{from} \quad f(\theta_2 \mid \theta_1^1, \theta_3^0, \ldots, \theta_p^0);$$
$$\theta_3^1 \quad \text{from} \quad f(\theta_3 \mid \theta_1^1, \theta_2^1, \theta_4^0, \ldots, \theta_p^0);$$
$$\vdots$$
$$\theta_p^1 \quad \text{from} \quad f(\theta_p \mid \theta_{-p}^1).$$

With θ^1 as the new starting point this iterative process is repeated r times.

Several practical issues have been ignored here. It can be shown under fairly general conditions that, if the Gibbs sampler is appropriate, convergence to the *target* distribution will be achieved after q iterations which can be discarded, but an appropriate value for q may not be readily determined. How large r should be is also an issue, arising from the fact that q is not known, and the correlation between values of θ^j after convergence. Rather than conducting one run, some practical strategies involve several runs. For discussion of these and other issues reference can be made to papers in Gilks *et al.* (1996) and to Brooks (1998), where other more complex MCMC methods are described.

In order for Gibbs sampling to be useful, the full conditional distributions need to be available in a form that is readily sampled. That this particular MCMC method is the one most usually found in the archaeological literature is because the currently most important routine application of Bayesian methods in archaeology, to radiocarbon calibration problems, can be approached via the Gibbs sampler and programmed in a way that is accessible to archaeologists, hiding the technical detail from them (Chapter 15).

Blocking of parameters is also possible. The Gibbs sampler is a special case of the more general Metropolis–Hastings algorithm, described in the references above, which does not result in the automatic acceptance of a candidate point at any stage of the simulation. There is currently considerable research interest in MCMC methods where the more straightforward (in principle) techniques such as the Gibbs sampler are not easily used. Several of these have been applied to archaeological data in the statistical literature (e.g., Buck and Litton, 1996; Buck and Sahu, 2000; Dellaportas, 1998; Brooks, 1998; Fan and Brooks, 2000). Some of these applications are described in Chapters 14 and 16.

13

Spatial analysis

13.1 Introduction

As Aldenderfer (1998: 101–2) and others (e.g., Orton, 2000c) have observed, spatial data analysis is widely regarded as an important topic in archaeology, but spatial statistics of the kind covered in statistical texts such as Cressie (1993) has had limited impact on archaeological practice. Hodder and Orton's (1976) text, *Spatial Analysis in Archaeology*, which was strongly influenced by developments in quantitative geography at the time, has not been superseded, and is still referenced in reviews of spatial analysis in archaeology (e.g., Aldenderfer, 1998; Wheatley and Gillings, 2002).

Several reasons can be suggested for this state of affairs. That spatial statistical methods can be mathematically complex, and are not readily accessible in the software typically available to archaeologists, is one reason. Aldenderfer (1998) implies as much, and Orton (1992) has suggested that archaeologists tend to prefer methods devised by archaeologists and have remained unaware of statistical developments that improve on them. Wandsnider (1996: 324) identifies the work of Whallon (1973, 1974) as having 'pioneered the subdiscipline of quantitative archaeological spatial analysis, which proposed methods and techniques of analysis compatible with the then-current analytic goals' in intrasite spatial analysis. It is argued that these goals have 'evolved dramatically' in the intervening years with an 'emerging consensus that the goal of spatial analysis lies in the inference of deposit formation history, which *subsumes* the earlier goal of identifying activity area assemblages and identifying site structure' (Wandsnider, 1996: 338; emphasis in the original). Extrapolating from this analysis, if the aim is to make archaeological inferences about the, often multiple, processes that give rise to the observed spatial pattern(s), sophisticated statistical spatial analysis may have a limited role to play. The nature of the processes may be poorly understood, and the problem of equifinality in which different causes (processes) potentially give rise to the same effect (spatial pattern) always needs to be borne in mind. Statistical analysis that, for example, demonstrates a departure from complete spatial randomness, and which may even identify the nature of the departure, does not *explain* the pattern. Similarly, models other than that of complete spatial randomness might or might not fit a particular archaeological distribution but, without linking archaeological theory, may offer little in the way of useful interpretation.

Another possible reason for the lack of impact of methods of spatial analysis, developed in the statistical literature, on archaeological spatial analysis is the impact that geographical information systems have had in archaeology since the early 1990s. Wheatley and Gillings (2002: 9) suggest that the label 'GIS' is a rather 'slippery' concept, but that GIS are 'computer systems whose main purpose is to store, manipulate, analyse and present information about geographic space' and, with some reservations, suggest that GIS may usefully be thought of as a 'spatial toolbox'. The impact of GIS in archaeology on 'computer-literate' archaeologists, since the early 1990s, can be traced in books and monographs such as Allen *et al.* (1990), Gaffney and Stančič (1991), Johnson and North (1997), Lock and Stančič (1995), Aldenderfer and Maschner (1996), Maschner (1996), Lock (2000), Westcott and Brandon (2000) and Wheatley and Gillings (2002), and review articles such as Kvamme (1999).

Arguably, GIS have attracted, and become the methodology of choice of, researchers who might otherwise have engaged with more 'traditional' statistical approaches to data analysis. This is coupled with the fact that GIS are not usually well articulated with statistical techniques other than at a basic level (Kvamme, 1999: 162; Fisher *et al.*, 1997). Shennan (quoted in Kvamme, 1996: 46) has wondered whether GIS reduce archaeological research and analysis to the inspection of 'pretty pictures'. Even practitioners, and enthusiasts about the potential of GIS in archaeology, have despaired at the way in which they have been used. Brandon, in the preface to Westcott and Brandon (2000), referring to practice in the mid-1990s, describes 'the current use of GIS in archaeology' as 'limited to including this catchy buzzword in the title of the presentations' and suggests that GIS are often used for 'nothing grander than "gee-whiz" visualization of data'. Scollar (1999) charts the relative rise of GIS and decline of statistics in papers given at the *Computer Applications in Archaeology* conference up to 1996, and since then mainstream statistical contributions have declined further.

GIS are clearly here to stay. A balanced account, which does address the way in which statistical methods may be used in GIS, is provided by Wheatley and Gillings (2002). It is possible that GIS answer a need that more formal spatial statistics cannot satisfy. The problem is that, because of the negative impact of GIS on the use of formal statistics, not enough has been done with the latter in recent years, and within the context of GIS, to allow a balanced judgment.

In the remainder of this chapter a variety of topics that have a spatial component are discussed, including spatial clustering, predictive modeling, point pattern analysis and spatial autocorrelation. The methods associated with these topics have been used for some time in archaeology though, in some cases, not much recently. The chapter concludes with a brief introduction to shape analysis. This has also been around for some time, but modern statistical developments have had little impact on the archaeological literature as yet.

13.2 Spatial clustering

Spatial k-means clustering, or pure locational clustering as Kintigh (1990) prefers to call it, was discussed and illustrated in Section 8.4 as an example of model-based cluster analysis. Part of the purpose of Section 8.4 was to point out that more modern statistical approaches to model-based clustering could avoid the main acknowledged limitation of spatial *k*-means clustering as used in archaeology, which is a tendency to produce circular clusters of similar size.

A complementary approach is Whallon's (1984) unconstrained clustering approach. Designed for the spatial analysis of an occupation floor, the word 'unconstrained' refers to Whallon's (1984: 243) requirement that characteristics such as the size, shape, density, composition, and internal patterns of association of clusters should be free to vary, and not imposed in advance of, or by, the clustering method. Kintigh (1990: 191) describes unconstrained clustering as an analytical strategy, rather than a specific method. Conceptually the idea is a simple one. Suppose there are *K* artifact types, and *n* units to be clustered. The units may be 'points' or grid squares, and are characterized by a vector of the *relative* density of the *K* artifacts. These relative densities form the input for a clustering algorithm, resulting in the identification of *G* clusters. Cluster labels are mapped onto their spatial locations and 'inspected for spatial integrity or interpretable spatial patterning' (Whallon, 1984: 244).

The most complex aspect of the procedure arises in the way relative densities are constructed. For each type of artifact a smoothed density surface is estimated, then for each unit to be clustered the smoothed estimate is divided by the sum of the estimates across all types. The smoothing can be carried out in various ways, and some possibilities are discussed in Kintigh (1990: 190–7), Gregg *et al.* (1991: 158–60) and Blankholm (1991: 78–80).

Potential problems in application include the fact that in areas of low absolute density quite small differences in the absolute densities can give rise to large differences in the relative density (Kintigh, 1990: 192). Another problem, noted by Ridings and Sampson (1990), is that the use of relative densities gives rise to fully compositional data of the kind discussed in Section 7.2.2. This can induce negative bias in the correlations between artifact types, and suggests that, in principle, some form of transformation may be desirable before cluster analysis. This is not usually done. Ridings and Sampson (1990) discuss the possibility of a log-ratio transformation in their spatial analysis of Bushman pottery decorations, but conclude there are too many zeros and small values in the data for this to be satisfactory. Orton (2000c: 587) has suggested that, despite the acclaim that welcomed the technique, 'it appears to have serious defects' in that it complicates rather than simplifies the data, can create spurious structure, and is sensitive to the smoothing used in creating density contours.

Notwithstanding such criticism, in practice applications of unconstrained clustering to ethnoarchaeological data have been judged to give good results. The Mask Site data of Table 2.6 were studied in Whallon (1984), Kintigh (1990) and Blankholm (1991), the last two comparing it with other methods, including spatial *k*-means analysis. Gregg *et al.* (1991: 194) applied both unconstrained clustering and spatial *k*-means analysis to Yellen's (1977) data on ethnographically recorded !Kung hunter-gatherer sites from Botswana, and concluded that the two methods in combination provided objective results that corresponded well with Yellen's subjective analysis. An interesting feature of the paper was the simulation of archaeological data by 'degrading' the observed data to different degrees. With sufficient degradation any analysis will, of course, fail to recapture the true structure of the site; however, it was concluded that the methods used were relatively robust to moderate degradation.

Unconstrained clustering was applied to archaeological data in Galanidou's studies of the spatial organization of the Upper Paleolithic sites of Klithi (Galanidou, 1993, 1998a) and Kastritsa (Galanidou, 1998b) in northwest Greece. She concluded that it was capable of uncovering robust patterns, or latent structure, not necessarily evident in excavation.

13.3 Predictive modeling

Suppose, in some region of study, archaeological sample surveys have identified a number of archaeological sites, with surveyed areas that do not contain archaeological sites designated as 'non-sites'. Idealizing sites and non-sites as points (i.e., very small in relation to the area surveyed) each may be associated with a vector, x_i, of variables potentially important in determining whether or not an archaeological site is located at that point. Alternatively, an area may be divided into units, such as grid squares, with units identified as containing archaeological sites or not. One aim of predictive models is to identify those variables that best discriminate between sites and non-sites (Section 5.4.2). Characteristically, the x_i consist of physical and environmental variables that can be measured from modern maps and survey. It is assumed that settlement decisions made by prehistoric people were strongly influenced or conditioned by characteristics of the natural environment that are reflected in the measurements that can be made today (Warren, 1990b: 202).

Predictive modeling developed in a big way in the United States in the 1980s. Surveys of the literature and reviews of methodology from that period include Parker (1985), Kohler and Parker (1986), Judge and Sebastian (1988), Warren (1990a) and Kvamme (1983, 1985, 1990b). Warren (1990a: 96) reviewed a number of statistical methods used in predictive modeling, including multiple linear regression, linear discriminant analysis and logistic regression, and concluded that 'considering the various limitations and benefits of these approaches, logistic regression analysis is clearly the method of choice for empirical predictive modeling of archaeological site locations'. Warren credits Kvamme (1983, 1985) with having developed this approach for archaeology (Section 5.4.2). Logistic discrimination is discussed in Section 9.3.

Applications of logistic regression include Kvamme (1992), Warren and Asch (2000), and Maschner and Stein (1995) whose application is described in Section 11.3.2. At a general level much predictive modeling has been criticized for being environmentally deterministic, in the sense that it treats past settlement patterns as being entirely determined by environmental factors, to the exclusion of cultural factors. Kvamme (1999: 181–3), who discusses the issues in the context of GIS, suggests that the argument is between those who view GIS as 'inherently atheoretical' and those who think that they are not theoretically neutral. Similar arguments have surrounded the use of statistics in archaeology (Section 1.2.1). That environmental variables are used in modeling is, presumably, because they are much easier to measure than relevant cultural variables. If the aim is simply to predict site locations, and if an 'environmentally determined' model works well, it might be argued that the charge of environmental determinism is irrelevant. It can be argued that modeling should have different and 'higher' aims, but this is a separate issue.

At a more specific level, the way in which logistic regression is used has been criticized by Woodman and Woodward (2002). They suggest that one problem is that the sampling mechanism used to generate the data is sometimes ignored, and different approaches are needed according to whether the data are generated by randomly sampling a landscape, or by selecting a known number of sites and non-sites from existing records. A second issue is that in many applications it is assumed that a linear model of the form (5.14) is appropriate, without testing. A third is that the possible existence of confounding variables and interactions is often ignored. These are essentially criticisms of modeling technique, rather than the logistic model itself.

A stronger criticism is that of Arbia and Espa (1996: 367), who note that no account is taken of the spatial dependence between observations and the possibility that 'archaeological sites tend to cluster in space and this fact results [in] higher probability of sites in the neighbourhood of existing sites'. Arbia and Espa conclude that failure to take this into account in the usual logistic regression model renders it 'statistically incorrect'. Developing ideas of Besag (1974), they propose what they term a logistic/autologistic model of the form

$$\pi_i = \frac{\exp(\sum_j \gamma_{ij} y_j + \mathbf{x}' \boldsymbol{\beta})}{1 + \exp(\sum_j \gamma_{ij} y_j + \mathbf{x}' \boldsymbol{\beta})}$$

where notation is as in Section 5.4.2, $y_j = 1$ if the jth neighbor of unit i contains a site, and summation is over those units considered to be neighbors. This is an interesting idea but, as Arbia and Espa (1996: 370) admit, is unfortunately difficult to estimate. They do not provide an illustrative application, and I am not aware of published applications of this model.

13.4 Point pattern analysis

13.4.1 Nearest neighbor analysis

Point pattern analysis has been undertaken at both the intrasite level, to investigate artifact distributions within sites (e.g., Dacey, 1973; Whallon, 1974; Gould and Yellen, 1987; Kintigh, 1990; Blankholm, 1991), and at the intersite level, to investigate the distribution of sites within a region (e.g., Hodder, 1971; Earle, 1976; Hodder and Orton, 1976; Grier and Savelle, 1994; Hill, 2000).

The starting point in many applications has been Clark and Evans's (1954) index of spatial randomness. Suppose n points of a spatial point pattern are observed in some study region of area A and perimeter length P. Suppose, also, that the null hypothesis that the points have a random distribution is of interest, and that the random distribution is generated by a Poisson process of intensity ρ per unit area. For the moment we take $\rho = n/A$. Let d_i be the distance from a point to its nearest neighbor, let $\bar{d}$ be the mean nearest neighbor distance, let $E(\bar{d})$ be its expected value, and let var($\bar{d}$) be its variance.

If data are non-random and clustered, most of the d_i will be 'small' and $\bar{d}$ can be expected to be less than $E(\bar{d})$. Conversely, if the data are regularly spaced then

$\bar{d}$ should be 'large' in relation to $E(\bar{d})$. This motivates the use of the statistic

$$R = \frac{\bar{d}}{E(\bar{d})}$$

as an index of departure from spatial randomness, with values close to 1 consistent with randomness, values less than 1 suggesting a clustered distribution and values greater than 1 suggesting a regular distribution. This is the Clark–Evans statistic.

Assuming known ρ, and ignoring the finite area of the study region, Clark and Evans derived the mean and variance of $\bar{d}$ as $E(\bar{d}) = 1/(2\sqrt{\rho}) = 0.5\sqrt{A/n}$ and $var(\bar{d}) = (4 - \pi)/4n\rho\pi = 0.068\,31/n\rho$, respectively. This gives rise to the form of the statistic often quoted,

$$R = \frac{2\bar{d}}{\sqrt{A/n}}, \qquad (13.1)$$

which should lie between 0 and 2.15. To test whether departure from randomness is significant or not, central limit theory may be invoked, in which

$$z = \frac{[\bar{d} - E(\bar{d})]}{\sqrt{var(\bar{d})}} \qquad (13.2)$$

is treated as a normally distributed, $N(0, 1)$, variable for large enough n (e.g., Whallon, 1974; Blankholm, 1991: 112). For smaller sample sizes other tests based on the Pearson type III distribution or chi-squared are available (Hodder, 1971; Whallon, 1974: Hodder and Orton, 1976: 40).

The Clark–Evans statistic, even for large n, is unsatisfactory as it stands, as it makes no allowance for the finite size of the area studied and associated boundary effects, and the dependencies among nearest neighbor distances (Diggle, 1983: 17). Boundary problems arise if the nearest neighbor to a point lies outside the study region and is unobserved. This means that the observed nearest neighbor will be further away than the true nearest neighbor, and $\bar{d}$ will be inflated.

One of the most satisfactory solutions to this problem seems to be that of Donnelly (1978) who used both integration and simulation methods to derive estimates of the mean and variance of the form

$$E(\bar{d}) = 0.5\sqrt{A/n} + 0.0514P/n + 0.0412P/n^{3/2}$$

and

$$var(\bar{d}) = 0.0703A/n^2 + 0.037P\sqrt{A/n^5}.$$

Used with these estimates, the statistic in equation (13.2) is 'for all practical purposes' normal (Ripley, 1979: 368). Simulations undertaken by Ripley (1979), for a square region and with $n = 25$ and 100, show that Donnelly's statistic performs well in comparison with several other tests of spatial randomness. Diggle (1983: 17) has observed that the approximation may break down in very convoluted regions, which can easily arise in archaeological applications, when recourse may be had to Monte Carlo methods to estimate the sampling distribution of $\bar{d}$. Other approaches to the boundary problem have been proposed in the archaeological literature but do not appear to have been used much.

In discussing the use of R, Graham (1980: 107–8) noted that the statistic 'gives no indication at all of the scale or structure of the clustering, or indeed of the number of clusters' and concluded that 'in view of the weakness of the simple model of randomness when applied to archaeology, it is not thought useful to discuss this topic further here'. That R is a global statistic that cannot reveal localized patterning of interest has also been noted by Kintigh (1990: 168) and Blankholm (1991: 109) as a limitation. More generally, a model of random spatial patterning generated by a Poisson process is unrealistic for many archaeological applications. Many phenomena of interest are likely to have been generated by multiple processes, possibly occurring over lengthy periods of time, and constrained by landscape or other features that may be internal to, as well as on the boundaries, of the region of study. More prosaically, data may be missing because of inadequate or incomplete sampling, or because sites or artifacts have disappeared from the archaeological record, or have simply never been recorded.

Kintigh (1990: 167–74) lists a variety of limitations of nearest neighbor analysis for archaeological problems. One is that it is frequently not clear how to define the limits of an area to be analyzed. Kintigh (1990) illustrates this for the Mask Site data of Table 2.6, where conclusions about the clustering or otherwise of tools depend on whether or not empty grid cells outside or on the edge of the artifact distribution are included in the analysis. This amounts to making a judgment about what constitutes a site, and militates against interpreting the Clark–Evans statistic in an absolute fashion. For intersite studies some boundaries, such as coastlines, may be sharply delineated, while others may be rather arbitrary and fixed by what area happens to have been surveyed thoroughly.

A related problem is that a lot of information is lost in reducing spatial data to a single statistic, R, and it is not always informative in the way one might hope. Hodder and Orton (1976: 43) provide one example where a value of R consistent with a random pattern is derived from a non-random arrangement of points associated with settlements located along three major transportation axes. The problem exhibited by their example is that the point pattern has a preferred orientation so is non-isotropic. Kintigh (1990: 169) provides another example where two rather different patterns of clustering give rise to identical values of R. Kintigh (1990: 174) concludes that, with caution, R might be used to investigate the *relative* patterning of different subsets of points, assessing significance of differences with Monte Carlo methods, but that it has little value as an absolute measure. The applications discussed below use R for comparative purposes.

Linear nearest neighbor analysis, where interest focuses on the randomness or otherwise of spacing along a 'linear' feature such as a waterway, has been reviewed, in an archaeological context, by Stark and Young (1981). I have seen few examples of this sort of analysis in the archaeological literature.

13.4.2 Applications of nearest neighbor analysis

Inuit winter residential sites
Grier and Savelle (1994) used the nearest neighbor statistic, among other tools, to investigate intrasite spatial patterning in 18 Thule Inuit winter residential sites located within one of three whaling zones, defined as core, intermediate and periphery. These zones were differentiated on the basis of whale abundance and inferred whaling intensity, it being hypothesized that there would be a greater degree of social organization and site structure in the core area, this decreasing

towards the periphery. This, in turn, led to the expectation that the mean nearest neighbor distance would be smallest for those sites showing the highest degree of organization. Analysis was based on R calculated for the first to third nearest neighbors, no attempt being made to test hypotheses based on R. The first nearest neighbors in the core, intermediate and periphery, with 10, 4 and 4 sites were (8.0, 6.7, 6.5, 8.2, 10.9, 9.4, 9.2, 7.4, 10.5, 7.9), (13.9, 12.8, 9.2, 5.2) and (12.7, 14.2, 10.2, 6.9). The means of 8.5, 10.3 and 11.0 were interpreted, qualitatively, to conform with expectations. Along with other analyses similarly interpreted, this was taken as evidence that social organization decreased away from the core.

It was acknowledged that the differences between zones were not necessarily statistically significant, and interpretation rested on the trends observed in a variety of analyses. I must admit to some unease about this kind of qualitative interpretation when the number of groups, three here, and sample sizes are small. The example discussed does, however, illustrate the use of the nearest neighbor statistic for descriptive and comparative purposes, in which significance testing is deliberately eschewed.

Environmental degradation in Jordan

Hill (2000) studied human-induced environmental degradation in the Wadi al Hasa, west-central Jordan, compiling a database of 339 sites classified by 10 periods in the process. He noted (Hill, 2000: 224) that maps of site locations for temporally adjacent periods were 'visually striking in many cases for the appearance of strong between-period segregation'. Following Kintigh's (1990) recommendation for using nearest neighbor statistics for comparative purposes, the value of R was calculated for sites from each period separately, and for each set of sites from temporally adjacent periods. For the single-period analyses the value of R was generally somewhat less than 1, which is to be expected if the total study area, A in equation (13.1), is used, but sites are concentrated in a subregion of this. For analyses involving pairs of periods R was, in most cases, greater than for either of the associated single-period analyses. Again, this is to be expected if there is between-period segregation. The analysis highlighted two pairs of temporally adjacent periods for which the expected relationship did not hold, and for which an explanation had to be sought.

13.4.3   Second-order methods

Second-order methods are concerned with local as opposed to global properties of spatial distributions. The second-order properties of a (stationary, isotropic) point process may be characterized by the K-function (Ripley, 1976, 1977, 1981, 1988) defined as

$$\rho K(t) = E[\text{number of events within distance } t \text{ of an arbitrary event}]. \quad (13.3)$$

If d_{ij} is the distance between two points, i and j, and $I_t(d_{ij}) = 1$ if $d_{ij} < t$ and 0 otherwise, an estimate of the expectation in equation (13.3) is $\sum_{i \neq j} I_t(d_{ij})/n$ so, if ρ is estimated by n/A, an estimate of $K(t)$, ignoring edge corrections, is

$$\hat{K}(t) = \frac{A}{n^2} \sum_{i \neq j} I_t(d_{ij}).$$

For complete spatial randomness $K(t) = \pi t^2$, so that a plot of

$$L(t) = \sqrt{\hat{K}(t)/\pi} \tag{13.4}$$

against t should be approximately linear for a random point pattern, with departures from linearity indicating the scale at which non-randomness is occurring (e.g., Venables and Ripley, 1999: 445). Alternatively, plots based on

$$L(t) = t - \sqrt{\hat{K}(t)/\pi}$$

against t, expected to be zero under complete spatial randomness, can be used.

Edge corrections are available. One simple one (Ripley, 1976) is to use

$$\hat{K}(t) = \frac{A}{n^2} \sum_{i \neq j} I_t(d_{ij})/w_{ij}$$

where w_{ij} is the proportion of the perimeter of a circle centered on point i and passing through point j that lies within the study region. In applications simulation may be used to generate envelopes within which the plot of $L(t)$ against t should lie if the point pattern is random at all scales.

For bivariate point patterns a K-function may be defined in an analogous way to equation (13.3), where the expectation is of the number of events of type 2 within a distance t of an arbitrarily chosen event of type 1. This leads to an estimate of the form

$$\hat{K}_{12}(t) = \frac{A}{n_1 n_2} \sum_{i}^{n_1} \sum_{j}^{n_2} I_t(d_{ij}),$$

where n_i is the number of points of type i, d_{ij} is now interpreted as the distance of the ith type 1 from jth type 2, and edge corrections have again been ignored. It is also possible to define a function $K_{21}(t)$ and, in theory, $K_{12}(t) = K_{21}(t)$. In practice, when these are estimated *with* edge corrections, $\hat{K}_{12}(t) \neq \hat{K}_{21}(t)$, but a weighted average of these may be used to estimate the theoretical function (Lotwick and Silverman, 1982). Upton and Fingleton (1985: 255) suggest plotting

$$L_{12}(t) = t - \sqrt{(\hat{K}_{12}(t)/\pi)}$$

against t and note that large positive values imply repulsion between two processes, while large negative values imply attraction, at the value of t involved.

K-functions have had little use in the archaeological literature (see below); however, they are closely related to the index of spatial association devised by Johnson (1984), for local density analysis, that has had some application. Specifically, the index devised by Johnson has the form

$$C_{12}(t) = \hat{K}_{12}(t)/\pi t^2$$

for bivariate point patterns, and

$$C(t) = \hat{K}(t)/\pi t^2$$

for a single point pattern. Some applications of this index are discussed shortly.

Blankholm's (1991) review included several other approaches to investigating spatial association, including dimensional analysis of variance (Whallon, 1973), Hodder and Okell's (1978) A-index and Carr's (1984) coefficient of polythetic association. He concluded that these were less satisfactory than Johnson's (1984) local density analysis. Earlier work is discussed in Hodder and Orton (1976: 198–223).

13.4.4   Applications of second-order methods

Grave distributions at Hallstatt
Graham (1980) compared three different methods: nearest neighbor analysis, spectral analysis and local density analysis. The source for the methodology on local density analysis was Johnson's unpublished 1976 thesis. The methods were applied to the distribution of graves within the cemetery of Hallstatt, Austria, and were also used for comparing the distributions of graves containing finds of different type and date. Plots of both $C(t)$ and $C_{12}(t)$ against t were used to detect departures from randomness, and evidence of a relationship between distributions. This is in the same spirit as plots based on $L(t)$ and $L_{12}(t)$, derived from the K-functions.

For 10 different types of finds, and fixed t, all possible values of C_{ij} were calculated, defining a similarity matrix between the distribution of types that was used as input to a non-metric multidimensional scaling routine. This analysis, unlike those using nearest neighbor and spectral methods, was unable to separate out late types from early types, so that local density analysis was the least satisfactory method in this respect. Graham (1980: 127) also expressed concern that local density analysis always treated structure as if it was circularly symmetric.

Artifact distributions at Pincevent and Mask Site
Johnson (1984) used local density analysis to investigate the distributions of 24 classes of artifact within hearth V105 at the Upper Magdalenian site of Pincevent, France. The index of association, C_{12}, was adapted to deal with cell frequency data. As in Graham's analysis, a similarity matrix for the distribution of types was generated, for five different values of t. These were subjected to correspondence analysis and a choice of t based on that value at which results appeared to stabilize (in terms of the correlation between components), with a reasonably high percentage of the inertia explained by the first factor. Correspondence and factor analyses at the chosen scale showed, among other things and most obviously, a clear distinction between faunal and lithic remains, corresponding to clearly segregated though overlapping spatial distributions (Johnson, 1984: 89). The factor analysis was based on a correlation matrix obtained by calculating correlation coefficients between the rows of the similarity matrix.

Blankholm (1991) followed Johnson closely in applying local density analysis, for coordinate data, to the Mask Site data of Table 2.6. Correspondence analysis and multidimensional scaling applied to the logged similarity matrix of the C_{ij} suggested that two of the five artifact categories had similar spatial distributions, the others being independent, leading to the identification of four spatially independent groups. Subsequent evaluation involved plotting and overlaying smoothed contour maps for each of these four groups, and comparing the interpretation derived from this with that of Binford (1978). It was suggested that, in some respects, local density analysis was unsatisfactory in being unable to reveal the 'three most clearly identifiable concentrations of bone splinters' (Blankholm, 1991: 107). This was regarded as a limitation of local density analysis, rather than the contouring method, presumably because the degree of smoothing was determined by the scale, t, suggested by the local density analysis. This may, in fact, equally well reflect limitations in the way t was selected, only five different scales being examined. Blankholm (1991: 109) concluded that local density analysis would work well if areas were relatively small, well defined, and/or had high densities, but might fail to detect clinal trends across an area, or transitory areas.

13.5 Spatial autocorrelation

13.5.1 Introduction

Spatial autocorrelation occurs when there is systematic spatial variation, or organized pattern, in the values of a variable across a map (Upton and Fingleton, 1985: 151). Another way of putting this is that measures of the variable over space are dependent. In some cases a demonstration of this dependence and a measure of its strength may be of direct interest, as in the discussion of the Classic Maya collapse below. In other cases, such dependence invalidates the use of conventional statistical tests and requires the development of special methods for dealing with spatially autocorrelated data. Kvamme (1993, 1996), for example, following Cliff and Ord (1975, 1981: 184–9), illustrated the use of the two-sample t-test, adjusted to deal with spatially autocorrelated data, to investigate the density of surface scatters of artifacts in quadrat data. Location may be defined by a point (e.g., the grid reference of a site on a map, where the site is small in relation to the scale of a map); an area, such as a quadrat (e.g., counts of the numbers of sites of a particular type within a quadrat); or volume (e.g., the number of artifacts of some kind in some 'unit' volume of a pit).

If there is a tendency for large measures of some variable at some location to be associated with large measures of the same variable at near locations then the data show positive spatial autocorrelation. An example might be the distribution of chipping debris in the manufacture of stone tools, measured as the density of debris in some unit, such as a quadrat, that is small in relation to the total spread of the debris. Quadrats with a high density are likely to be close to other quadrats with high density. Negative autocorrelation occurs if high measures at a location tend to be associated with low values of the same variable at near locations.

Some early archaeological applications involving spatial autocorrelation were discussed in Hodder and Orton (1976: 174–83), and much of what was done then and subsequently draws on the quantitative geographical literature, such as Cliff and Ord (1973, 1981), that was then fashionable within archaeology. Moran's (1950) I statistic, which may be adapted to point or quadrat data and which is defined in equation (13.5) below, is that most widely used in the archaeological literature to detect spatial autocorrelation (Read, 1989: 60–3; Wheatley and Gillings, 2002: 131–4).

13.5.2 The Classic Maya collapse

Measuring and interpreting spatial autocorrelation
Bove (1981) investigated the relationship between time and distance for 47 sites in the Lowland Maya region of Mesoamerica in the Late Classic period around

AD 750–900. The end of monument construction activity at these sites can be dated, at least approximately, by the dates associated with the most recent monuments, given in Bove's Table 1. Cessation of construction activity, as measured by these dates, occurred over a relatively short timespan, hence the term 'collapse'. Bove's analysis of these data, using trend-surface analysis, is discussed in Section 5.3.3.

Subsequently several papers were published using Bove's data and measures of spatial autocorrelation, to investigate the existence of spatial patterning in the data (Whitley and Clark, 1985; Kvamme, 1990a; Williams, 1993). All the papers make use of some form of Moran's I statistic as generalized by Cliff and Ord (1973, 1981), which can be written as

$$I = \frac{n \sum_{(2)} w_{ij}(x_i - \bar{x})(x_j - \bar{x})}{W \sum_i (x_i - \bar{x})^2}, \qquad (13.5)$$

where

$$W = \sum_{(2)} w_{ij}$$

with $\sum_{(2)} = \sum_{ij, i \neq j}$, and the w_{ij} are weights. The null hypothesis of no spatial autocorrelation can be tested using a standard normal deviate of the form

$$z = [I - E(I)]/\sqrt{\mathrm{var}(I)}.$$

The expectation and variance can be evaluated assuming either that the data are independently sampled from a normally distributed population, or by using randomization arguments (Cliff and Ord, 1981: 42). Assuming normality gives $E(I) = -(n-1)^{-1}$ and

$$\mathrm{var}(I) = \frac{n^2 S_1 - n S_2 + 3 W^2}{W^2(n^2 - 1)},$$

where

$$S_1 = 2 \sum_{(2)} w_{ij}^2$$

and

$$S_2 = 4 \sum_i \left(\sum_j w_{ij} \right)^2.$$

For the Mayan dates the randomization argument leads to similar numerical results, so formulas are not given here. The precise form of equation (13.5) depends on how data are recorded. For point data, as used in Kvamme (1990a), and in the present context, x_i is the terminal date of site i and $w_{ij} = d_{ij}^{-\alpha}$, where d_{ij} is the distance between sites i and j and α is a positive constant. Kvamme (1990a) uses $\alpha = 1$, while Williams (1993) presents results for $\alpha = 1$ and 2.

For gridded data, as used in Whitley and Clark (1985), x_i is the average date of sites within cell i, and $w_{ij} = 1$ or 0 according to whether or not cell j is contiguous with cell i. Commonly, contiguity is defined by requiring cells to have a common edge (the Rook's case) or a common edge or corner (the Queen's case), though Whitley and Clark (1985) also experimented with a more general directed (binary) weight matrix. For the Queen's case, Whitley and Clark (1985: 388) reported a standard normal deviate of 'less than 1.7' that was 'not statistically significant at the 0.05 level of probability'. In other words, they concluded that spatial autocorrelation was 'minimal'.

Kvamme's paper was prompted by disbelief in this conclusion, deemed 'surprising' in the light of previous analyses and the 'considerable pattern' evident in a 'simple plot of the raw data' (Kvamme, 1990a: 198). Kvamme (1990a: 201) was critical of the use of gridded data and a binary weight matrix that he thought threw away a lot of 'critical' information. His own analysis of the point data using both the normality assumption and randomization, and with $\alpha = 1$, gave rise to values of $z = 2.8$ with (two-sided) p-values of about 0.0045, leading to the conclusion that there was strong evidence of spatial autocorrelation.

Williams (1993: 706), in turn, was critical of Kvamme's critique of Whitley and Clark, arguing that the use of gridded data when point data were available 'emphatically' did not 'invalidate' the analysis, that the differences in the methodologies used were 'minor', and that the results obtained in the two analyses were in 'complete agreement'. Williams's case was possibly overstated. For the Queen's case analysis of Whitley and Clark, and the analysis of Kvamme with $\alpha = 1$, the p-values for the I statistic, calculated under the normality assumption, were 0.0724 and 0.0045. While Williams (1993: 708) was right to criticize slavish adherence to 5% levels of significance (or p-values of 0.05) it is stretching things to claim that these results are in 'complete agreement'.

Of perhaps more interest are the results reported for gridded data and the Rook's case, with a p-value of 0.1557, and point data for $\alpha = 2$, with a p-value of 0.0595, that did agree reasonably with Whitley and Clark's original results. These analyses point to the general lesson that results can be sensitive to the aggregation, or otherwise, of point data and choice of weight matrix. The particular lesson seems to be that spatial autocorrelation in the Mayan data is not so strong as to be unambiguously detected by the different methods and weighting schemes used.

Methodological extensions
Neiman (1997) took Bove's (1981) data, updated to account for more recent research and with the addition of 22 sites, as the starting point for his analysis of spatial patterning. The paper contains a number of relatively novel statistical features. One of these is the use of the lowess smoother (Section 5.6) to estimate the broad geographical trend. A local quadratic function was used, with the local neighborhood, α, chosen by cross-validation. Specifically, there were 69 data points and, for fixed α, 69 surfaces were fitted, omitting each point in turn. Each omitted point was predicted from the surface derived from the other 68, and the prediction error obtained by comparison with the true value. The sum of squared errors was used as a measure of the overall 'badness of fit'. The neighborhood, α, was then varied, and that value selected which minimized the badness of fit. This procedure resulted in the identification of clear global trends in the data (Neiman, 1997: 279).

Neiman (1997: 281) was investigating a model, one of whose implications was that the residuals from the fitted surface would exhibit dependence, and that the scale over which this dependence occurred would 'indicate the size of catchments within which Maya elites competed for resources'. To identify this scale a *(semi-)variogram* of the residuals was used. In the present context this is a plot of

$$\gamma(d) = \frac{1}{2n(d)} \sum (\hat{\varepsilon}_i - \hat{\varepsilon}_j)^2,$$

against the fixed distance d, where $\hat{\varepsilon}_i$ is the ith residual from the fitted surface, $n(d)$ is the number of pairs of points at distance d apart, and summation is over these pairs. This form assumes isotropy (Bailey and Gatrell, 1995: 164). The variogram is a tool, borrowed from geostatistics, that has been advocated by Ebert (2002), who illustrated it with data on the distribution of struck flints, and is described by Wheatley and Gillings (2002: 196–8). It has not been used much in archaeology.

Given a model that predicts that residuals will be correlated for small d, the expectation is that the plot will initially rise and then level out at the value of d that defines the size of the catchment of interest. Neiman's (1997: 280) variograms, calculated with and without four outliers, which were smoothed using the lowess smoother, exhibited an unexpected pattern, with an increase up to about 65 km but then a decline to values at 150 km similar to those seen at about 35 km. These results were interpreted by Neiman (1997: 283) as implying upper (65 km) and lower (35 km) limits on the scale of political competition, following a series of arguments linked to a Darwinian theory of wasteful advertising that underpins his paper. These results are compatible with epigraphic evidence that spheres of political competition averaged about 50 km (Neiman, 1997: 284).

13.5.3 Spatial patterns among blood types

Sokal *et al.* (1987) studied the spatial patterns of blood types in medieval Hungarian cemeteries with a view to seeing if there were patterns that could support inferences concerning familial relationships among contiguous graves. Paleoserological methods were used to determine ABO serotypes of skeletal remains, resulting in the characterization of serotypes as A, B, AB or O.

For any pair of graves the type of *join* may be characterized as homotypic (A/A, B/B, AB/AB or O/O) or heterotypic (A/B, A/AB, etc.). There are 10 possible joins in all. For some fixed distance, d, define a symmetric matrix, $\mathbf{X}$, with zero diagonal elements, whose off-diagonal elements $x_{ij} = 1$ if graves i and j are within distance d of each other and 0 otherwise. For any two serotypes, S1 and S2 say, define a symmetric matrix, $\mathbf{Y}_{S1/S2}$, with zero diagonal elements, whose off-diagonal elements $y_{ij} = 1$ if the join for graves i and j is S1/S2 and 0 otherwise. The matrix $\mathbf{X}$ can be defined for different values of d, and there are 10 possible $\mathbf{Y}$ matrices. To measure spatial autocorrelation for fixed d and a particular join a Mantel statistic (see Section 12.4.3) of the form

$$Z = \sum_{(2)} x_{ij} y_{ij}$$

can be defined, which has expectation

$$\sum_{(2)} x_{ij} \sum_{(2)} y_{ij}$$

if there is no spatial autocorrelation for the join type, and for the value of d, used.

To assess the statistical significance of an observed Z, Monte Carlo methods (Chapter 12) rather than theoretical results were used. The rows and columns of **X** may be randomly permuted and Z recalculated. This was repeated a large number of times, and the significance of the observed Z assessed by reference to the distribution of values thus generated.

This methodology was applied to material from three medieval Hungarian cemeteries for between 10 and 13 values of d for each of the different join types. Sample sizes varied from 57 to 73. For two of the cemeteries, archaeological evidence suggested the presence of different ethnic groups, and the Mantel statistics calculated tended to support this. As a generalization, a relatively small number of the statistics were significantly different from expectation under the hypothesis of no autocorrelation, but there was some fairly clear patterning, with a tendency for homotypic joins to show positive autocorrelation at short distances and negative autocorrelation at longer distances, with the opposite being the case for heterotypic joins. Subdividing the sample into different ethnic groups produced samples too small to investigate possible familial relationships with any certainty. The third cemetery, which on archaeological evidence was ethnically more homogeneous, showed no evidence of spatial structure.

13.6 Shape analysis

13.6.1 Introduction

Statistical shape analysis is a subject that underwent considerable development in the 1990s, resulting in a number of books such as Bookstein (1991), Small (1996), Dryden and Mardia (1998), Kendall *et al.* (1999) and Lele and Richtsmeier (2001). This recent work has had a limited impact on the mainstream archaeological literature as yet, though some of these publications mention archaeology as a potentially fruitful field for application, and illustrate methods using archaeological data. This section briefly reviews some of these applications, including earlier archaeological work on identifying patterns in post-holes that pre-dates recent developments.

13.6.2 Post-hole patterns

Circular and rectangular structures can frequently be recognized in the archaeological record from post-holes identified in the course of excavation, but such recognition is problematic when the density of post-holes is high. Several papers dealing with this problem were published, mostly in the 1980s, some of which were reviewed in Read (1989: 36–7). These papers include Fletcher and Lock (1981, 1984a,b, 1990), Litton and Restorick (1983), Bradley and Small (1985) and Bell *et al.* (1990). The most sustained attack on the problem was that of Fletcher and Lock (co-authors of Bell *et al.*, 1990) who were motivated by the problem of recognizing structures in the post-hole distribution from excavations of the Iron Age hillfort of Danebury in Hampshire, England. The most accessible account of their approach (in terms of where it is published) is Fletcher and Lock (1984a), but it was superseded in certain respects by the 1990 papers which drew on ideas from Litton and Restorick (1983).

Two main stages are involved in the approach. The first is to identify possible structures in the post-hole distribution. This is a non-statistical problem, though it can be automated by computerized search, allowing for a certain tolerance in the precision with which structures were erected. Once the number of possible structures has been determined, the problem then is to see if this could have occurred by chance, given the observed density of the post-holes. Fletcher and Lock (1984a,b) gave some theoretical results (reproduced in Read, 1989: 36) based on the null model of a random distribution of points over an area A, that agreed well with results using simulated data assuming randomness. The methodology of the 1990 papers, which use a different approach to simulation than that of the earlier papers, is described in Section 12.4 as an example of the use of randomization methodology in archaeology.

A different approach to detecting circular structures in post-hole patterns was adopted by Bradley and Small (1985), and is discussed in Read (1989: 36). Small (1996: 190–3) reanalyzed the data from Bradley and Small (1985). There are some differences between the two analyses, so the later account is described here.

The median distance from a point to its nearest neighbor, in a Poisson process of intensity ρ, is $\sqrt{(\log 2)/\pi\rho}$, so if the observed median nearest neighbor distance between post-holes is m the intensity may be estimated by $\hat{\rho} = (\log 2)/\pi m^2$, where natural logs are used. The area of the 'hypothetical region of post-hole activity' may be estimated as $\hat{A} = n/\hat{\rho}$ where n is the number of post-holes, and an estimate is used, rather than the actual area excavated, because post-hole activity may not occur in all parts of the excavation.

If, on a map of the post-hole distribution, all post-holes are linked to those within 3.0 m, a count may be made of n_i, the number of clusters of post-holes containing $i = 1, 2, \ldots$ post-holes. In Bradley and Small (1985) the distribution of n_i was compared with that expected of a Poisson process resulting, in the examples presented, in an apparent overrepresentation of clusters with small i and some larger values of i. Subsets of the larger clusters so identified were, in some cases, candidates for intentionally erected circular structures of size $s \leq n_i$, though archaeological judgment may also suggest candidate structures that are not completely linked.

An annulus, or circular ring, may be defined by all points whose distance, d_{ij}^2, from some center, satisfies the criterion

$$r^2 \leq d_{ij}^2 \leq r^2(1 + \varepsilon)^2$$

for some inner circle of radius r and outer circle of radius $r(1 + \varepsilon)$, where ε is expressed as a proportion of r. If an annulus containing s post-holes can be found it can be asked if this is an unusual occurrence for a Poisson process of intensity ρ in a region of area A. Small (1996: 189) shows that the expected number of circular arrangements is given by

$$E(s) = \frac{(2\pi)^{s-1}}{(s-2)!}\rho^s r^{(2s-2)}A\varepsilon^{s-2}$$

where, in application, ρ and A are replaced by their estimates.

For the later Bronze Age site of Aldermaston Wharf, Berkshire, England, two possible six-post structures were identified with $r = 3.66$ m, $\varepsilon = 0.792$ and

$r = 3.15$ m, $\varepsilon = 0.1492$, respectively. Calculations for $\hat{\rho} = 0.076$ and $\hat{A} = 803$ m^2 gave values of $E(s)$ of 1.07 and 3.03, suggesting, on the statistical evidence only, that these possible structures could have arisen by chance. Similar analyses, for possible seven- and eight-post structures from the site of South Lodge Camp, Wiltshire, produced expected values of 0.001 and 0.15, lending support to the idea that the configurations represented real structures. Note that the expected values for all these analyses differ somewhat from those in Bradley and Small (1985), but do not affect the conclusions.

13.6.3 Comparing shapes

Small (1996), Le and Small (1999) and Dryden (2001) have all used the data on Iron Age brooches from Münsingen in Switzerland, published in Hodson *et al.* (1966), to illustrate aspects of shape analysis. Shape may be defined as 'all the geometrical information that remains when locational, scale and rotational effects are filtered out from an object' (Dryden and Mardia, 1998: 1). For a drawing of an archaeological object, such as the profile of a brooch (i.e., a two-dimensional construct), *landmarks* are selected points on the object that, taken together, define (approximately) the shape of the brooch.

Without going into detail, after removing locational, scale and rotational effects, a measure of the (dis)similarity between pairs of brooches can be defined that may be used as input for some 'standard' multivariate procedure such as principal coordinate analysis, principal component analysis or multidimensional scaling. The measure of (dis)similarity can be defined in more than one way. Small (1996: 92–4), Le and Small (1999) and Dryden (2001) used a Procrustes distance, which is a symmetrical measure of the 'fit' of one shape to another, for four, four and five landmarks, respectively. In their multivariate analyses a plot of the first two axes, labeled by age of the brooches, suggested a change in shape over time, summarized by Small (1996: 94) as 'with the passing of time, the brooches at Münsingen became larger and more elongated'.

14

Bayesian methods

[Bayesians] have a view of statistics that is in many ways closer to the archaeologists': that probability is subjective, and that statistics is about the 'orderly influencing of opinions by data'. Orton (1999: 29)

14.1 Bayes' theorem

Bayes' theorem, which first appeared in a paper by Bayes (1763) published posthumously, is at the heart of a particular approach to statistical thinking and analysis. As such it embraces the whole of statistics, and is not readily summarized in a short chapter. The main aim here is to try and explain the (potential) appeal of Bayesian statistics for archaeological data analysis, and to illustrate aspects of the methodology and examine some of its strengths and limitations through a discussion of the way it has been applied to archaeological problems. Book-length treatments of Bayesian inference include Bernardo and Smith (1994) and O'Hagan (1994), while Lindley's (2000) paper on the philosophy of statistics, and the subsequent discussion, outlines a variety of views on the subject. Buck *et al.* (1996) is intended specifically for archaeologists.

Bayes' theorem may be expressed in various ways. Buck *et al.* (1996: 21) begin with

$$P(\text{parameters} \mid \text{data}) \propto l(\text{parameters; data}) \times P(\text{parameters}),$$

which says that the posterior probability is proportional to the likelihood times the prior probability, where $P(\cdot)$ is probability and $l(\cdot)$ likelihood. For discrete data

$$P(\theta_i \mid \mathbf{x}) = \frac{P(\mathbf{x} \mid \theta_i) P(\theta_i)}{\sum_i P(\mathbf{x} \mid \theta_i) P(\theta_i)} \qquad (14.1)$$

for parameters θ_i, and where $\mathbf{x} = (x_1, x_2, \ldots, x_n)$ is the data, and proportionality rather than equality arises if the denominator is omitted. For continuous data

$$f(\theta \mid \mathbf{x}) = \frac{f(\mathbf{x} \mid \theta) f(\theta)}{\int f(\mathbf{x} \mid \theta) f(\theta) \, d\theta} = c^{-1} f(\mathbf{x} \mid \theta) f(\theta) \qquad (14.2)$$

where $\theta = (\theta_1 \; \theta_2 \; \cdots \; \theta_p)'$ is a p-dimensional parameter vector. In equation (14.2) $f(\mathbf{x} \mid \boldsymbol{\theta})$ is the probability of the data, given a probability model dependent on the unknown $\boldsymbol{\theta}$ and, viewed as a function of $\boldsymbol{\theta}$, is sometimes called the *likelihood function* and written as $l(\boldsymbol{\theta} \mid \mathbf{x})$. The term $f(\boldsymbol{\theta})$ is the *prior distribution* for $\boldsymbol{\theta}$ and expresses in probabilistic form knowledge or beliefs about the true value of $\boldsymbol{\theta}$. The term $f(\boldsymbol{\theta} \mid \mathbf{x})$ is the *posterior distribution* of $\boldsymbol{\theta}$ given $\mathbf{x}$ and describes how one's original prior beliefs about $\boldsymbol{\theta}$ have been modified in the light of the data. The normalizing constant $c = \int f(\mathbf{x} \mid \boldsymbol{\theta}) f(\boldsymbol{\theta}) \, d\boldsymbol{\theta}$ ensures that the posterior distribution integrates to 1, and is often omitted in presentations so that

$$f(\boldsymbol{\theta} \mid \mathbf{x}) \propto f(\mathbf{x} \mid \boldsymbol{\theta}) f(\boldsymbol{\theta}).$$

One way of viewing this is that Bayes' theorem provides a mechanism for updating one's beliefs or knowledge about a set of parameters in the light of the data.

The application of Bayes' theorem to statistical inference has proved controversial. A central point of contention is the role played by the prior distribution. There are circumstances in which this can be 'objectively' defined but, equally, it can be used to express subjective beliefs about the true value of $\boldsymbol{\theta}$. Many statisticians have been unhappy about this, partly because it is seen as allowing subjective considerations to influence scientific analyses that ought to be objective.

In contrast to 'classical' statistics, the Bayesian approach also allows statements about parameters to be expressed in probabilistic form. As an example, a 95% confidence interval in the classical approach, for a single parameter θ, has the strict interpretation that in an infinite number of repeated experiments 95% of the intervals would cover the true value of θ. As far as the one interval usually to hand goes, it either does include the true value or it does not, and you do not know which. It is tempting to interpret a 95% confidence interval as saying that there is a 95% probability that the true θ is included in the interval. This may seem 'natural', but it is not true. The Bayesian analog of a confidence interval, by contrast, allows this natural interpretation.

For a long period arguments about the relative merits of the classical, Bayesian and other approaches to inference raged mainly at the philosophical level. The reason for this is that, until the advent of modern computing power, Bayesian formulations of problems proved relatively intractable unless great care was taken in the formulation of the prior. Essentially these were defined on the basis of mathematical convenience, rather than in the light of true beliefs, often leading to analyses that, in practical terms, led to the same conclusions as a 'classical' analysis.

Among the mechanisms that allow tractable results to be obtained are *conjugate* prior distributions, *improper* prior distributions and *vague* prior distributions. Conjugate priors are defined in such a way that, on combination with the model for the data, a posterior distribution having the same form as the prior distribution is obtained. This, in effect, allows the use of 'standard' results to be applied in situations where the use of a conjugate prior is deemed appropriate or convenient. Improper priors are not genuine probability distributions and are often used, convenience aside, to express the fact that one has limited knowledge about $\boldsymbol{\theta}$. They are a particular case of vague priors that are also defined to represent prior ignorance about the parameters.

For the philosophically indolent and pragmatic statistician, having usually been trained in the classical tradition, the case for adopting Bayesian methods did not necessarily seem compelling. A methodology that laid great emphasis on the use of priors to represent existing knowledge or beliefs, and then devoted much effort to defining and justifying prior ignorance, could seem perverse.

The situation caricatured above has altered drastically in recent years, because of developments in computational power and associated theory. This has allowed applications of Bayesian inference to flower, untrammeled by the need to model problems in ways that are mathematically convenient. Archaeology has benefited from this development, with most of the applications to be described dating from the late 1980s. A brief account of some of these developments, in particular Markov chain Monte Carlo methods, is given in Section 12.5.

14.2   Bayesian inference in archaeology – an overview

Proponents of Bayesian ideas, not just in archaeology, can display an evangelical fervor in their enthusiasm for the cause that sometimes overstates the power of the methodology in particular instances. Reviews of Bayesian applications in archaeology (e.g., Litton and Buck, 1995: 2; Buck *et al.*, 1996: 1) tend to cite the work of scholars such as Doran and Hodson (1975: 308), Orton (1980: 220; 1992: 139), Ruggles (1986) and Cowgill (1989b: 79; 1993) in support of the Bayesian view. In fact, most of these references simply note the possibility of using Bayesian ideas and provide no instances of application. Buck *et al.* (1996: 1) quote Orton (1992: 139) to the effect that 'I have for many years advocated the ... Bayesian approach' with approval, but omit the equally telling continuation 'but must confess to have done very little about it'. That 'very little' had been done is at least partly a function of the computational difficulties described earlier. Applications of Bayesian methods to archaeological problems pre-dating the late 1980s are scarce. The paper by Freeman (1976), which investigates the existence or otherwise of the megalithic yard, is one such paper, and is applied to a problem described by Fieller (1993: 283) as being 'of negligible interest to archaeologists' (Chapter 19). More recent applications, particularly to radiocarbon calibration (Chapter 15) have, however, led to valuable developments in an important area of application. Although the success of the Bayesian approach in specific applications has sometimes been exaggerated, an apologia in terms of the approbation of distinguished archaeologists is no longer required, given the achievements that have been made.

Many of these achievements are the consequence of collaboration between archaeologists and statisticians, working initially at the University of Nottingham, UK, in the late 1980s and 1990s. About three-fifths of the references to Bayesian applications in archaeology that I have been able to trace have been produced by members of this group and their collaborators, and a substantial proportion of the remainder directly stem from problems originally investigated by the Nottingham group. Given that the mathematics involved in a Bayesian analysis can be challenging, collaboration between archaeologists and statisticians is arguably more important than in the application of more 'standard' statistical methods. Another interesting body of work involving Bayesian methods has arisen from collaboration between archaeological scientists and statisticians at the Universities of Bradford and Leeds in the UK (Allum *et al.*, 1999; Aykroyd *et al.*, 1996, 1997, 1999, 2001; Lucy *et al.*, 1996, 2002).

Cowgill (2001) and others (e.g., Buck *et al.*, 1996) have noted that the Bayesian as opposed to classical approach to statistics has a strong conceptual appeal. Cowgill has also noted the tension between this appeal and the undoubted complexity (at least for archaeologists) of much Bayesian analysis. In the next section some applications of Bayesian analysis in archaeology are discussed in detail. The intention here is to illustrate applications at rather different levels of complexity and, in the last case, to raise some issues about possible limitations of the use of Bayesian methods in archaeology. Other applications, to problems as diverse as radiocarbon calibration, seriation, and investigating the existence of the megalithic yard are discussed in the relevant chapters elsewhere in the book.

14.3 Bayesian inference in archaeology – examples

14.3.1 Estimating proportions

Artifact types at Teotihuacan
In this subsection some work of Robertson (1999) that, in purely statistical terms, amounts to estimating a proportion is described. More technical detail is provided than that given by Robertson. This is at the (mathematically) simple end of Bayesian applications in archaeology. A related application of Orton (2000b) is also described.

The beta probability distribution with parameters α and β has the form

$$f(\theta) = \frac{\Gamma(\alpha + \beta)}{\Gamma(\alpha)\Gamma(\beta)}\theta^{\alpha-1}(1-\theta)^{\beta-1} \tag{14.3}$$

for $0 \leq \theta \leq 1$, where $\Gamma(\alpha)$ is the gamma function, and $\Gamma(\alpha) = (\alpha - 1)!$ for integer α. Since it is a probability distribution and integrates to 1, it follows that

$$B(\alpha, \beta) = \frac{\Gamma(\alpha)\Gamma(\beta)}{\Gamma(\alpha + \beta)} = \int_0^1 \theta^{\alpha-1}(1-\theta)^{\beta-1}\, d\theta,$$

where $B(\alpha, \beta)$ is the beta function. The mean of the distribution is $\mu = \alpha/(\alpha + \beta)$ and the variance is $\sigma^2 = \mu(1 - \mu)/(\alpha + \beta + 1)$, from which it follows that

$$\alpha = \mu\left[\frac{\mu(1-\mu)}{\sigma^2} - 1\right] \tag{14.4}$$

and

$$\beta = (1 - \mu)\left[\frac{\mu(1-\mu)}{\sigma^2} - 1\right]. \tag{14.5}$$

Now let x be the number of successes in n trials so that x has a binomial distribution,

$$f(x \mid \theta) = \frac{n!}{x!(n-x)!}\theta^x(1-\theta)^{n-x}, \tag{14.6}$$

and $\hat{\theta} = x/n$ is an estimate of θ. If, in a Bayesian analysis, interest centers on making inferences about θ in the binomial distribution (14.6), and if a beta prior distribution (14.3) is assumed for θ, Bayes' theorem in the form of equation (14.2) may be used to derive the posterior distribution $f(\theta \,|\, x)$. The denominator, $\int f(x \,|\, \theta) f(\theta) \, d\theta$, is readily shown (O'Hagan, 1994: 5) to have the form

$$\frac{n!}{x!(n-x)!} \frac{B(x+\alpha, n-x+\beta-1)}{B(\alpha, \beta)},$$

so that with the numerator of $f(x \,|\, \theta) f(\theta)$ the posterior distribution for θ is

$$f(\theta \,|\, x) = \frac{1}{B(x+\alpha, n-x+\beta-1)} \theta^{x+\alpha-1} (1-\theta)^{n-x+\beta-1}, \qquad (14.7)$$

a beta distribution with parameters $x + \alpha$ and $n - x + \beta$. It follows from this that the mean of the posterior distribution is given by $\tilde{\mu} = (x+\alpha)/(n+\alpha+\beta)$. Unless $x/n = \mu$, this shows that $\hat{\theta} < \tilde{\mu} < \mu$; that is, the posterior mean lies between the observed proportion of successes and the mean of the prior.

This theory has been applied to an archaeological problem as follows. Teotihuacan, about 30 miles to the northeast of Mexico City, is a Classic period (AD 300–900) Mesoamerican site, of about 100,000 inhabitants at its peak, that has attracted considerable archaeological attention. Robertson (1999) describes a mapping project involving the systematic recovery of approximately 1 million artifacts from the site. Superimposing a set of collection tracts on the site and mapping the proportion of an artifact type within each tract provides one approach to studying spatial variability across the site. For example, it is of interest to try and identify areas having an unusually low or high relative proportion or density of some artifact of interest, with a view to investigating further why this might be so.

Notwithstanding the large number of artifacts collected, for any particular investigation the relevant population may be much smaller than the total, so that the sample sizes within tracts may be both highly variable and often small. As an example, Robertson (1999: 142) focuses on San Martín Orange Ware, manufactured over about two centuries mostly in the form of large cooking and storage vessels. The relevant population, used to define the proportion of San Martín Orange Ware present within a tract, was taken to be 25 wares (including San Martín Orange) broadly relating to the period when San Martín Orange was in common use.

Of particular concern was the fact that the observed proportions had very different variances because of the differing sample sizes within tracts, with proportions based on small samples sizes, including zeros, being particularly unreliable. Essentially Robertson (1999) uses Bayesian analysis to produce a smoother map of the proportions of a type, based on posterior estimates of θ given by $\tilde{\mu}$ rather than $\hat{\theta}$.

The argument used is that the mean and standard deviation of $\hat{\theta}$ across all collections with $n > 50$ provide reasonable estimates of μ and σ^2, and hence α and β from equations (14.4) and (14.5), that are little affected or not at all by any individual collection tract. This gives rise to values $\alpha = 1.6$ and $\beta = 16.4$ that can be used in conjunction with equation (14.7) to derive posterior estimates, $\tilde{\mu}$, of the proportion within each collection tract. It is worth observing that only equations (14.4) and (14.5), and knowing the form of the mean and variance of a beta distribution, are needed to apply this theory. Knowledge of the mathematics leading to the results is not essential.

The general effect of this procedure is to 'shrink' posterior estimates of θ to the prior mean. The observed data will dominate calculations for large samples. For samples where $x = 0$ and hence $\hat{\theta} = 0$ the posterior mean will be $1.6/(n + 18) > 0$, while at the other extreme where $x = n$ and $\hat{\theta} = 1$ the posterior mean is $(n + 1.6)/(n + 18) < 1$. The overall effect is therefore to reduce variability compared to the observed proportions, and produce a more interpretable and smoother map on applying mapping procedures to the posterior estimates (Robertson, 1999: 144). A related application is reported in Ortman (2000: 635).

Sampling for 'nothing' – the Bayesian approach
Related theory is used in Orton's (2000b) discussion of problems that arise in archaeological site evaluation, centered on how confident one can be that failure to observe archaeological remains in a sample survey of a site reflects the true situation. The classical approach to this problem was discussed in Section 4.5.2. Assume there are n sampling points, where the true proportion of archaeological remains on the site is θ, and the observed proportion is $\hat{\theta} = 0$. In the Bayesian approach, as developed in Nicholson and Barry (1995), $f(x \mid \theta) = (1 - \theta)^n$, where $x = 0$ and a prior beta distribution for θ, with $\alpha = 1$, is assumed. For integer β this leads to the posterior distribution (14.7) of the form

$$f(\theta \mid x) = (1 - \theta)^{(n+\beta-1)}(n + \beta)$$

which can be rearranged to give

$$n + \beta = \frac{\log(1 - p)}{\log(1 - \theta_{post})}$$

where θ_{post} is an upper $100p\%$ posterior confidence limit for θ.

Nicholson and Barry (1995) and Orton (2000b) discuss the implication of these results and the choice of β in the prior specification. It is argued that, from a Bayesian perspective, the classical result for sample size n amounts to choosing $\beta = 0$ corresponding to a prior belief that $\theta = 1$. If this were the case it might be argued that survey is unnecessary. A value of $\beta = 1$ corresponds to the belief that all values of θ in the range 0 to 1 are equally likely, and will have a negligible effect on n. As β increases the implied values of θ that one believes are plausible become smaller. Another way of looking at this is to take $n = 0$, and write

$$\beta = \frac{\log(1 - p)}{\log(1 - \theta_{prior})}$$

where θ_{prior} is the prior confidence limit before data collection. Thus if a prior belief is expressed as being 95% ($100p$) sure that θ is less than 0.1 this gives $\beta \approx 29$, rounding up to the nearest integer.

Orton (2000b) discusses some of the practicalities of applying the theory, using a hypothetical worked example with $p = 0.9$ and $\theta_{post} = 0.01$, that gives rise to $n = 229$ using a classical analysis that assumes point samples. Large values of β that correspond to fairly strong beliefs that θ is relatively small are needed to reduce this significantly. Complications arise from the fact that it is usually more effective to take areal samples than (approximate) point samples, and that archaeological features differ in shape and areal extent. Depending on what is assumed about the nature of the features, the value of n may be reduced, but this can actually increase the amount of excavation required because each sample now corresponds to an area. To the best of my knowledge, the ideas discussed in Orton (2000b) have yet to be applied in anger to an archaeological problem.

14.3.2   Bayesian clustering

The 'standard' model
Applications of Bayesian clustering methods to archaeological problems have been reported in Buck and Litton (1991), Buck (1993), Buck and Litton (1996), Buck *et al.* (1996), Dellaportas (1998) and Baxter and Buck (2000). The 'standard Bayesian formulation' of the mixture model for clustering (Dellaportas, 1998) starts with the mixture model of equation (8.2). Conjugate prior distributions are assumed as follows. The prior for $\mu_g \mid \Sigma_g$ is $N(\xi_g, \Sigma_g/c_g)$ for some means ξ_g and precision parameters $c_g > 0$; the prior for Σ_g is the inverse Wishart distribution $W^{-1}(\zeta_g, Z_g)$, with $\zeta_g > 0$ degrees of freedom and scale matrix Z_g; and the prior for π is the Dirichlet distribution, $D(\alpha)$, with parameters $\alpha = (\alpha_1 \, \alpha_2 \cdots \alpha_G)$. The subsequent mathematical formulation, needed for the MCMC methodology, takes about a page to write down and is not reproduced here (see Dellaportas, 1998). Dellaportas adapted the basic model to deal with the fact that two of his four variables were binary, one with missing data, and with the problem that one of his continuous variables was measured, in some cases, with error. Technical details are given in his paper.

This standard formulation is complex. A general question of interest is whether the extra complexity, compared to exploratory or model-based approaches based on maximum likelihood estimation (Section 8.4), is justified in terms of the extra insights gained in application. This is a difficult question to answer, and the analysis of Dellaportas' (1998) application that follows is intended to raise a number of issues of general interest, rather than as a criticism of a specific paper.

Example: Classifying Neolithic stone tools
Dellaportas (1998) analyses data on Neolithic ground stone tools from Thessaly in Greece, with the aim of classifying them into tool types. There are four variables, two continuous and two binary. The technical achievement of Dellaportas' paper is to develop a framework within which such mixed-mode data, including measurement error and missing data, can be analyzed. The following discussion focuses more on the outcome than the technique, however.

The continuous variables are *widening*, defined as the ratio of width to thickness, and *length*. For the latter variable some tools are fragmented so the true length is not known. The binary variables are *hardness*, coded as hard or soft and for which some values are missing, and *symmetry*, a dichotomous variable relating to the shape of the working edge of the tool. Data are given for 147 tools, with complete information available for 93 of these, and complete information on widening and length available for 117 tools. It is assumed at the outset that two tool types, *axes* and *adzes*, are represented in the data, and that the four variables used are relevant to classifying tools into one or other of these types.

The plot of length against widening is shown in Figure 14.1, with labeling to be explained shortly. The following observations may be made about the analysis in Dellaportas (1998).

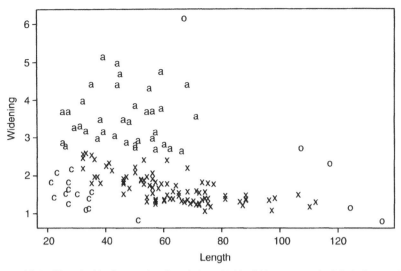

Figure 14.1   Plot of widening against length for 147 Neolithic stone tools, labeled according to the four-cluster solution obtained using classification maximum likelihood methods on logged data. Cases labeled 'a' correspond very closely to the group defined as adzes in Dellaportas (1998); cases labeled 'c' and 'x' correspond closely to his axe group; cases labeled 'o' are outliers put into one group by the clustering method.

The assumption that there are two classes of tool is restrictive. The possibility of a third tool type, *chisel*, is discussed but not investigated, though it is observed that they might fit into the bottom left corner in the figure. Expectations are that fragmentation will reduce the length of the tools by 1–3 mm (Dellaportas, 1998: 290). The mean recorded length is about 57 mm and the minimum is 21 mm. The magnitude of the errors is small in relation to a typical length and seems unlikely to affect general patterns in the data, so that modeling it is possibly unnecessary from a practical view. What is done in the paper, when fragmentation occurs, is to model the observed lengths as normally distributed about the true lengths, $N(x_T - 2, 0.25)$, where x_T is the true value. From this the conditional density of the unobserved true values given other parameters in the model, needed for the Gibbs sampling procedure, can be derived (Dellaportas, 1998: 291).

Despite the sophisticated nature of the statistical machinery developed, the results obtained can be summarized simply, along the lines that if widening is greater than about 2.6 the tool is an adze (see Figure 3 in Dellaportas, 1998). A slightly more complex rule would be to apply this criterion after noting that anything longer than 75 mm is an axe. Hardness and symmetry did not appear to affect the classification.

Graphical inspection of density estimates for length and widening suggests that the assumption of bivariate normality is unlikely to be true and that a log transformation may help matters. This, of course, ignores group structure in the data. Once the groups are defined by the analysis their properties can be investigated, and it is evident that they are not bivariate normal. This is particularly the case for those tools classified as axes, which correspond closely to those labeled c and x in Figure 14.1. Thus, a fundamental modeling assumption of the Bayesian analysis is clearly untrue. Taking logs improves matters, but does not eliminate the problem.

A non-Bayesian cluster analysis using classification maximum likelihood (Chapter 8) produces essentially the same (or better) results, as a comparison of Figure 14.1 with Figure 3 in Dellaportas (1998) will confirm. The S^* method of Banfield and Raftery (1993), as implemented in the S-Plus package using the mclust function, has been used. At the four-cluster level, apart from outliers, this recovers, with a few differences near the boundary, the adze group of Dellaportas (1998). His axe group is split into two, the smaller of which, in the bottom left corner of Figure 14.1, might tentatively be defined as *chisels*. Some comment on the fourth dispersed group is needed. The cases involved appear visually to be outliers – a phenomenon not explicitly allowed for in the analysis of Dellaportas (1998). Previous experience in applying classification maximum likelihood methods to archaeological data suggests that such outliers are often grouped together in a single incoherent cluster (Papageorgiou *et al.*, 2001), leaving an otherwise sensible clustering, and this is what appears to be happening here. The individual cases involved could be classified as axes or adzes on common-sense, non-statistical grounds, given the other results and if such a classification is needed.

Despite the impressive statistical machinery assembled, the archaeological analysis does not convince one of the need for it. The classification developed can be derived more simply using only the two continuous variables, and the Bayesian analysis is overelaborate, demonstrating technique but using data that clearly violate the assumptions on which it is based. The general methodology developed in Dellaportas (1998) might well work with other data sets where binary or categorical variables do have a stronger influence on the outcome, but given its complexity more convincing applications are needed to persuade one to use it in preference to simpler methods.

Other papers

An early application using the same basic framework was provided in Buck and Litton (1991) who provided additional structure in the sense that ξ_g in the prior for μ_g was specified in terms of hyperparameters $\mathbf{d}$ and $\mathbf{D}$ through the assumption $\xi_g \sim N(\mathbf{d}, \mathbf{D})$. Their application was to clustering 72 clay pipes, on the basis of 13 measurements descriptive of size and shape. The results of the subsequent Bayesian analysis were very similar to the conclusions derived from a non-Bayesian maximum likelihood analysis reported in the same paper, resulting in the conclusion that there were four or five clusters in the data. In other words, the substantive interpretation of the data did not appear to differ much from what could be achieved by more standard and accessible methods.

Buck and Litton (1996: 502) described as 'undoubtedly complex' another approach that followed Müller *et al.* (1996). The analysis was of 15 log-transformed trace elements in 150 specimens of medieval tin-glazed wares. These were known to come from three distinct kiln sites, two in Spain and one in Italy. Apart from two or three outliers (relative to the kiln site of origin), the Italian group showed up as quite distinct in an analysis based on the first two principal components. One of the Spanish groups also appeared as a concentrated group, largely distinct from the other more dispersed Spanish group. Even without knowledge of the kiln groups, an analysis based on the first two principal components alone suggested three main groups in the data that, apart from a few outliers, corresponded to the kiln groups. The Bayesian analysis did not obviously add to this. In other words, for a relatively 'easy' clustering problem it is illustrated that this particular Bayesian approach 'works', but the substantive conclusions to be drawn from the analysis can be derived far more easily and quickly using less complex methods.

Elsewhere Buck *et al.* (1996: 313) observed that work on Bayesian cluster analysis of archaeological data 'is still in the developmental stage' and that the analyses they presented, including that described above, were 'simply to illustrate the methodology'. Similar comments are made in the earlier papers intended for an archaeological audience, and the same is arguably true of Dellaportas (1998). The methods developed are complex and, in principle, provide an approach for dealing with complex data structures not otherwise easily handled. The machinery developed is impressive, but the archaeological applications to date have been largely demonstrations of technique. Applications that convince one that it is worth embracing the extra complexity because of the additional insight into the data, compared to what can be achieved by more standard methods, do not yet exist.

14.3.3 A miscellany

The intention here is to note, with little discussion, other applications of Bayesian analysis to archaeological problems that have appeared, mostly since the publication of Buck *et al.* (1996), and which are not referenced in other chapters.

Geophysical prospection
In geophysical surveys of archaeological sites, magnetometers may be used to record the magnetic field strength at points across a surface. The strengths may be mapped to form an image, which approximates the underground magnetic susceptibilities and may reflect the presence of archaeological features. Bayesian methods of image analysis (Aykroyd *et al.*, 2001; Aykroyd and Al-Gezeri, 2002) can be used to produce a 'sharper' or smoother image by estimating the true magnetic susceptibilities, θ say, on the basis of a model for the observed susceptibilities, x, of the form $f(x \mid \theta)$. The estimation may be based on the posterior distribution $f(\theta \mid x)$ calculated via Bayes' theorem using a prior that expresses the assumed relationship between θ and the true susceptibilities at surrounding points. Mathematical and computational details are complex. Related, and earlier, work that examines alternatives to the Bayesian approach, and includes investigation of the analysis of core samples (i.e., measurements are determined vertically rather than horizontally), includes Allum *et al.* (1995, 1996, 1997, 1999). These papers can be viewed as examples of Bayesian methodology applied to archaeological spatial analysis. Buck *et al.* (1996: 253–91) discuss earlier applications of such analyses.

Dendrochronology
The transverse section of the trunk of a felled tree will reveal a pattern of growth layers, or tree rings, of varying width, that reflect the environmental conditions that affected the growth of the tree. In certain environments each ring relates to a year of growth, and trees of the same species, age and felling date, growing within that environment, will tend to exhibit the same pattern. Dendrochronology involves the study of such patterns for dating purposes. Given trees of known felling date, and others whose lifespan overlaps with those of known age, it is possible to match

patterns and construct an absolute chronology that may be used, for example, to date timbers found in excavation or in standing buildings. Absolute chronologies so derived may be used to calibrate dates arrived at by other methods, such as radiocarbon dating discussed in the next chapter and Section 5.6, that are subject to biases and errors of various sorts. This is a simplified account; for a longer but reasonably concise account of the background, Buck *et al.* (1996: 334–52) can be consulted.

Rather than using absolute dates derived from tree-ring dating to calibrate dates derived from other methods, the problem of 'wiggle-matching' arises when a 'floating' sequence of tree-ring widths is available, for which dates are unknown, and for which a date is sought on the basis of radiocarbon dating of a sample of the rings. This gives rise to a set of radiocarbon dates. It is known which is earliest, and the difference between dates, in terms of calendar years, is known. The problem then is to match the sequence of radiocarbon determinations as closely as possible to a sequence on the radiocarbon calibration curve, thus providing an absolute date for the floating tree-ring sequence.

Bayesian approaches to this and other dendrochronological problems are discussed in Buck *et al.* (1996: 233–48, 334–52) who summarize earlier work of Litton and Zainodin (1991), Christen (1994a) and Christen and Litton (1995). Later treatments include Bronk Ramsey *et al.* (2001). Christen's (1994a) approach to some of the problems involved requires that uncertainties in the calibration curve itself be taken into account. A later Bayesian treatment of this problem is Gómez Portugal Aguilar *et al.* (2002). Another Bayesian application in dendrochronology, that of estimating the number of sapwood rings in oak trees, and felling dates, is given in Millard (2002).

14.4 Discussion

Other applications of Bayesian methodology to archaeological problems are discussed in Chapter 15 and Sections 16.3.3 and 19.2.4. The most successful and influential application has been to radiocarbon calibration problems (Chapter 15). The essential problem is to convert radiocarbon dates to calendar dates, and informative prior information about the relative chronology of events is usually available. For most users the archaeological problem is recognizably important, and useful results can be obtained without a deep understanding of the underlying theory because freely available and user-friendly software exists (e.g., OxCal (Bronk-Ramsey, 1995) and BCal (Buck *et al.*, 1999)).

This cannot be said about many other applications of Bayesian methods to archaeological problems, particularly those that appear in the statistical literature. That methods are mathematically and computationally complex will deter archaeological users, but is not in itself a good reason for not using them. Future software developments may, in any case, remove some of these difficulties. For those committed to the Bayesian view, however, the challenge is not simply to convince archaeologists of the philosophical merits of their approach, but to multiply instances where its power is fully exploited and results superior to those achievable by more 'conventional' means are obtained.

15

Absolute dating – radiocarbon calibration

Radiocarbon dating is really the dating of the moment of death of living things, because it is at this point that the flow of carbon, from biosphere to dying organism, stops. (Hedges, 2000: 467)

15.1 Introduction

As Buck *et al.* (1996: 2) put it, 'dating is vitally important to archaeologists, because without a time-scale there is no history or prehistory'. In this chapter, statistical aspects of absolute dating, in the treatment of radiocarbon dates, are examined. In Chapter 16 methods of relative dating are considered. As noted in Chapter 14, Bayesian approaches to the problem of radiocarbon calibration, when multiple dates are available, represents one of the most interesting and successful applications of Bayesian statistics in archaeology. Before describing this, a little background is needed.

Carbon occurs naturally in the form of three isotopes, ^{12}C, ^{13}C and ^{14}C, the last of which is radioactive and decays with time. Formation of ^{14}C occurs in the upper atmosphere where it combines with oxygen, mixes rapidly and, through photosynthesis, enters plant life. Animals eat plants, and consequently all members of the living animal and vegetable world (the biosphere) contain ^{14}C (Aitken, 1990: 57). In a living organism the ratio of ^{14}C to non-radioactive carbon is approximately constant, but the former decays after death.

A simple model for the process is to take t to be the time since death, with $N(0)$ the number of atoms of ^{14}C present at death, and $N(t)$ the number of atoms present when measured. Assuming the rate of decay is proportional to the number of atoms present leads to the result that

$$N(t) = N(0)e^{-\lambda t}$$

and

$$t = -\frac{1}{\lambda} \ln \frac{N(t)}{N(0)},$$

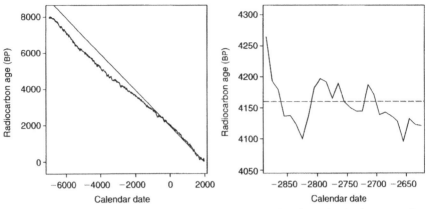

Figure 15.1   The left-hand plot shows radiocarbon ages plotted against calendar dates, negative values representing dates BC. The reference line shows the ideal that would result if radiocarbon ages corresponded directly to calendar dates. The right-hand curve shows a section of the plot with the horizontal line corresponding to a radiocarbon age of 4160 BP.

where λ is the decay constant. In principle λ is known, and $N(t)$ but not $N(0)$ can be determined experimentally. If, however, $N(t)$ and $N(0)$ are measured as *proportions* relative to one of the other isotopes of carbon, and if it is assumed that the proportion in living animals and plants remains constant over time, then an estimate from modern material can be made and t estimated.

This is a simplified model that is not true for a variety of reasons discussed in Aitken (1990), Bowman (1990) and Hedges (2000). In particular, the assumption of a constant ratio of ^{14}C to other isotopes is untrue, with the consequence that radiocarbon dates cannot be equated with calendar dates in a simple manner. This can be demonstrated by comparing radiocarbon dates estimated from samples of known age. The left-hand panel of Figure 15.1 shows a plot of radiocarbon ages against calendar dates determined from tree-ring dating (dendrochronological age – see Section 14.3.3) and a reference line showing the ideal relationship that would be obtained if radiocarbon and calendar dates were equivalent. It can be seen that for the period BC the radiocarbon ages tend to be too 'young', so that some correction or calibration is needed. The non-linear form of the radiocarbon curve, discussed in Section 5.6, is also evident.

Conventionally radiocarbon dates are reported as BP (before the present, defined as before AD 1950) in the form

$$x \pm \sigma,$$

where σ is the standard deviation based, ideally, on an assessment of all the contributions to the error in the laboratory measurement (Hedges, 2000: 484). In statistical analysis σ is typically treated as known.

A radiocarbon determination dates the *event* that caused the death of the organism sampled. The usual assumption in radiocarbon dating is that this can be equated, at least to a good approximation, with an archaeological event. Ensuring this equivalence is non-trivial and the subject of repeated warnings in the radiocarbon dating literature (e.g., Bowman, 1990; Buck and Litton, 1996). It is quite common, for example, for an archaeological context to contain material that pre-dates the formation of the context, possibly by a considerable time, and a matter of archaeological expertise to select suitable samples for dating.

A *phase* can be defined as a period of time during which a period of archaeo-logical activity took place (Litton and Buck, 1996: 470). A timespan, such as the period of occupation of a site, will frequently be divided into phases, which may be disjoint (possibly abutting) or overlapping. Given the concept of an event and phase, a variety of problems, by no means exhaustive, arise in which the general problem is to make inferences about the calendar dates of events on the basis of one or more radiocarbon dates:

1. A single radiocarbon date associated with an archaeological event of interest is available; there may or may not be known limits for the date of the event.
2. A set of radiocarbon dates associated with an event is available. Two situations can be distinguished: (i) the true date of each event is assumed to be identical; (ii) the true date for each event is assumed to be similar, but not necessarily identical.
3. A set of radiocarbon dates within the same phase is available, but the true dates are assumed to differ (the difference between this and 2(ii) is one of degree).
4. As (3) but with radiocarbon dates available for several phases, and with some archaeological knowledge about the relationship between phases.
5. A set of radiocarbon dates is available for which a (partial) ordering of the true dates is known, arising from knowledge of the archaeological stratigraphy. Such ordering can occur within and between phases.

It is generally agreed that radiocarbon dating is most valuable when multiple dates are available, and a major contribution of the Bayesian approach to radiocarbon calibration has been to develop a framework within which the interpretation of such dates can be integrated with prior archaeological information.

After a date has been calibrated it may variously be expressed as a calibrated BP (Cal BP) date, or as a calendar date (e.g., Cal BC or simply BC) by subtracting 1950 from the Cal BP results. Other conventions are, or have been, used for reporting.

15.2 Combining dates

15.2.1 Theory

When several radiocarbon dates pertain to the same event, or to events assumed to have identical dates, they can be combined into a single date (with higher precision than any individual date) before further analysis. If this is to be done it is necessary to check whether the radiocarbon dates are consistent with the assumption that the true dates are the same. Ward and Wilson (1978) distinguish between a Case I situation where it is *known* that the true dates are the same, by virtue of determinations having been made on the same object for example, and Case II situations, where it is required to test the *assumption* that the true dates are the same. Their Case I methodology, which is that most widely used, is described below.

There are n observations of the form $x_i \pm \sigma_i$, and the x_i are viewed as realiza-tions of random variables $X_i = \mu + \varepsilon_i$, where μ is the true radiocarbon date. It is assumed that $\varepsilon_i \sim N(0, \sigma_i^2)$.

The weighted mean of the observations, $\bar{x}$, is

$$\bar{x} = \left(\sum_{i=1}^{n} x_i / \sigma_i^2 \right) \Big/ \left(\sum_{i=1}^{n} 1 / \sigma_i^2 \right). \qquad (15.1)$$

To test the hypothesis that the series of determinations are consistent (i.e., have the same true radiocarbon date) the statistic

$$T = \sum_{i=1}^{n} (x_i - \bar{x})^2 / \sigma_i^2 \qquad (15.2)$$

is used, which has the chi-squared distribution with $n - 1$ degrees of freedom under the null hypothesis. If the null hypothesis is not rejected the data can be combined, the pooled age being $\bar{x}$ with variance

$$\left(\sum_{i=1}^{n} 1 / \sigma_i^2 \right)^{-1}.$$

Dolukhanov *et al.* (2001: 701–2), who do not cite Ward and Wilson (1978), derive equation (15.1) as a weighted least squares (or minimum chi-squared) estimate by replacing $\bar{x}$ by μ in equation (15.2), from which $\hat{\mu} = \bar{x}$ emerges as the estimate that minimizes T.

15.2.2   Application

To illustrate, the data of Table 15.1, given in Bayliss *et al.* (1997: 44) and Bronk Ramsey and Bayliss (2000), will be used. The dates are for nine antler picks found at the base of a ditch excavated at Stonehenge, and interpreted to be those which were used to dig the ditch and placed at the base as soon as the ditch was dug (Bayliss *et al.*, 1997: 42). The assumption is that the picks were deposited in the ditch at the same time, but it is possible that the ages of the picks themselves may differ, though the expectation is that any difference will not be great. The first seven determinations are of high precision relative to the remaining two.

Table 15.1   Radiocarbon dates for nine antler picks found in the base of a ditch excavated at Stonehenge (Bayliss *et al.*, 1997)

Ref. No.	Radiocarbon age
UB-3787	4375 ± 19
UB-3788	4381 ± 18
UB-3789	4330 ± 18
UB-3790	4367 ± 18
UB-3792	4365 ± 18
UB-3793	4393 ± 18
UB-3794	4432 ± 22
BM-1583	4410 ± 60
BM-1617	4390 ± 60

Using all nine determinations $T = 15.0$ with a p-value of 0.060, while using just the high-precision data gives $T = 14.6$ with a p-value of 0.024. The pooled standard deviation is approximately 7 in both cases. One of these statistics is significant at the 5% level, and because it is not certain that antlers are of the same date, but could represent different growing seasons, Bayliss *et al.* (1997) chose to err on the side of caution and not combine the dates.

15.3 Some Bayesian solutions

15.3.1 Calibration of a single date

A Bayesian formulation for calibrating a single date is as follows. Let θ be the *unique* but unknown calendar date, and $\mu(\theta)$ the unique but unobserved true radiocarbon date associated with it. Let x be a realization of the random variable $X = \mu(\theta) + \varepsilon$ where $\varepsilon \sim N(0, \sigma^2)$ so $X \sim N(\mu(\theta), \sigma^2)$. The problem is to make inferences about θ on the basis of observation of x and σ. That this is not easy in general is because of the 'wiggly' nature of the radiocarbon calibration curve, since even if there is no error ($\sigma = 0$) and $x = \mu(\theta)$ it may be multi-valued. This is illustrated in the right-hand panel of Figure 15.1, where a radiocarbon age of 4160 BP with no error is seen to correspond to five possible calendar dates separated by about 160 years.

In the figure, piecewise linear interpolation has been used which can be modeled as (Naylor and Smith, 1988; Buck *et al.*, 1992, 1996)

$$\mu(\theta) = \begin{cases} a_1 + b_1\theta, & (\theta \leq \theta_0), \\ a_k + b_k\theta, & (\theta_{k-1} < \theta \leq \theta_k, \quad k = 1, 2, \ldots, K), \\ a_K + b_K\theta, & (\theta > \theta_K), \end{cases}$$

where $K + 1$ is the number of knots used to define the calibration curve, and the a_i and b_i are known. This leads to a likelihood function of the form

$$l(x_i|\theta_i) \propto \exp\left\{-\frac{[x_i - \mu(\theta_i)]^2}{2\sigma_i^2}\right\} = z_i, \tag{15.3}$$

where subscripting is introduced to accommodate the later discussion of multiple dates.

If an improper prior $f(\theta) \propto 1$ is assumed then the likelihood and posterior distribution, $f(\theta \mid x)$, upon which inference is based, are equivalent. It is possible to modify the above theory to take into account other forms of interpolation than piecewise linear and/or to allow for error limits associated with the calibration curve. This is done in the OxCal package (Bronk Ramsey, 1995, 2001) used to generate Figure 15.2 where, for illustration, a radiocarbon age of 2450 ± 20 BP has been assumed. It is clear from the form of the posterior that there is no simple summary of what the radiocarbon age means in terms of the calendar date and no simple summary, such as a posterior mean or mode, makes sense. Approximate 95% and 68% highest posterior density intervals are indicated at the bottom of the plot, and the limits of the former have been printed out; however, the figure itself is probably the best summary.

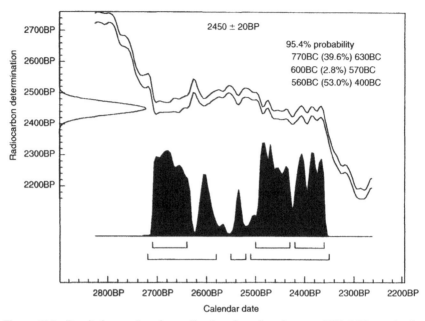

Figure 15.2   Results from radiocarbon calibration of a radiocarbon age 2450 ± 20 BP using the `OxCal` package.

15.3.2   Multiple dates – single phase

Here we consider the situation where there are n radiocarbon dates, $\mathbf{x} = (x_1, x_2, \ldots, x_n)$, with associated errors, corresponding to true dates $\boldsymbol{\theta} = (\theta_1 \theta_2 \cdots \theta_n)$ that are different, and for which there is no prior information about their ordering. The dates are treated as belonging to a single phase, with boundaries given by β and α, with $\alpha < \beta$ (i.e., α is later than β).

Bayes' theorem may be written in the form

$$f(\boldsymbol{\theta}, \alpha, \beta \mid \mathbf{x}) \propto l(\mathbf{x} \mid \boldsymbol{\theta}, \alpha, \beta) f(\boldsymbol{\theta}, \alpha, \beta),$$

and it is assumed that

$$f(\boldsymbol{\theta}, \alpha, \beta) = f(\boldsymbol{\theta} \mid \alpha, \beta) f(\alpha, \beta).$$

Using the notation of equation (15.3),

$$l(\mathbf{x} \mid \boldsymbol{\theta}, \alpha, \beta) \propto \prod_{i=1}^{n} z_i$$

and, assuming that θ_i is uniformly distributed between α and β, so $f(\theta_i \mid \alpha, \beta) = (\beta - \alpha)^{-1}$, and the θ_i are distributed independently of each other,

$$f(\boldsymbol{\theta} \mid \alpha, \beta) = (\beta - \alpha)^{-n},$$

leading to

$$f(\boldsymbol{\theta}, \alpha, \beta \mid \mathbf{x}) \propto (\beta - \alpha)^{-n} f(\alpha, \beta) \prod_{i=1}^{n} z_i.$$

It remains to specify $f(\alpha, \beta)$. One possibility is to have

$$f(\alpha, \beta) = 1 \quad \text{for } \theta_0 > \beta > \alpha > \theta_\infty, \tag{15.4}$$

where θ_0 and θ_∞ are dates that bound α and β and that can be chosen to be informative or uninformative about them. This is essentially the form given in Buck *et al.* (1996: 249). Christen's (1994b) original formulation is slightly different. He specifies priors for α and β which take them to be uniformly distributed in some interval; that is

$$\alpha \sim U(a_1, b_1) \quad \text{and} \quad \beta \sim U(a_2, b_2).$$

The a_i and b_i are defined so that the intervals $b_i - a_i$ are wide and do not overlap (Christen, 1994b: 491). Given the lack of overlap Christen then takes $f(\alpha, \beta) = f(\alpha) f(\beta)$. The Gibbs sampling scheme for the MCMC analysis is also given by Christen. Applications of the general approach can be found in Christen (1994b), Buck *et al.* (1994a) and Buck *et al.* (1996: 245–52).

15.3.3 Multiple dates – ordered

Let $\mathbf{x} = (x_1, x_2, \ldots, x_n)$ be a set of radiocarbon dates, associated with events whose true dates $\boldsymbol{\theta} = (\theta_1 \theta_2 \cdots \theta_n)$ are subject to order restrictions of the form $\theta_i < \theta_j$ or $\theta_i \leq \theta_j$, $i \neq j$, and are bounded by θ_0 and θ_∞. Remembering that Bayes' theorem can be written in the form

$$f(\boldsymbol{\theta} \mid \mathbf{x}) \propto l(\mathbf{x} \mid \boldsymbol{\theta}) f(\boldsymbol{\theta}), \tag{15.5}$$

and assuming that the dates are determined independently, the likelihood can be written as

$$l(\mathbf{x} \mid \boldsymbol{\theta}) \propto \prod_{i=1}^{n} z_i \tag{15.6}$$

with

$$f(\boldsymbol{\theta}) = I_A(\boldsymbol{\theta}), \tag{15.7}$$

where

$$I_A(\boldsymbol{\theta}) = \begin{cases} 1, & \boldsymbol{\theta} \in A, \\ 0, & \boldsymbol{\theta} \notin A, \end{cases}$$

and A is the set of values for which the inequalities governing the θ_i hold. For example, if $\theta_0 > \theta_1 > \theta_2 > \cdots > \theta_n > \theta_\infty$ then $I_A(\boldsymbol{\theta}) = 1$ for any set of values for which this holds, and is 0 otherwise. This choice of prior is intended to express prior ignorance, in the sense that, in the absence of other knowledge and subject to the constraints, any combination of dates is equally probable. This formulation is essentially that of Buck *et al.* (1991) as modified in Buck *et al.* (1992, 1996).

The precise form of the Gibbs sampling scheme necessary to generate samples from the posterior distribution depends on the nature of the inequalities governing

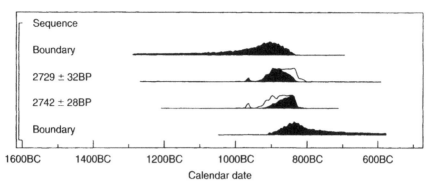

Figure 15.3 Results from radiocarbon calibration of two radiocarbon ages 2729 ± 32 BP and 2742 ± 28 BP using the prior information that $\theta_1 > \theta_2$.

the θ_i. An example is given in Buck *et al.* (1994b), also used for illustration in Buck *et al.* (1996) and Litton and Buck (1996).

For illustration a subset of data from Buck *et al.* (1992) is used. This involves two combined radiocarbon dates from the Late Bronze Age phase at Runnymede Bridge (Needham, 1991), given as 2729 ± 32 BP and 2742 ± 28 BP. Although, on the basis of the BP date, it would appear that the second event is the older, it is known from archaeological evidence that it should post-date the first event. Thus the prior information is that $\theta_1 > \theta_2$. If the dates are calibrated individually without taking into account this knowledge the distributions for the true calendar dates are virtually indistinguishable. A Bayesian analysis, using OxCal, that takes account of the prior knowledge produces the central two posterior distributions shown in Figure 15.3.

The shaded distributions are those obtained when prior knowledge is incorporated, and may be contrasted with the outlines showing the shape of the posteriors ignoring the prior information. Relative to the latter analysis the probability mass is shifted to older dates in the upper plot and to younger dates in the lower plot, reflecting what is known (or believed) to be true about the relative chronological ages of the associated events. Approximate 68% highest posterior density intervals are 900–830 BC and 905–830 BC if prior knowledge is not used, and 910–850 BC and 876–832 BC using this knowledge.

15.3.4　Multiple dates – multiple phases

One model used in phasing problems is as follows. Let there be M phases with boundaries $(\psi_1, \psi_2), (\psi_3, \psi_4), \ldots, (\psi_{2M-1}, \psi_{2M})$, so that for the jth phase $\psi_{2j-1} > \psi_{2j}$ and let $\boldsymbol{\psi} = (\psi_1 \psi_2 \cdots \psi_{2m})$. As in the case of ordered dates, bounds to the set of phase dates, ψ_0 and ψ_∞, can be defined. Bayes' theorem may now be written as

$$f(\boldsymbol{\theta}, \boldsymbol{\psi} \mid \mathbf{x}) \propto l(\mathbf{x} \mid \boldsymbol{\theta}, \boldsymbol{\psi}) f(\boldsymbol{\theta}, \boldsymbol{\psi}), \qquad (15.8)$$

where it is assumed that

$$f(\boldsymbol{\theta}, \boldsymbol{\psi}) = f(\boldsymbol{\theta} \mid \boldsymbol{\psi}) f(\boldsymbol{\psi}), \qquad (15.9)$$
$$f(\boldsymbol{\psi}) = I_A(\boldsymbol{\psi}) \qquad (15.10)$$

and

$$I_A(\boldsymbol{\psi}) = \begin{cases} 1, & \psi \in A, \\ 0, & \psi \notin A, \end{cases}$$

where A is now the set of values for which inequalities governing the ψ_i hold. This assumes that any set of dates, ψ_i, consistent with the imposed prior constraints (i.e., any set of legal or allowable dates) is equally probable, and is intended as an uninformative prior. Apart from the constraints $\psi_{2j-1} > \psi_{2j}$, the relationship between pairs of phases may or may not be known. If relationships are known, for example that one phase precedes another without overlap, additional constraints will be imposed.

The likelihood is as in equation (15.6). If θ_{ij}, for $j = 1, 2, \ldots, n_j$, are the true dates for the events within phase j, and if $\boldsymbol{\theta}_{\cdot j}$ is the set of these events then an uninformative prior, assuming a uniform distribution, for a single date is that

$$f(\theta_{ij} \mid \psi_{2j-1}, \psi_{2j}) = \begin{cases} (\psi_{2j-1} - \psi_{2j})^{-1}, & \psi_{2j-1} > \theta_{ij} > \psi_{2j}, \\ 0, & \text{otherwise}, \end{cases}$$

and for the set of dates within the phase

$$f(\boldsymbol{\theta}_{\cdot j} \mid \psi_{2j-1}, \psi_{2j}) = (\psi_{2j-1} - \psi_{2j})^{-n_j}$$

and

$$f(\boldsymbol{\theta} \mid \boldsymbol{\psi}) = \prod_{j=1}^{m} f(\boldsymbol{\theta}_{\cdot j} \mid \psi_{2j-1}, \psi_{2j}) = \prod_{j=1}^{m} (\psi_{2j-1} - \psi_{2j})^{-n_j}.$$

This particular formulation is based on the (unpublished) appendix to Zeidler *et al.* (1998) where the Gibbs sampling scheme for the constraints involved in the problem studied is also given. More accessible but less detailed accounts of the mathematics involved (as opposed to the archaeological problem) are given in Buck *et al.* (1996) and Litton and Buck (1996), and related models are discussed in Nicholls and Jones (2001).

As an illustrative example a subset of the data from Zeidler *et al.* (1998) will be used. That paper studied a seven-phase cultural sequence for the Jama River Valley in western Ecuador using 37 radiocarbon dates. Only the nine dates for the first two ceramic phases are used here. These are shown in Table 15.2.

Table 15.2 Radiocarbon dates for nine contexts from two ceramic phases from the Jama River Valley, Ecuador (Zeidler *et al.*, 1998)

Ref. No.	Radiocarbon age (BP)	Phase
ISGS-1221	3630 ± 70	I
ISGS-1222	3620 ± 70	I
ISGS-1223	3560 ± 70	I
PITT-426	3545 ± 135	I
ISGS-1220	3500 ± 70	I
ISGS-2336	3030 ± 80	II
AA-4140	2845 ± 95	II
HO-1307	2800 ± 115	II
ISGS-2377	2500 ± 160	II

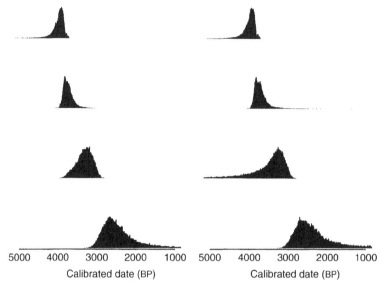

Figure 15.4   From the top down, the four rows of figures show posterior distributions for phase boundaries ψ_1 to ψ_4. The left-hand set of figures uses the prior information that the first phase, defined by ψ_1 and ψ_2, precedes the second phase. The right-hand set of figures ignores this prior information. The data used are from Table 2 of Zeidler *et al.* (1998), for the first two phases only.

There is strong archaeological evidence to suggest that phase I precedes phase II, so that $\psi_1 > \psi_2 \geq \psi_3 > \psi_4$, bearing in mind that larger values correspond to earlier dates. The left-hand side of Figure 15.4 shows the outcome of a Bayesian analysis, based on output from the BCal package (Buck *et al.*, 1999), that takes this prior knowledge into account. The right-hand side of the figure is based on an analysis that ignores the relationship between phases and assumes only that $\psi_1 > \psi_2$ and $\psi_3 > \psi_4$. A comparison of the two analyses shows that they give broadly similar results, but that there is greater overlap between the posterior distributions for ψ_2 and ψ_3 when the relationship between phases is ignored. For example, the 95% highest posterior density regions for ψ_2 and ψ_3 when prior knowledge of their relationship is used are (1980, 1580) BC and (1730, 1030) BC, an overlap of 150 years. The corresponding intervals when the prior knowledge is not used are (2010, 1520) and (2360, 980), so that the latter interval embraces the former.

15.3.5 Prior assumptions

In schemes such as those represented by equations (15.5) to (15.7) and (15.8) to (15.10) the prior specifications (15.7) and (15.10) are intended to be uninformative. It has been pointed out in a number of papers that, in certain circumstances, these priors are rather more informative than intended (Bronk Ramsey, 1999, 2000; Steier and Rom, 2000; Steier *et al.*, 2001; Nicholls and Jones, 2001). The general problem is that if the *span* of a set of dates is of interest then the supposedly uninformative priors noted above favor longer spans. In the context defined by equations (15.5) to (15.7) the span is $s(\theta) = \theta_1 - \theta_n$, where θ_1 and θ_n are the earliest and latest dates. In the context defined by equations (15.8) to (15.10) the span is $s(\psi) = \psi_1 - \psi_{2M}$. Let $R = \theta_0 - \theta_\infty$ or $R = \psi_0 - \psi_\infty$ and, in what follows and according to context, assume $R \gg s(\theta)$ or $R \gg s(\psi)$.

Bronk Ramsey (1999, 2000) notes that given n dates, whether ordered or unordered, the number of possible combinations for a given span, $s(\theta)$, is proportional to $s(\theta)^{n-2}$. This implies a prior in which longer spans are regarded as more likely than shorter spans. For example, a span of ks, $k > 1$, is favored over a span of s by a factor of k^{n-2}. This is a consequence of using the prior distribution (15.7). One way to avoid this effect is to replace this prior with one of the form $s(\theta)^{2-n}$ for 'legal' combinations and 0 otherwise.

An alternative approach, and that used in OxCal, is to treat bounding dates, θ_0 and θ_∞, as dates additional to the original n and use a prior of the form $s(\theta)^{-n}$. This is equivalent to assuming a uniform prior for R. The prior (15.7) can be motivated by assuming that the sampled events are derived from a continuum of events that have a Poisson distribution, chosen from an infinite timeslice. The imposition of boundaries and use of uniform prior for R retains the Poisson assumption but imposes the restriction that events are chosen from a finite timeslice. Intuitively the latter procedure should result in estimates of dates less spread out than in the former procedure (Bronk Ramsey, 2000).

Nicholls and Jones (2001) give an equivalent analysis for what can be regarded as a very specific phase model, where phases abut, so that $\psi_{2j} = \psi_{2j+1}$ for $j = 1, \ldots, M - 1$. Their actual context envisages abutting layers of earth in which vertical mixing takes place within but not between layers, and for which they provide a model of the depositional process. They observe that the prior given by (15.10) favors longer spans and propose, instead, a prior of the form

$$f(\psi) = \frac{s(\psi)^{1-M}}{R - s(\psi)} \tag{15.11}$$

which, if $R \gg s(\psi)$, has a similar form to that proposed by Bronk Ramsey (1999) for dates in a single-phase model.

The prior (15.11) can be justified by a model of the deposition and recovery process in which datable artifacts are generated by a Poisson process of rate $\lambda(t)$, in the interval $[\psi_1, \psi_{2M}]$, where the phase boundaries, $\psi_{2j} = \psi_{2j+1}$, mark change-points in $\lambda(t)$. The change-points are modeled as the realization of a Poisson point process of constant intensity in the interval $[\psi_1, \psi_{2M}]$ (Nicholls and Jones, 2001: 512). Nicholls and Jones provide an example contrasting the use of priors (15.11) and (15.10) where the posterior density for the former favors shorter spans much more than the latter. A Bayesian model comparison is undertaken that shows that the former model is much to be preferred.

For more general phase models, where phases do not necessarily abut, Nicholls and Jones (2001) suggest that a prior similar to (15.11), replacing $(1 - M)$ with $(2 - d)$, where d is the number of phase parameters, may be better than (15.10). They also note that, in this more general setting, a justification for (15.11) in terms of some process model is lacking, and that the development of such models is desirable.

The literature discussed above is, at the time of writing, recent, and many previous published analyses have used priors of the form (15.7) and (15.10). To what extent is this a problem? Nicholls and Jones (2001: 516–17) suggest that replacing (15.10) by (15.11) 'makes no significant difference for data sets of the kind considered in the Bayesian radiocarbon literature up to the present'. Where the true span is short, and there are many dates and/or phases, problems may arise with spans that are too long being favored. This was a specific concern in Nicholls and Jones (2001: 504) who were concerned with New Zealand prehistory, which is relatively short, and where the true span for a set of dates may have a similar order of magnitude to the standard deviation of the observation model.

15.4   Other applications

15.4.1   Outlier detection

Christen (1994b) addressed the problem of detecting outliers in radiocarbon determinations, a problem deemed to be 'relatively common' (Christen, 1994b: 495). His approach was to define x_i as an outlier if it needed a shift, δ_i, on the radiocarbon scale 'for it to be consistent with the rest of the sample' (Christen, 1994b: 496). This leads to a modified likelihood for a single observation of the form

$$l(x_i \mid \theta_i) \propto \exp\left\{ -\frac{[x_i - \{\mu(\theta_i) + \phi_i\delta_i\}]^2}{2\sigma_i^2} \right\} = z_i,$$

where $\phi_i = 1$ if a shift is needed and 0 otherwise. A vague prior was assumed for the δ_i and a prior for the ϕ_i of the form

$$f(\phi_i = 1) = p_i \quad \text{for } 0 \le p_i \le 1.$$

Christen (1994b: 497) emphasized that the p_i should be specified before the data are seen, and in his two examples used $p_i = 0.1$ for all the i.

The Gibbs sampler for obtaining the posterior distribution of the ϕ_i is given in Christen (1994b). Other than the examples given there, one of which is repeated with minor variations in Buck *et al.* (1994a), the methodology seems not to have been used much, though its availability in the BCal package (Buck *et al.*, 1999) may change this.

15.4.2   Sample selection

Buck and Christen (1998) and Christen and Buck (1998) investigated the selection of samples for radiocarbon dating, motivated by the following problem. Suppose N potentially datable samples have been collected of which n have actually been dated, $\mathbf{x}_n = (x_1, x_2, \ldots, x_n)$, giving rise to estimated calendar dates $\boldsymbol{\theta} = (\theta_1\theta_2 \cdots \theta_n)$. The chronological ordering of the θ_i is known, as is their relationship to phase boundaries, ψ_j, with $\boldsymbol{\psi} = (\psi_1\psi_2 \cdots \psi_J)$. If funds become available for further radiocarbon dating, what samples should be selected to improve the precision of inferences about $\boldsymbol{\psi}$?

Having stated the problem in general terms, it needs to be made more specific. Let n' be the number of new samples selected for dating, with $\mathbf{x}_{n'}$ the associated radiocarbon dates. From the samples originally collected the posterior distributions of the ψ_i can be estimated, along with their associated variances. Let $V(\boldsymbol{\psi} \mid \mathbf{x}_n)$ be the *sum* of these variances, with $V(\boldsymbol{\psi} \mid \mathbf{x}_{n'}, \mathbf{x}_n)$ the corresponding quantity given the additional samples. The theoretical analysis given in

Christen and Buck (1998) leads to the definition of a risk function defined as the expectation, given $\mathbf{x}_n$, of

$$\frac{V(\boldsymbol{\psi} \mid \mathbf{x}_{n'}, \mathbf{x}_n)}{V(\boldsymbol{\psi} \mid \mathbf{x}_n)} + C(n'), \tag{15.12}$$

where $C(n')$ is a cost function defined, in the papers, as the cost of obtaining the new determinations relative to the overall cost. Essentially equation (15.12) balances the improved precision obtained from the new samples against the cost of obtaining them. The core of the MCMC methodology described in Christen and Buck (1998) is aimed at approximating the expectation, given $\mathbf{x}_n$, of $V(\boldsymbol{\psi} \mid \mathbf{x}_{n'}, \mathbf{x}_n)$. The expectation of equation (15.12) may be calculated for all possible selections of size $1, 2, \ldots, N - n$, and the recommended selection is that which gives the minimum risk.

16

Relative dating – seriation

... seriation is a unidimensional ordering technique used to arrange items in a series such that the position of an item, relative to other items, reflects its similarity to those other items. Although it need not necessarily be a chronological ordering technique, archaeologists have made frequent use of seriation to arrange units along a temporal dimension. (Marquardt, 1978: 266)

16.1 Introduction

In the previous chapter we saw how the Bayesian approach to absolute dating, via radiocarbon calibration, used information on *relative* dates to sharpen inferences about the absolute dates of events. Knowledge of relative dates might be derived from the stratigraphy of a site, where the material dated is associated with events whose relative order is inferred from the relationship between the contexts in which they occur. Relative dating may be of interest in its own right, in circumstances where stratigraphic evidence and radiocarbon dates are unavailable. An example would be a cemetery site, where there are no intercutting burials, but where one is prepared to make the assumption that the similarity in date of pairs of graves may be reflected in the similarity of the grave goods within them. A *seriation* of such data might involve an attempt to produce an ordering of the graves, so that graves similar to each other in terms of content are close to each other in the ordering, and such that the sequence as a whole can be interpreted as a relative chronological ordering.

Let $\mathbf{X}$ be an $n \times p$ data matrix whose rows correspond to contexts or units, such as graves, and whose columns correspond to artifact types. Let x_{ij} be either a measure of the relative abundance of type j within grave i (i.e., $\sum_j x_{ij} = 100\%$), or a measure of the presence ($x_{ij} = 1$) or absence ($x_{ij} = 0$) of the type within the grave. In the former case $\mathbf{X}$ is a relative abundance matrix and in the latter an *incidence* matrix. The matrix $\mathbf{X}$ satisfies what is sometimes called the *ideal model* for seriation if its rows can be permuted to obtain another matrix, $\mathbf{A}$, such that in each column of $\mathbf{A}$ the elements increase to a maximum and then decrease, or the elements increase, or the elements decrease. The terms 'increase' and 'decrease' are understood in a weak sense, so that adjacent elements within a column may be identical (e.g., Kendall, 1971a; Laxton, 1990).

A matrix **A** that satisfies the above properties will be called a *Q-matrix* if relative abundance data are used, and a *P-matrix* if incidence data are used. A P-matrix has the property that all the 1s in each column are brought together. If **X** can be permuted so that a P- or Q- matrix results, the ordering of the rows that achieves this is a seriation of the data. It is usually to be hoped that this ordering can be given an archaeological interpretation, often in terms of the relative chronology of the units.

For concrete illustration and later application a small data set, given originally in Jacobi *et al.* (1980), and reproduced in Buck and Litton (1991) (where it is described as highly simplified) and Buck and Sahu (2000), is shown in Table 16.1. For six sites this shows the percentages, by site, of seven types of Mesolithic flint tool found. The matrix can be permuted into a Q-matrix, as is shown in Table 16.2. If the non-zero percentages in the latter table were replaced by 1s this would be a P-matrix. If the seriation shown is judged to be a chronological one additional, archaeological, information is needed to determine which is the beginning and which is the end of the sequence, since the reverse ordering also produces a Q-matrix.

It is an archaeological problem to select the units to be seriated and the variables to be used. Given such a selection, and given that one accepts that the ideal model is what one requires of a seriation, it is then a mathematical problem to determine whether the resultant data matrix can be permuted into the ideal form and, if so, what that permutation is. If such a permutation does not exist the problem is to

Table 16.1 Percentages of seven different types of Mesolithic flint tools from six sites in southern England (Jacobi *et al.*, 1980)

Site							
1	20	3	4	42	18	0	13
2	85	3	12	0	0	0	0
3	26	40	8	0	0	26	0
4	20	1	4	13	58	0	4
5	67	10	23	0	0	0	0
6	26	29	8	3	0	33	1

Table 16.2 Percentages of seven different types of Mesolithic flint tools from six sites in southern England. This is Table 16.1 with permuted rows and now in the form of a Q-matrix (Jacobi *et al.*, 1980)

Site							
2	85	3	12	0	0	0	0
5	67	10	23	0	0	0	0
3	26	40	8	0	0	26	0
6	26	29	8	3	0	33	1
1	20	3	4	42	18	0	13
4	20	1	4	13	58	0	4

find a permutation that approximates the ideal in some sense, and to determine how good the approximation is. This can be viewed as a statistical problem. Interpretation of any ordering obtained is, once again, an archaeological problem.

16.2   A brief review

Seriation has a long history and has attracted a considerable literature. Flinders Petrie (1899), who dealt with a problem similar to that described above, for about 900 Predynastic Egyptian graves and 800 varieties of pottery, has been credited by one commentator as the inventor of statistical archaeology (Kendall, 1969a: 68). Collaboration between a statistician (Robinson, 1951) and archaeologist (Brainerd, 1951) on a mathematical formulation of the seriation problem stimulated a considerable amount of work from the 1950s on reviewed, up to about the mid-1970s, by Doran and Hodson (1975: 268–84) and Marquardt (1978). A more recent technical review is Djindjian (1990), while Wilcock (1999) includes a brief bibliographic review.

Wilcock (1999: 38) suggests that seriation 'is one of the extremely few quantitative analytical methods which can be said to have been developed by archaeologists for archaeological application', but notes 'an explosion of methodology' from the late 1960s on led by mathematicians and statisticians. Prior to this 'explosion' an important manual approach to seriation was that of Ford (1962). Relative abundances are represented as horizontal bars on long narrow strips of paper that are then arranged in relation to one another until something approaching the ideal model emerges. Ford's (1962) much reproduced illustration of the method can be found in Doran and Hodson (1975: 279), Marquardt (1978: 262), Djindjian (1990: 80) and O'Brien and Lyman (2000: 125).

Prominent among mathematicians and statisticians taking an interest in seriation was Kendall (1963, 1969a,b, 1970, 1971a,b) whose work can be seen as the inspiration for Wilkinson (1971, 1974), Laxton (1976, 1987, 1990) and Laxton and Restorick (1989). The work of Ihm (1981, 1990) was important in demonstrating the connection between different seriation algorithms. The appeal of the seriation problem, for mathematicians, stems at least partly from the fact that it can be modeled in a mathematical manner amenable, in the ideal case, to theoretical solution. That such solutions were not always readily implemented led to the development of a variety of algorithms for the rapid seriation of a set of data (e.g., Renfrew and Sterud, 1969; Gelfand, 1971a,b), possibly useful as a means of providing a good starting position for more complex, iterative algorithms. What is now the WinBASP package, software developed specifically for archaeological applications, had it origins in the 1970s as a means of making seriation methods developed by Wilkinson (1974) more widely available (Graham *et al.*, 1976; Scollar *et al.*, 1993).

Statistical considerations enter when the ideal model is not attainable, either because of 'noise' in the data or because a good seriation does not exist (possibly because of poorly chosen variables). This leads to the search for good approximations to an ideal seriation, and begs the question both of what counts as an approximate solution and how should its quality be measured.

Kendall (1971a) proposed the use of non-metric multidimensional scaling techniques (Section 7.4), and MDS methods were also exploited by Cowgill (1968,

1972), LeBlanc (1975) and others (Marquardt, 1978: 279). More recently, correspondence analysis (which can be regarded as an MDS technique) seems to have supplanted other MDS approaches as the method of choice. Wilcock (1999: 38) suggested that MDS had 'gone out of favour in the 1990s'. Scott (1993) developed a parametric approach to seriation in which the true dates were treated as missing values and estimated using an EM algorithm (Section 10.1.4). His model makes strong distributional assumptions, and I am not aware of applications of it.

The advance of correspondence analysis can be traced from the paper by Hill (1974), which applied the technique to an archaeological seriation problem involving an incidence matrix. This paper appeared in the statistical literature and had little immediate impact on archaeological practice. Likewise, the use of correspondence analysis for seriation by French scholars from the mid-1970s, reviewed in Djindjian (1989), was little noticed in the English-language literature. Ihm (1981), among others, recognized that correspondence analysis was mathematically equivalent to earlier algorithms, such as that of Wilkinson (1974) based on reciprocal averaging (Djindjian, 1990: 83–4). As noted in Chapters 1 and 11, the paper by Bølviken *et al.* (1982) is regarded by many as having put correspondence analysis on the 'archaeological map'. The collection of Madsen (1988a) contains several papers that use correspondence analysis successfully for the purposes of seriation. In his 1990 review Djindjian (1990: 79) observed that the main approaches to seriation had been reciprocal averaging, rapid seriation, proximity graphs, operational research and data analysis. This classification might be queried, as reciprocal averaging is the same as correspondence analysis and can be regarded as a data analysis method, while proximity graphs (Renfrew and Sterud, 1969) are a form of rapid seriation. Djindjian (1985) stated that correspondence analysis 'is today the most popular method of seriation' and in the 1990 paper concluded that correspondence analysis could 'replace all the heterogeneous techniques' previously used for different data types such as percentage data or presence/absence data (Djindjian, 1990: 88).

Not everyone would agree with Djindjian's assessment, but it is clear that correspondence analysis is a popular method for seriation. Lucy (2000) suggested that the use of correspondence analysis in burial studies was popular in Scandinavia but less so in Britain. She noted that for correspondence analysis to work well the types used in an analysis needed to change quickly and should be uniform over the area analyzed. She suggested that this was reasonable for some areas, such as Iron Age Denmark, but had produced unsatisfactory results for English cemetery material. Edited collections that include several examples of the use of correspondence analysis for seriation, mostly in burial studies, appearing since 1990, include Jørgensen (1992), Jensen and Nielsen (1997) and Müller and Zimmerman (1997). McHugh (1999: 82–3) includes a brief review of the use of correspondence analysis for seriation in burial studies.

16.3   Practical seriation

16.3.1   Similarity matrices

Many seriation algorithms operate, explicitly or implicitly, on a symmetric similarity matrix, S. Robinson (1951) defined a similarity coefficient, suitable for relative abundance matrices, of the form

$$s_{ij} = 200 - \sum_{k=1}^{p} |x_{ik} - x_{jk}| \qquad (16.1)$$

assuming that the measurements are percentages. This is a transformation, to a similarity coefficient, of the Manhattan distance between rows (see Section 6.4.1). Kendall (1971a,b) defined an operator, the circle product, of the form

$$s_{ij} = (\mathbf{X} \circ \mathbf{X}')_{ij} = \sum_{k=1}^{p} \min(x_{ik}, x_{jk}), \qquad (16.2)$$

which results in a similarity coefficient proportional to (16.1) for relative abundance data (assuming percentages sum to exactly 100). For incidence data, equation (16.2) is just the number of types common to the ith and jth units. The similarity matrix in this case can also be written as $\mathbf{S} = \mathbf{XX}'$.

Some earlier work on seriation was concerned with the relationship between $\mathbf{X}$ and $\mathbf{S}$ in the ideal case (Kendall, 1969a; Wilkinson, 1971). This had relatively limited practical application, but did suggest the use of $\mathbf{S}$ as input into an ordination algorithm Kendall, 1971a). Some of the possibilities are now considered.

16.3.2 Correspondence analysis and MDS

The horseshoe effect
As discussed in Section 16.2, correspondence analysis and non-metric MDS have been widely used as practical methods for seriating large and noisy bodies of data. Although seriation involves an attempt to recover a one-dimensional ordering of the data, from Kendall (1971a) on it has been common to present results in the form of two-dimensional plots based on the two leading axes or components extracted in the analysis. For successful seriations this typically results in a 'horseshoe'-shaped plot and, informally, the presence of the horseshoe effect is often taken to indicate a successful seriation. The required one-dimensional seriation is obtained by reading round the horseshoe, archaeological criteria being used to determine the early and late ends of the sequence. An example was given in Figure 11.1.

The reason for the horseshoe effect is discussed, for example, in Kendall (1971a: 225). Informally, it occurs because, with many types, units at either end of the sequence will appear to be 'similar' in the sense that there will be many types absent from both units. The two-dimensional configurations produced by CA and MDS may need to 'bend round', resulting in the horseshoe effect, to reflect this 'similarity'. In some areas of study the horseshoe effect is regarded as a nuisance, and techniques such as detrended CA exist to unbend it (Section 11.4.2). Archaeologists have been largely untroubled by this, regarding a clear horseshoe as evidence of a good seriation; but see Lockyear (2000b) for an exception.

One reason for using a two-dimensional representation, and one emphasized by Kendall (1971a), is that it allows the method to 'fail'. Serious departures from a horseshoe shape may indicate that a good seriation is not possible, or that factors other than chronological ones may be operating. Units at some distance from what is otherwise an acceptable horseshoe may be associated with 'outliers' that may require separate consideration. Djindjian (1985: 126) has suggested that

Table 16.3 An artificial data set, from Baxter (1994a: 121), in Q-matrix form, used in the text to illustrate when correspondence analysis fails to recover the ideal form

Site					
1	4	0	95	0	1
2	6	0	75	0	19
3	8	0	56	0	36
4	10	40	50	0	0
5	26	54	20	0	0
6	43	51	5	1	0
7	44	31	0	25	0
8	65	30	0	5	0
9	66	29	0	5	0
10	88	12	0	0	0

correspondence analysis may best be used in an iterative fashion where units and types may be successively removed from analysis until a satisfactory seriation is achieved.

CA or MDS

As discussed in Section 16.2, CA appears to be the method of choice of many practitioners and to have supplanted the earlier preference for (non-metric) MDS methods. Claims that CA is identical to or clearly better than competing methods are, however, sometimes exaggerated. Djindjian (1985: 121–3), for example, states that correspondence analysis is equivalent to the Robinson seriation method, to Petrie's method of direct seriation, and 'most important of all' to multidimensional seriation equated, in part, with MDS. Equivalent claims are made in Djindjian (1990), in a paper described as a 'major contribution in bringing the various methods [of seriation] under a single mathematical formulation' (Ammerman, 1992: 240).

In comparing the merits of CA and other methods, including MDS, it is perhaps most helpful to think of CA as a form of MDS (e.g., Cox and Cox, 2001). Whereas MDS typically operates on a similarity or *proximity* matrix **S**, CA operates on the data matrix **X**. There is, however, an implicit underlying proximity measure involved in CA, which is the chi-squared distance between rows (Djindjian, 1985). Recall from Section 11.5 that the Euclidean distances between cases in a typical CA row plot are intended to approximate chi-squared distances. Note also that CA can be applied directly to an abundance matrix. A first step in the CA algorithm is to convert this to a relative abundance matrix. Subsequent operations differentially weight rows according to the original absolute abundance.

From this perspective the various forms of MDS and CA that have been used for seriation can be viewed as different algorithms applied to different proximity measures (that are sometimes implicit). One approach to judging the merits of an algorithm is that if an ordering satisfying the ideal model exists the algorithm should be able to recover it (Laxton, 1990). Table 16.3 is an artificial data set taken from Baxter (1994a: 121) that was inspired by Laxton (1990). Figure 16.1 shows the outcome of non-metric MDS and CA on these data.

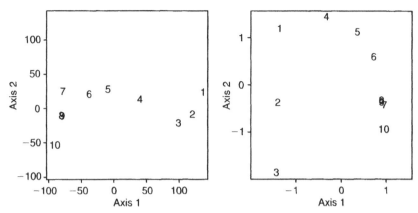

Figure 16.1 The left-hand panel shows the result of a non-metric MDS analysis of the data of Table 16.3. The right-hand panel shows the outcome of a correspondence analysis of the same data.

The MDS recovers the true order of the data, while the CA, however it is read, fails to do so. The conditions under which CA may fail are discussed in Baxter (1994a: 122). In archaeological terms, the true order for a section of the data matrix may be inverted for abundance matrices consisting of common types showing limited change in use over longish periods of time, and uncommon types peaking sharply in popularity in one period and possibly short-lived. It might reasonably be argued that this example is pathological and, in practice, one should not consider seriating such a matrix for archaeological reasons (Section 16.4). The point of the example is merely to demonstrate that it is possible to get differences between CA and MDS, and that CA is not always the superior technique.

If an incidence matrix can be permuted to a P-matrix it is petrifiable. If the same is true of its transpose it is two-way petrifiable. In this case CA will recover the correct row and column order (Hill, 1974). An analogous result holds for relative abundance matrices (Laxton, 1990). Thus, if one imposes stronger criteria for a 'perfect' seriation than is required for the ideal model then CA will recover the seriation order if it exists. Laxton (1990: 41) suggests, however, that the requirement of two-way petrification, or its equivalent for a relative abundance matrix, is an 'extremely restrictive assumption'.

As another example, the data of Table 16.1 are analysed using non-metric MDS and CA in Figure 16.2. This has been used in Buck and Sahu (2000) to illustrate that CA fails to recover the true order, and is outperformed by alternative methods of analysis. The true order (2, 5, 3, 6, 1, 4) (or its reverse) is not recovered by either method if one reads the seriation order off the first axis. The ordering is different for the two methods. If, as many users would do, one attempts to read round a horseshoe, then the true order is obtained. It is, admittedly, rather difficult to discern a horseshoe effect in the plots, and the real message is probably that one has no business trying to seriate these data, notwithstanding the fact that they can be ordered in the form of a Q-matrix. There are three pairs of similar sites, weakly linked to other pairs and, arguably, not enough data to merit an attempt at seriation.

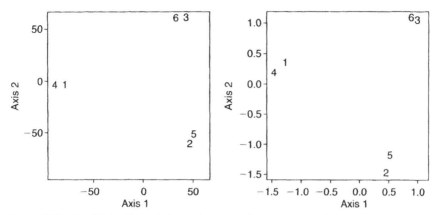

Figure 16.2 The left-hand panel shows the result of a non-metric MDS analysis of the data of Table 16.1. The right-hand panel shows the outcome of a correspondence analysis of the same data

16.3.3 Bayesian methods

Laxton (1990: 41) concluded his paper by noting that questions concerning orderings that approximated the ideal model, and concerning the merits of different orderings in relation to the true model, had been little addressed. He suggested that a stochastic formulation of the model was needed, together with a Bayesian approach to analysis. This is attempted in Buck and Litton (1991) and Buck and Sahu (2000) for abundance data, and Halekoh and Vach (1999) for incidence data. For the most part the models developed, including computational details, are complex and have not yet been much used. Their practical utility is as yet unclear.

Buck and Sahu's (2000) approach is illustrated on two small data sets, one of which is that of Table 16.1. Bayesian model-selection methods are used to evaluate the relative performance of what is called an extended Robinson–Kendall model, that approximates the ideal model, and CA, interpreted as a model-based technique. In both examples the former approach is judged to be the more successful. One example uses five different types of painted pottery from seven layers from a refuse mound at Awatovi, Arizona, where the true ordering is known from the stratigraphy – the data are also given, in fuller 10 × 5 form, in Laxton and Restorick (1989) and Laxton (1990). Both the Robinson–Kendall and CA models lead to the same seriation, but the former is judged superior, essentially because the goodness of fit to the original data is better. If the aim is simply to seriate the data the relevance of goodness of fit could be queried. It has previously been remarked that the data of Table 16.1 are, arguably, unsuited for a seriation exercise. Given this, more experience of Bayesian seriation methods, with larger and realistic data sets, is needed before their merits, relative to the much simpler exploratory methods, can be properly assessed.

16.4 Archaeological considerations

Publications in the archaeological literature are often more concerned with what Cowgill (1972: 381) called 'method and assumptions' than techniques. An extended discussion, with an emphasis on seriation in Americanist archaeology, is given in O'Brien and Lyman (2000). This references very little of the technical material referred to in the previous sections (Marquardt, 1978, is one exception) and, for example, does not mention CA or the work of Kendall, and those influenced by him, that mainly appears in the European literature. What follows is largely a comment on work that has appeared in the Americanist literature concerning seriation as 'method'.

Dunnell (1970, 2000) discussed the considerations involved in constructing a seriation, and distinguished between seriation as a 'method' and seriation as a 'technique'. In terms of this distinction, CA, MDS and the Bayesian approaches discussed would all be regarded as techniques. 'Method' was defined as 'a set of assumptions, their corollaries and relations, organized for the solution of a particular class of problem (e.g. chronology)' (Dunnell, 1970: 306).

Dunnell draws a distinction between (relative) frequency seriation (e.g., of assemblages characterized by frequencies of different ceramic types) and occurrence seriation (e.g., where objects are characterized by the presence or absence of attributes). The latter is equivalent to analysis based on incidence data, as described earlier. The 'single theoretical principle' upon which occurrence seriation rests is that '*the distribution of any historical or temporal class is continuous through time*' (Dunnell, 1970: 308; emphasis in the original). Frequency seriation adds to this the assumption that a class (carefully defined in Dunnell, 1970, and for which the shorthand 'type' is being used here) 'exhibits a unimodal curve through time'. That is, it has an inception; increases in popularity to a peak; and then declines to extinction (Dunnell, 1970: 309). This is just the 'ideal model' discussed earlier. Note that this can be viewed as a model for the population. Samples of material, on which analysis is based, are likely to depart from the ideal model (even if it is true) simply because of sampling error, and quite apart from further complications caused by the vagaries of survival, recovery, and the way material is quantified.

Both CA and MDS, as often used, produce a representation of the data from which one may, if one wishes, derive a relative ordering of assemblages/objects by reading off the scores on the first axis, or reading round a horseshoe. The extent to which the pattern departs from the expected horseshoe provides an informal means of assessing the quality of the ordering, and practitioners are perfectly at liberty to iterate to a solution, by omitting assemblages and/or types that seriously disturb the expected pattern. Dunnell (1970, 2000) variously characterizes this as a 'statistical' or 'similarity' approach to seriation, since manipulation of a (sometimes implicit) similarity matrix is needed to get results. Interpretation of any order that is accepted is exclusively an archaeological problem.

This statistical approach is contrasted with graphic techniques, such as that of Ford (1962) described earlier, characterized by Dunnell (1970: 310) as a 'visual' kind of frequency seriation. Although this and the statistical approach are viewed as simply different techniques for achieving the same end, a preference seems to be expressed for the visual technique, since it 'presents rather more information than do the comparable statistical techniques' (Dunnell, 1970: 306). From the perspective of the early 21st century this distinction between the two approaches seems overstated. It is easy, with modern software, to rapidly seriate frequency data using a 'statistical' technique, and there is nothing to stop the analyst subsequently displaying the results in any preferred 'visual' fashion, modifying the order if this seems justified.

Dunnell (2000: 549) also distinguishes between deterministic and probabilistic solutions, the former being those in which the ideal model is satisfied exactly. This leads to the identification of eight different families of seriation techniques, according to whether frequency or occurrence data are used; the ordering principle used; and whether a deterministic solution is sought or not. From a practical point of view this could easily be collapsed into two main approaches. A 'statistical' ordering can be presented and modified using any method of display that an investigator wishes; and deterministic solutions can be obtained from a statistical ordering by omitting rows and columns of the data matrix that cause departures from it. This approach, or something like it, has been advocated in the European literature for several years (Djindjian, 1985).

The distinction between 'statistical' and more 'traditional' approaches to seriation has possibly been exaggerated, and the flexibility of statistical approaches not fully appreciated. All that is needed is to view the outcome of a 'statistical' seriation as the start rather than the end point of an analysis. This is not to deny that any method that seeks a chronological ordering in a set of data depends on assumptions. Simplistically, for example, a type that remains at the same level of popularity throughout the period for which a seriation is sought is useless for analysis, as may be assemblages or types for which the sample size is small.

Dunnell (1970, 2000) and O'Brien and Lyman (2000: 117–30) discuss the conditions under which it is reasonable to infer that a seriation represents a chronology. There seems to be general agreement that the units (or groups or assemblages) to be seriated should have a comparable duration in time, and belong to the same cultural tradition. Failure to meet these conditions may be recognized by a noticeable departure of the seriated data from the ideal model. That units should also come from the same 'local area' has occasioned more debate. The idea here is to avoid confounding temporal with spatial variation. Dunnell (1970) discussed the difficulties of defining a local area, since it is impossible to eliminate the effects of spatial variation, and suggested undertaking multiple seriations, using different materials, to try and identify orderings common to all seriations. More recently ideas from evolutionary theory have been harnessed to account for 'what produced the unique, historically unrepetitive sequence of forms on which [seriation] depended and also accounted for the unimodal distributions' so that, now 'the underlying theory is understood, the liability imposed by spatial variation for seriation as an empirical method promises to become a major tool for investigating spatial organization of human communities, as well as a dating method' (Dunnell, 2000: 550; see also Neiman, 1995; Lipo *et al.*, 1997; O'Brien and Lyman, 2000).

Statisticians should be aware that these later developments are grounded in an approach to archaeological theory, evolutionary archaeology, that not all archaeologists share (see Hodder, 2001, for what is on offer). The later Americanist publications cited in this section are largely silent on more recent practice in Europe and, in particular, the quite widespread – and apparently successful – use of correspondence analysis for seriation is unremarked on. Whether this is because the extent of such use is unknown, or simply considered irrelevant and not worthy of mention, is difficult to tell. My feeling is that techniques, such as correspondence analysis, used with modern software, and within the context of a 'method', are rather more flexible and useful than has sometimes been allowed.

17

Quantification

Informally, quantification is the answer to the question 'how much?'; more formally, we can define it as the process of measuring the amounts of pottery of different categories in one or more assemblages. For many classes of finds, the question does not pose a problem: one can simply count the objects. For pottery, the question of how one 'counts' objects, which are almost always broken and incomplete, is far from trivial. (Orton, 1993a: 169)

17.1 Introduction

Statistical analysis in archaeology is frequently concerned with the analysis and/or comparison of assemblages. In seriation studies, for example, an attempt is made to order assemblages on the basis of their similarity with a view to recovering an ordering that may have a chronological interpretation (Chapter 16). Among the techniques that have been used to aid such analyses are multidimensional scaling (Section 7.4) and correspondence analysis (Section 11.4). In studies of vertebrate faunal remains, bone counts may be used to infer the function of a site – for example, whether it was a 'kill site' where animals were slaughtered with subsequent selective removal of body parts rich in meat; or a 'home base' to which such parts were removed for consumption; or some combination of these or other depositional agencies (e.g., Rogers, 2000a). The aim of these and similar studies may be to infer or compare the functions of sites and the human agency (or lack of it) that led to their formation.

The archaeological problems here are often difficult. A variety of statistical techniques are available as aids to answering such problems, and these techniques operate on tables of numbers generated by archaeological investigation. The present chapter deals with some aspects of how such numbers are generated, or the problem of *quantification*. This is a non-trivial problem in the case of finds (e.g., bone, pot, glass, stone) that fragment, and which, for a variety of reasons, may not survive in the archaeological record to be recovered and counted, or may survive differentially.

In statistical (and general) terms, problems arise simply because what is sampled may be a very non-random sample from a population of interest. The population of interest, or the population of which the sample is a part, may also be difficult to define (see Chapter 3 of Orton, 2000a, for a good discussion of this). As a simple but stark example, by no means unrealistic, how should one compare two faunal assemblages formed at the same time in different environmental conditions, when one is excavated complete and the other only survives in the form of some teeth, or not at all?

Initial discussion concerns the quantification of vertebrate faunal remains (e.g., Ringrose, 1993a; Lyman, 1994; O'Connor, 2000: 54–67, 2001; Orton, 2000a: 53–7; Rogers, 2000a,b). Following this, research on the quantification of pottery will be described. This has had ramifications for the study of other categories of find, such as glass, stone and bone, that are also discussed. A general review of the quantification problem, with some emphasis on stone tools, is provided by Shott (2001).

17.2 Quantification of vertebrate faunal remains

17.2.1 Simple approaches to quantification

The following account leans on Ringrose (1993a). He distinguishes between *living, death, deposited, fossil* and *sample* assemblages. The living assemblage is the community of live animals whose carcasses form the death assemblage, a subset of which is deposited at a site, as the result of a variety of depositional agents. The fossil assemblage is that part of the deposited assemblage surviving at the time of recovery. The sample assemblage is the part of this that has been recovered. Ringrose (1993a: 123–4) distinguishes between paleozoological approaches, where interest is in the living assemblage, and zooarchaeological approaches, where the focus of archaeological inference is on the death and deposited assemblages and the processes that relate them. *Taphonomy* is 'the study of the transition between past live communities and excavated remains'.

From a statistical point of view the starting point for analysis is the sample assemblage. This sample is a sample of a sample of a sample of a sample (Ringrose, 1993a). The samples are usually non-random, and the processes that link a sample at one level to that at a higher level are often poorly or imperfectly understood. An obvious first step is to count what is in the sample assemblage. This apparently elementary process is not straightforward. The main problem is that the simplest ways of counting produce results that can be of limited use in addressing the archaeological questions of interest. Ringrose (1993a: 122) distinguishes between an *element*, which is an anatomical part, a *bone*, which is an element from a particular animal, and a *specimen*, which is a recovered bone or fragment of bone.

The number of identified specimens (NISP) is a simple method of quantification and can be used to describe the sample assemblage. It is subject to numerous biases if it is to be used for inferences about the life, death or deposited assemblages, and has been dismissed by several commentators (Grayson, 1984; Ringrose, 1993a: 125–6; O'Connor, 2000: 54–7). Among the problems are that some animals have more bones than others, and many specimens in a sample may come from the same animal, so that it is of little use for estimating taxonomic abundance. The minimum number of elements (MNE) is the NISP calculated for each skeletal part, possibly making some allowance for fragmentation, and the minimal animal units (MAU) is the MNE divided by the number of copies of the skeletal part in a living animal.

The minimum number of individuals (MNI) is the smallest number of animals needed to account for the specimens of a taxon found in a sample. The MNI may be determined for each element, and the maximum across elements then gives the MNI for the sample as a whole. O'Connor (2001: 706) suggests that 'it is one of the more remarkable features of zooarchaeological methodology that MNI estimates continue to be used, despite serious inherent flaws, and two decades of detailed criticism'. Criticisms include the fact that it is not clear what MNI measures, that it can be correlated with sample size (Grayson, 1981, 1984), and that it cannot be subjected to mathematical manipulation to obtain estimates of relative abundance (O'Connor, 2001: 706). If separate contexts from a site can (legitimately) be amalgamated the MNI will depend on whether such amalgamation does, in fact, take place, or whether it is calculated separately for each context and summed to get an overall MNI. Finally, account needs to be taken of the fact that elements in a skeleton may be multiple. Ringrose (1993a) gives a critical account, describing several methods that have been proposed for dealing with paired bones. Some of these require an ability to recognize pairs as from the same animal. O'Connor (2000: 60) notes that his 'own attempts to match pairs of modern bones in which the true pairs are known have been abysmally unsuccessful, prompting a swift retreat from MNI estimation'.

The NISP and MNI measures may be regarded as descriptive of a sample assemblage, but whether usefully so has been a matter for debate. Ringrose (1993a: 132) argues that the NISP might be viewed as an estimate of the number of bones of 'which at least a part has survived to be in the fossil assemblage', while the MNI is an estimate, usually conservative, of the number of animals, some of whose bones have survived to be in the fossil assemblage. This is of limited value if the aim is to say something useful about the deposited, death or living assemblages, and to do this models that in some way link the sample assemblage to these assemblages are needed.

17.2.2   Models – the Lincoln/Peterson index

If counts based on the sample assemblage are to be used to make inferences about the deposited, death or living assemblage, some form of modeling is essential. One approach to estimating numbers in the death assemblage is the Lincoln or Petersen index, based on ideas derived from capture–recapture studies in animal ecology, and proposed independently for archaeological applications by Poplin (1976), Fieller and Turner (1982) and Wild and Nichol (1983) (Ringrose, 1993a). The idea is simple. If n_1 animals are captured, tagged and released, and at a later stage n_2 animals from the same population are captured and it is observed that m are tagged, then

$$\hat{N} = n_1 n_2 / m$$

estimates the population size. This is biased for small samples. An unbiased version is available (Seber, 1982: 60; Wild and Nichol, 1983), of the form

$$\hat{N} = \frac{(n_1 + 1)(n_2 + 1)}{m + 1}.$$

This can be applied to bone counts using elements that occur in pairs, such as the left and right femur, by equating n_1 and n_2 with the number of identified specimens from the left- and right-hand sides respectively, and m with the number of matched pairs. Confidence intervals can be calculated for $\hat{N}$, which is an estimate of the number of animals in the death assemblage (Fieller and Turner, 1982). Different estimates may be made for each element that occurs in pairs.

Not everyone has been convinced of the merits of the Lincoln/Petersen index (e.g., Horton, 1984). Ringrose (1993a: 129) lists several problems with its use. One is that left and right bones from the same animal may not be treated independently, in a statistical sense, as required by theory; a second is that m needs to be 'fairly high' for estimates of $\hat{N}$ to be useful; a third is that the estimate is sensitive to changes in the value of m. This places a premium on the correct matching of pairs. Those who are skeptical about the possibility of correct matching regard the estimates produced as 'unreliable' (O'Connor, 2000: 63). O'Connor also makes the point that if estimates for different taxa in an assemblage are to be compared, informal modeling of the depositional process that guarantees that comparison is valid needs to be done, as otherwise estimates will be 'devoid of context'. Further discussion of the use of capture–recapture statistics in archaeozoology is provided by Winder (1991, 1993, 1997).

17.2.3 Maximum likelihood estimation of bone counts

An ambitious method for modeling bone counts in the deposited (or death) assemblage, using maximum likelihood estimation, has been developed by Rogers (2000a). Apart from the method itself, this and related papers are interesting for the way they deploy a variety of statistical techniques, some rarely found in the archaeological literature. In what follows we shall assume that interest lies in drawing inferences about the deposited assemblage.

It is assumed that a deposited assemblage is formed by k depositional agents, each of which contributes a fraction, α_k, of the assemblage. The model aims to estimate these parameters and the expected number of animals in the assemblage, taking into account 'attritional' processes that cause the fossil and sample assemblages to differ from the deposited assemblage.

The sample assemblage of bones is described by a vector $\mathbf{y} = (y_1 \ y_2 \ \cdots \ y_P)$, where P is the number of skeletal parts in the assemblage and y_i is the count for skeletal part i. Let $\mathbf{a}_i$ be one possible way in which a single animal can contribute to the total count, and let M be the total number of possible ways (configurations) in which some contribution can be made; then

$$\mathbf{y} = \sum_{i=0}^{M} x_i \mathbf{a}_i,$$

where x_i is the number of animals represented by configuration i. For $i = 0$, $\mathbf{a}_0 = (0 \ 0 \cdots 0)$ represents the absence of bones from an animal that formed part of the deposited assemblage but which is not represented in the sample assemblage.

It is assumed that $\mathbf{x} = (x_0 \ x_1 \cdots x_M)$ has a multinomial distribution with parameters $\mathbf{p}$ and index T, where T is the number of animals brought to the site of the deposited assemblage and p_i, $i = 0, \ldots, M$, is the probability of configuration i.

If it is now assumed that T has a Poisson distribution with mean κ, $T \sim P(\kappa)$, then $x_i \sim P(\kappa p_i)$ and

$$\mathbf{y} \sim MVN(\boldsymbol{\mu}, \boldsymbol{\Sigma}),$$

the multivariate normal distribution with mean vector $\boldsymbol{\mu}$ and covariance matrix $\boldsymbol{\Sigma}$, assuming either that κ is large, or that M is large and a sufficiently large number of the p_i are appreciably greater than zero.

This statistical model is integrated with a site formation process model so that the elements of $\boldsymbol{\mu}$ and $\boldsymbol{\Sigma}$ are reexpressed in terms of a small number of parameters of archaeological interest. It is assumed that there are k agents of deposition, such as natural death, kills or transportation, with α_k the fraction of the deposited assemblage contributed by agent k, and that the assemblage is subject to attrition with intensity β.

To implement the model several quantities are assumed known without error. These are m_{ki}, the mean number of skeletal part i transported to the site per animal by agent k; f_{kij}, the mean product of the numbers of skeletal parts i and j transported per animal by agent k; and s_i, the sensitivity of the ith skeletal part to attrition. It is assumed that the ith skeletal part survives attrition with probability $\exp(-\beta s_i)$, where the s_i are scaled so that on average half the bones in a complete skeleton are lost when $\beta = 1$. With these assumptions Rogers (2000a) shows that

$$\mu_i = \kappa e^{-\beta s_i} \sum_k \alpha_k m_{ki}$$

and

$$\Sigma_{ij} = \kappa e^{-\beta(s_i + s_j)} \sum_k \alpha_k f_{kij},$$

The log-likelihood, which is proportional to

$$-[\log |\boldsymbol{\Sigma}| + (\mathbf{y} - \boldsymbol{\mu})' \boldsymbol{\Sigma}^{-1} (\mathbf{y} - \boldsymbol{\mu})],$$

is thus a function of κ, β and the α_k, and must be maximized subject to the constraints that $\kappa > 0$, $\sum_k \alpha_k = 1$, and $\alpha_k \geq 0$ for all k. A practical problem with large P is that it may not be possible to invert $\boldsymbol{\Sigma}$ because it is singular or nearly so. When this arises Rogers (2000a: 124) suggests replacing $\mathbf{y}$, $\boldsymbol{\mu}$ and $\boldsymbol{\Sigma}$ by $\mathbf{E}'\mathbf{y}$, $\mathbf{E}'\boldsymbol{\mu}$ and $\mathbf{E}'\boldsymbol{\Sigma}\mathbf{E}$, where the columns of $\mathbf{E}$ are the normalized eigenvectors of $\boldsymbol{\Sigma}$ associated with the leading eigenvalues in a principal component analysis of $\boldsymbol{\Sigma}$.

Much of Rogers (2000a) is concerned to demonstrate and evaluate this model, on both simulated and archaeological data, using a package abcml (Analysis of Bone Counts by Maximum Likelihood) developed to undertake the maximum likelihood estimation. For the simulation, studies of the Hadza, a population of human foragers in Tanzania, are used to inform the calculation of those quantities assumed known that are necessary for the model. Among other things, 1000 simulated data sets are generated for $\kappa = 50$, $\beta = 0$ or 1, and two agents of deposition with $\alpha_1 = 0.4$ and $\alpha_2 = 0.6$. Kernel density estimates (Chapter 3) of the distribution of the parameter estimates obtained suggest that they are reasonably symmetrically and tightly distributed about the true values, with rather more variance in the case of assemblages subject to attrition. It is shown that the MNI statistic seriously underestimates the number of animals.

In analysis of archaeological data the Hadza pattern of transporting skeletal parts was used to determine the quantities associated with different agents of deposition. In particular 'the hypothesis that the bighorn sheep assemblages from two caves in the western U.S.A. were deposited by hunters who behave like modern Hadza' (Rogers, 2000a: 118) was tested. Aspects of the analysis were problematic, not least the fact that bone counts used, taken from the original publications, were biased and not ideal for abcml. The dimensionality of the data was reduced from 24 to 7, using principal components analysis, for the reasons already discussed. Assuming two possible agents of deposition, a kill-site pattern and a home-base pattern, both for small animals, the analysis produced results much more consistent with the former and estimated a somewhat larger numbers of animals than in the original publications. Analysis of residuals showed, however, that many observed counts were badly predicted by the model. A chi-squared-like goodness-of-fit statistic also led to rejection of the hypothesis that the 'assemblages were deposited by hunters who behave in accordance to the Hadza small-animal transport model' (Rogers, 2000a: 122). A feature of the chi-squared analysis was the use of bootstrapping methodology (Chapter 12) to generate critical values for the statistic.

A lot of statistical and archaeological assumptions are necessary to make the model operational. In another application Rogers and Broughton (2001: 764) describe the assumptions as 'heroic'. The need to characterize the nature of the depositing agents through specification of m_{ki} and f_{kij} is particularly demanding. In a concluding discussion of his method, Rogers (2000a: 122) acknowledges this and concludes that, at least in the short run, it may be limited for archaeological inference and of more value in ethno- and experimental archaeology. Other applications of the abcml methodology are to be found in Rogers (2000b,c) and Rogers and Broughton (2001).

17.3 Pottery quantification

17.3.1 Introduction

The focus of this section is on pottery quantification, and is heavily influenced by the work of Clive Orton and his colleagues, over a lengthy period, that resulted in what is known as the *pie-slice* method of pottery quantification (Orton, 1975, 1982, 1993a,b; Orton and Tyers, 1990, 1991, 1992, 1993; Orton *et al.*, 1993). This work has subsequently influenced research on the quantification of bone (Moreno-García *et al.*, 1996; Orton, 1996), glass vessels (Cool, 1994; Cool and Baxter, 1996) and stone tools (Shott, 2000, 2001) and this is discussed in a separate section.

Orton (1993a: 177–8) distinguishes between a target population or *life assemblage*, defined as the pots in use at a certain point at a certain time, and the *death assemblage*, defined as the pots disposed of from a certain place over a certain period. Pots are categorized by type, such as their form or function. The life and death assemblages will usually differ. Orton has argued that for the purposes of interpretation it is necessary to estimate parent populations (which may be the life or death assemblages) rather than simply characterize the sample assemblage (e.g., Orton, 2000a: 51).

The sample assemblage is typically a biased sample from the population of interest, but in effecting a comparison between assemblages one may be prepared to assume, as a reasonable approximation, that similar biases are operating (Orton, 1975). Assemblage size cannot be used for comparative purposes, because sampling fractions are unknown, and the composition of a single assemblage is of limited inferential value because of the numerous sources of bias that can cause it to differ from the parent population. If, however, it can be assumed that similar biases affect two or more assemblages then it is reasonable to try and compare them in a statistically rigorous fashion. Thus, early work focused on the comparison of different and competing measures of the amount of pot in an assemblage, and later work sought to develop this in a direction that allowed the statistical significance of differences between assemblages to be evaluated.

17.3.2 Simple measures and estimated vessel equivalents

Up to about 1960, *sherd count* was the predominant method of pot quantification. A problem with this is that different types of vessel may break differentially so that, other things being equal, a type that is prone to fragmentation will be overrepresented compared to one which is not. That this issue is not trivial can be seen with reference to the work of Byrd and Owens (1997). They were concerned to compare the use of *surface area*, as opposed to sherd count, as a measure of quantification. Surface area can be tedious to measure exactly, and Byrd and Owens (1997) described a sieving technique to determine the *effective area* of a sherd, the size of a sherd being equated with that of the mesh size at which it came to rest. Regression analysis was used to show a good relation between total effective area and actual surface area. Using experimental data, where a known number of pots of different type were broken differentially, it was shown that effective area and sherd count produced markedly different estimates of the amount of pot, with the former much closer to the truth. Using real data, in a seriation problem where the ordering of contexts was known, it was shown that the use of effective area but not sherd count produced the correct ordering.

Surface area is related to *weight*, which has also been used as a measure of the amount of pot, and other possibilities are *vessels represented* and *vessel equivalent* (Orton, 1993b). To understand these last two the concept of a *sherd family*, defined as all the sherds in one context that belong to the same vessel (Moreno-García *et al.*, 1996: 453), is needed. Vessels represented is equivalent to the number of sherd families present, whereas vessel equivalent is obtained by summing the *proportions* of the vessels present in the sherd families. The first measure requires an ability to recognize and sort sherds into sherd families, and this will often be impossible to do with certainty. The second measure also requires sorting into sherd families if it is to be used as the basis for the method of quantification, via pottery information equivalents, discussed in the next subsection. Variants of vessels represented that try to allow for identification problems include the minimum and maximum numbers of vessels represented (Orton, 1982). The solution to the difficulty in the case of vessel equivalents is to use *estimated vessel equivalents* (EVEs). One common approach is to group rim sherds according to their sherd family and record the proportion of the rim represented. Summing these proportions across a type give the rim-EVE for that type.

Orton (1975) was a theoretical comparison of the properties of sherd count, weight, vessels represented and EVEs as a measure of pottery. It is posed, without loss of generality, in terms of assemblages that consist of just two types, and assumed that interest focused on estimating R_i, the ratio or relative proportion of the two types in assemblage i, and R_1/R_2, the ratio of these relative proportions for two assemblages. Using standard statistical theory associated with the estimation of ratios (described by Cochran, 1977), and assumptions about the breakage and recovery processes that lead from the target population to the sample, Orton (1975) investigated the bias in estimates of these quantities assuming large samples. The assumptions were that the number of sherds into which a vessel breaks is a random variable, whose mean depends on the vessel type and context, and that all sherds derived from the target population have an equal probability of recovery from a context that depends only on the context. Under these assumptions it was shown that, in general, only EVEs will produce unbiased estimates of R_1 and R_1/R_2. Weight and sherd count provided unbiased estimates of R_1/R_2, with the former requiring fewer of the assumptions, while vessels represented was biased for both quantities.

17.3.3 Pottery information equivalents

Orton's advocacy of EVEs has influenced the practice of pottery quantification in Britain. The motivation for moving beyond the EVE (Orton and Tyers, 1990: 88) was to develop theory that allowed confidence intervals to be set for proportions of types within an assemblage and to allow an assessment of the statistical significance of the differences between assemblages. Following Orton and Tyers (1990), the following notation will be used. It is assumed that pottery is recorded by EVEs, though this is not essential for the general theory.

Let there be m records for an assemblage, with m_j the number of records for the jth type, w_i the measure for the ith record, W_j the total measure for the jth type, and W the overall total for an assemblage. It is assumed that records are independent which, in practice, means that records refer to sherd families. Let $\sum_j$ indicate summation over the jth type, so $W_j = \sum_j w_i$, for example, and let $S_j^2 = \sum_j w_i^2$. The notation $\sim j$ refers to all types except the jth.

The proportion, p_j, of the jth type in an assemblage may be estimated as

$$\hat{p}_j = \frac{W_j}{W} = \frac{\sum_{i=1}^m \delta_{ij} w_i}{\sum_{i=1}^m w_i},$$

where $\delta_{ij} = 1$ if the ith record relates to the jth type and 0 otherwise. This is a ratio estimate, and standard theory (Cochran, 1977: 32–3) leads to the result that the variance of $\hat{p}_j$ is approximately

$$\frac{m}{(m-1)W^4}(W_{\sim j}^2 S_j^2 + W_j^2 S_{\sim j}^2). \tag{17.1}$$

Now suppose that the estimate of $\hat{p}_j$ had been obtained by sampling from a notional set of n_j complete vessels, so that the estimated variance, using a binomial model, is

$$\hat{p}_j(1 - \hat{p}_j)/n_j.$$

Equating this with equation (17.1) and solving for n_j gives the number of whole vessels that would give rise to the same estimated variance as the m measurable records:

$$n_j = \frac{(m-1)}{m} \frac{W_j W_{\sim j} W^2}{W_{\sim j}^2 S_j^2 + W_j^2 S_{\sim j}^2}. \tag{17.2}$$

If there are J types in an assemblage, calculations based on equation (17.2) lead to J different estimates of n_j. In the absence of complications these should not differ significantly, and assuming that all types have the same mean and variance for w, and using pooled estimates of the mean and sum of squares of w, W/m and S^2, leads to a pooled estimate

$$n = \frac{(m-1)}{m} \frac{W^2}{S^2}. \tag{17.3}$$

Before pursuing the implications of these results it may be helpful to provide a small illustrative example for the data of Table 17.1. This is based on that given in Baxter and Cool (1995: 92) where EVEs were used for five types of glass found in a single context from excavations in Roman Chester. For a pottery vessel, an EVE value is typically recorded as the proportion of the rim that survives, and can vary continuously between 0 and 1 (or 0 and 100%). Other kinds of artifact, including glass vessels, may be characterized by recognizable (diagnostic) zones. For a single vessel, the EVE value is recorded as the proportion of the maximum number of zones that are recognized in the surviving fragments. Here, for example, $W_1 = 40 + 40 + 40 + 40 = 160$, where 40 indicates that two-fifths of the diagnostic zones are present in the sample.

If pooling is justified only the totals are needed, which on substitution into equation (17.3) gives an estimate for n of $(83/84) \times (1694^2/44\,924) = 63.12$. Orton and Tyers (1992) coin the term *pottery information equivalent* (PIE) for this quantity. Multiplying the estimated PIE total by the EVE proportion for a type gives the estimated PIE for the type; thus that for type 1 is $63.12 \times 160/1694 = 5.96$. The estimated *proportion* of a type will be the same, whether EVEs or PIEs are used, but as the PIE total will differ from assemblage to assemblage similar proportions will transform into different numbers.

The PIE is an abstract quantity. Moreno-García *et al.* (1996: 452) state that one PIE of pottery contains as much statistical information as one complete vessel,

Table 17.1 The number of records (m_j) and EVE totals (W_j) for five types of glass found in a single context in excavations of Roman Chester; other notation is explained in the text (Baxter and Cool, 1995)

Type	m_j	W_j	S_j^2	S_j^2/W_j	$\bar{W}_j$	n_j
1	4	160	6400	40.00	40.00	43.37
2	15	422	15780	37.39	28.13	49.54
3	5	98	2156	22.00	19.60	75.14
4	5	104	2360	22.69	20.80	73.00
5	55	910	18228	20.03	16.55	60.70
Total	84	1694	44924			

and note that it is best thought of as a statistical device to aid analysis rather than an entity in its own right. A result established in Orton and Tyers (1990: 91) is that the PIEs for an assemblage behave as if they are samples from a multinomial distribution, so that a set of assemblages can be organized in the form of a contingency table and treated using statistical methods appropriate to such tables. In particular, log-linear modeling and correspondence analysis (Chapter 11) are available as ways of analyzing and displaying the data.

The foregoing methodology is associated with the pie-slice package of Orton and Tyers (1993) and was developed specifically with applications to pottery in mind. The pot is classified according to fabric, form and context, resulting in a three-way table of PIEs. Using the notation of Chapter 11 the fully saturated model, [123], is expected to be unnecessarily complex, whereas the independence model, [1][2][3], is expected to be too simple for describing the data. Model-selection strategies such as those discussed in Chapter 11 or in Orton and Tyers (1990) can be used to select a model. If, for example, the model [12][3] is accepted, this would suggest a fabric by form interaction that was completely independent of context. Once the important two-way interactions are identified the appropriate two-way marginals of the three-way table can be displayed graphically using correspondence analysis, in order to understand the nature of the interaction. Examples are to be found in Orton and Tyers (1990, 1992, 1993), Orton *et al.* (1993) and Tyers (1996).

Before applying the pie-slice methodology a number of practicalities have to be addressed. Among them, the theory requires that S_j^2/W_j should be similar for each type, as should the mean EVE totals, $\bar{W}_j$. This is often not the case. Data sparsity may also be a problem, requiring either the amalgamation of forms, fabrics or types, or the omission of categories for which such amalgamation cannot be justified. For a discussion of these and related problems, and their resolution, reference may be made to Orton and Tyers (1990, 1991, 1992) and Baxter and Cool (1995).

17.4 Other find types and the pie-slice approach

17.4.1 Bone

Bone quantification was discussed in an earlier section of this chapter. The pie-slice approach was developed with pottery in mind but it was realized that it could also be applied to animal bone data, provided they were suitably quantified and recorded. This is discussed, with a detailed application, in Moreno-García *et al.* (1996). The form and fabric classifications of pot are analogous to the use of species and bone element for animal bones. For each bone, diagnostic zones characteristic of the bone can be recognized, varying in number from 2 to 10 (Rackham, 1986). Provided bones are recorded in the equivalent of sherd families, the equivalent of an EVE for a bone can be measured as the *completeness*, which is the proportion of the maximum number of diagnostic zones present.

The example presented in Moreno-García *et al.* (1996) used data from sites mostly in the Greater London area of England which, before any reduction, involved 17 species and 24 bone elements. The contexts were divided into Roman, Saxon and medieval, and post-medieval. The analysis of Orton (1996) concentrated on the Roman data for the four most common species. Both papers provide good illustrations of the iterative strategy of analysis, in which the most obvious structure in the data is discovered and then removed in order to reveal subtler structure on reanalysis. In Orton (1996), for example, the initial correspondence analysis showed a strong contrast between horses and their associated contexts and other species. Removing horses and the contexts they dominated from the data set allowed a better contrast between other species, cattle, pig and sheep/goat, and their associated contexts to be seen.

There are some differences between the application of the pie-slice approach to pot and to bones. One is that the quantifying measure is continuous in the former case and discrete in the latter. With a small number of records this means that outliers may cause more problems with the discrete data. The theory developed for pots requires that records be independent, with the implication that recording should be by sherd family. Equivalent recording for animal bones less obviously satisfies the requirement of independence, and Moreno-García *et al.* (1996: 442) query the use of bones from articulated skeletons, which they omit from their analysis. Another difference identified is that animal bone data are far more likely than pot data to display a three-way interaction, which can be identified but is not so readily explored. An implication of this is that it is worth looking at the correspondence analysis for the three possible two-way marginal tables as a matter of course, but these may not tell the full story.

17.4.2 Glass

The major obstacle to the use of quantitative methods for assemblage comparison is often the lack of a suitable method of quantification. It was this problem that led Cool (1994) to devise a system of quantifying vessel glass, reported more accessibly in Cool and Baxter (1996). Glass generally survives in the archaeological record in much smaller quantities than pot, and the measures of quantity that have been proposed for pot are less obviously suited to glass. The solution, inspired by sight of the work that led to Moreno-García *et al.* (1996), was to use diagnostic zones, the number of which typically varies between five and eight, according to vessel type.

A practical problem with this approach to quantification is that diagnostic zones such as rims may themselves fragment, and if this is not recognized 'double-counting' may occur. As with other materials the success in quantification is thus dependent on the expertise of the specialist and the distinctiveness of the material. The analogous difficulty in pot studies is less serious, since a sherd family may be spread over more than one record, but the EVE total is unaffected.

Apart from the papers noted above, Cool and Baxter (1999) provide an illustration of an application to the study of Roman vessel-glass assemblages. Clear differences between assemblages of the 4th and earlier centuries AD were demonstrable (Figure 11.2), and within the earlier periods, assemblages of the 1st to 2nd centuries were largely distinct from those of the 2nd to 3rd centuries. A general trend was an increase over time in the proportion of drinking vessels in the assemblages, relative to glass vessels serving other functions. One analysis of late 1st century drinking vessels classified by type showed some evidence of regional differences, with Northern/Midland sites favoring low cups and Southern sites favoring tall beakers. English beer-drinkers might recognize in this an early manifestation of current regional preferences for different styles of pint glass.

A possible criticism of this analysis is that it was based on small EVE numbers, so interpretation needs to be undertaken with caution. This is unavoidable with glass, which survives in far smaller quantities than pot, but the strong patterns in the data, rather than the positioning of individual points, justified the interpretation, which at the very least establishes a hypothesis capable of further investigation given suitable future quantified data.

17.4.3 Stone tools

Shott (2000, 2001) has investigated the quantification of stone tool assemblages using ideas derived from the work of Orton and colleagues. There are important differences between pot and stone tool assemblages that Shott (2000: 727) summarizes as follows:

> In sum, pottery was used as relatively few irreducible wholes but is recovered in many small pieces. Tools were used as reducible wholes and are recovered whole or in fewer, larger pieces. A method designed to quantify pottery may not work well on tools.

To the forest of acronyms already met in this chapter Shott adds MNT, ETE and TIE, the tool equivalents of MNI, EVE and PIE. As with glass and bone quantification, most of Shott's examples employed diagnostic zones to determine the ETE. Typically, fewer diagnostic zones are available, though this depends on the tool type. In three of four examples only three diagnostic zones were used, and there was a higher proportion of complete tools than is typical of assemblages of other finds types. These were also the examples in which the MNT and TIE estimates agreed reasonably well, raising the possibility that 'the statistical exertions demanded by Orton's method may not be necessary in lithic assemblages' (Shott, 2000: 734). This suggests that minimum number statistics descriptive of the sample assemblage, but of limited value for inference about, say, the death assemblage for other finds types, may be of more value for stone tools where the scope for severe fragmentation is smaller.

The exception to this generalization was an example for a tool type where rather more diagnostic zones were available (eight), and there was rather more fragmentation, in the sense that only 6/183 tools survived intact, than in other examples. In this example there was a rather larger discrepancy between the MNT and TIE than was found elsewhere.

18

Lead isotope analysis

18.1 Introduction

An example of lead isotope ratio data was introduced in Section 2.2.1, where some of the statistical issues that have surrounded its application in archaeology were noted, and this should be read as background. In this chapter the statistical issues are examined in more detail.

To recapitulate, specimens from ore bodies mined in antiquity for copper can be characterized by the measurement of three lead isotope ratios. A sample of n specimens may be used to estimate the extent of the lead isotope field for the ore body. If ore bodies have distinct fields there is some hope that the lead isotope signature of an artifact can be used to identify potential sources of origin or, at least, exclude many ore bodies as possible sources.

As noted in Section 2.2.1, application of this methodology has proved controversial. Some of the scientific and archaeological aspects of this debate, in the context of the trade of metals in the Mediterranean Bronze Age, are discussed in Knapp (2000) and Gale (2001). The statistical argument has centered on whether or not multivariate methods are useful in the analysis of lead isotope data, but issues concerning the treatment of outliers, data transformation, and appropriate sample sizes have also arisen.

Supporters of the use of multivariate methods include Sayre *et al.* (1992a, 1993). Their position seems to have been largely uninfluenced by strong criticism of the methodology used in the 1992 paper (Sayre *et al.*, 2001). They use (Sayre *et al.*, 1992a: 82)

> the multivariate probability calculations of specimens matching groups as our primary statistical tool for evaluation [of whether or not an artifact could have a particular field as a source] because it is a very straightforward statistic that can be calculated for each specimen and because it is the only such statistic that takes into account sample size.

The probability calculations referred are based on Hotelling's T statistic, in turn based on Mahalanobis distance calculations (see Section 6.4.3), and assume that lead isotope fields have a trivariate normal distribution. Results are often displayed in the form of bivariate plots for pairs of ratios with particular fields delineated by ellipsoidal confidence intervals. Figure 6.1 illustrated this for data from the islands of Seriphos and Kea in the Cyclades, using data published in Stos-Gale *et al.* (1996). The calculation of the ellipsoid requires the normality assumption. Notwithstanding the reference to sample size, Sayre *et al.* (1992a, 2001) also present probability calculations that assume the covariance matrix of the data is determined without error, equivalent to assuming an infinite sample size. Given that $n < 20$ for many fields studied, and is sometimes in single figures, there seems little justification for this. Cases that appear to be statistical outliers are omitted in the ultimate definition of a field and subsequent probability calculations.

To check normality the principal components of a set of data are calculated, and histograms formed for the component scores on each component. If the histograms approximate normality, which it is claimed they mostly do, it is concluded that the data are normal (Sayre *et al.*, 1992a: 79). There are several problems with this, discussed more fully below. One is sample size. Work by Reedy and Reedy (1988) that recommends a sample size of at least 20 is noted, along with the remark that this is 'too restrictive because of the availability of Hotelling's T statistic for multivariate statistical treatments' (Sayre *et al.*, 1992a: 79).

Initial discussion of the 1992 paper broadly welcomed it. Leese (1992: 321), for example, observed that a consensus seemed 'to be emerging in which multivariate methods, making use of all the available data, should be used for lead isotope studies'. At the same time she noted that it might be more appropriate to transform the data logarithmically. Any such consensus was challenged by Budd *et al.* (1993: 241), based in the Department of Archaeological Sciences in the University of Bradford, UK, who took the discussants to task for failing 'to address what would appear to be fundamental flaws in the methodology of lead isotope studies and the statistical analysis of data'. Among other things, they were critical of the procedures used to identify outliers, of the use of confidence ellipsoids to delineate fields, and of the use of small samples. Elsewhere, the Bradford group has argued that simple three-dimensional graphical analysis will often be sufficient for delineating fields and assessing whether artifacts could come from them. Pernicka (1993: 259), in the discussion of Budd *et al.* (1993), agreed, suggesting that there was 'hardly any need for multivariate statistical methods at all', given the low dimensionality of the data.

Subsequent debate can be traced in the papers listed in Section 2.2.1. Most of the statistical issues of interest were raised in the papers noted above and discussion of them, and to these we now turn.

18.2 Statistical issues

18.2.1 Data transformation

Presentations on the subject of lead isotope ratio analysis to a statistical audience almost invariably provoke the suggestion that the data should be transformed logarithmically (Leese, 1992). The measurements are strictly positive and variables differ by an order of magnitude so this seems sensible. In fact, it is rarely done and seems to make very little difference when it is attempted (Sayre *et al.*, 1992b).

The issue of data transformation is probably the least important of those that have been raised in connection with the statistical analysis of lead isotope data.

18.2.2   Outliers

Reedy and Reedy (1992) are critical of the approach adopted by Sayre *et al.* (1992a) for outlier detection. Budd *et al.* (1993) and Scaife *et al.* (1996) have been similarly critical. One criticism is that cases are sometimes identified as outliers because they lie outside the limits of 90% confidence intervals on bivariate plots for pairs of ratios, whereas some such cases are to be expected. This problem is exacerbated if small samples are used. The apparently elliptical scatter of cases from a field, on bivariate plots, may be a function of the fact that cases that do not conform to such a scatter are designated as outliers.

This last point is of general interest in provenance studies using other forms of material. It might be reexpressed by noting that one starts by assuming trivariate normality of a field and then designates as an outlier any case not in conformity with this assumption. This results in the discovery of a field that is normal, and is arguably a 'self-fulfilling prophecy'.

In the context of lead isotope studies it is arguable that the detection and treatment of outliers is not, or should not be, a statistical problem. Budd *et al.* (1993) and Scaife *et al.* (1996) argue that if a case is selected from a field, and there are no known problems with it, it should be used in characterizing the field. Rejection as an outlier simply because it does not conform to expectations of normality involves a circular argument. In other words, if the aim of an analysis is to delineate the extent of a field then the rejection of cases simply because they are *statistical* outliers is unacceptable. The logic of this position seems to me to be unassailable, though it has been resisted (Sayre *et al.*, 1993, 2001).

Gale and Stos-Gale (1993: 253) observe that 'it cannot be stressed too highly that, if a few lead isotope analyses of ores from a given ore body appear statistically to be outliers from the main distribution, the first step must be to repeat the analyses more carefully to determine whether the original analyses were false'. They report instances where such reanalysis resulted in previously identified outliers conforming to the field from which they originated. Thus, observation of statistical outliers may provide an indication of measurement problems but, unless these are confirmed, does not in itself justify removing a case from an analysis.

18.2.3   Normality

As has been intimated, not everyone sees the need for using multivariate methods to analyse lead isotope data or the need for the assumption of normality of lead isotope fields. The assumption is, nevertheless, a central plank in one approach to analysis that is still current (Sayre *et al.*, 2001), and merits investigation.

Early attempts to investigate the assumption suffered from several problems. Using histograms based on principal component scores (Sayre *et al.*, 1992a) suffers from the problem of small sample sizes and the inefficiency of histograms (particularly with small samples) as a means of detecting non-normality. The use of univariate statistical tests of the three ratios (Gale and Stos-Gale, 1993) suffers from the problem that marginal normality of the ratios does not establish trivariate normality.

Part of the problem in the past has been the small sample sizes typically available, which do not permit the use of tests of multivariate normality with any substantial power. This problem was partially alleviated by the publication by Stos-Gale *et al.* (1996) of several data sets sufficiently large in size to allow the use of formal tests of normality. The largest of these, with $n = 62$, was for the Kea data, used in Figure 6.1.

For these data, Baxter and Gale (1998) applied seven univariate tests of normality to the individual ratios and found no evidence of non-normality. They then transformed bivariate pairs of ratios to principal components and tested for normality of the components using the univariate tests and a multivariate test due to Royston (1982, 1983) that is a function of the Shapiro–Wilk statistic for the individual marginals. Royston's test suggested non-normality at the 2% level of significance for the $^{208}Pb/^{206}Pb$ and $^{207}Pb/^{206}Pb$ ratios, and the Shapiro–Wilk and modified Anderson–Darling statistics (D'Agostino, 1986) and QH^* statistic of Chen and Shapiro (1995) were found to be significant for the second principal component at the 1% level or better.

The motivation for transformation to principal components was the observation that in some of the bivariate plots of the ratios there was an apparent 'edge' to the scatter, indicative of non-normality. This can be seen in the plot of the $^{208}Pb/^{206}Pb$ and $^{207}Pb/^{206}Pb$ ratios for Kea in Figure 6.1. Transformation to principal components approximately aligns this edge with one of the new axes and improves the chance of detecting non-normality. Because of this 'data-snooping', allied to the multiple testing of different components, and the use of a variety of tests of normality the significance levels quoted above must be treated with considerable circumspection. Another difficulty is that Royston's (1982, 1983) test, which can be applied to combinations of ratios as well as their principal components, is coordinate-dependent.

To avoid these difficulties Baxter (1999a) investigated the Kea and other data sets using a variety of coordinate-free tests of multivariate normality applied directly to the three ratios. These included tests of multivariate skewness and kurtosis, and an omnibus statistic based on these due to Mardia and Kent (1991), equivalent to tests developed by Koziol (1986, 1987), the multivariate Shapiro–Wilk statistic of Malkovich and Afifi (1973), the test of Baringhaus and Henze (1988), and its later generalization in Henze and Zirkler (1990) that is equivalent to the test of Bowman and Foster (1993b) based on multivariate kernel density estimates. Other than the test for kurtosis, all these tests led to the rejection of the assumption of multivariate normality at the 5% or 1% level. Similar results were obtained for other data sets for which adequate data were available.

The multivariate Shapiro–Wilk test can be regarded as a form of projection pursuit (Section 7.5) that identifies the most non-normal projection of the data. Figure 18.1 shows kernel density estimates for this projection and the three univariate ratios, and illustrates the multimodal nature of the non-normality, undetected by the marginal analysis.

Baxter (1999a) concluded that non-normality was the rule rather than the exception, and that a previous failure to demonstrate this was largely attributable to the small sample sizes typically available. Other work also suggests this. For example, until 1997, discussion of lead isotope data from Cyprus often proceeded under the assumption that these data, based on sites from around the island, could be treated as a sample from a multivariate normal distribution. This was on the basis

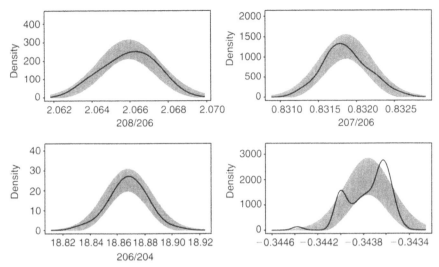

Figure 18.1 Kernel density estimates for the three lead isotope ratios for the Kea field are shown, using the Sheather–Jones method of bandwidth estimation (Sheather and Jones, 1991). The final figure shows the KDE for the most non-normal combination of the ratios identified by the multivariate Shapiro–Wilk statistic (Malkovich and Afifi, 1973). The normal reference bands suggest this latter KDE is indicative of non-normality, unlike the individual ratios.

of 43 observations. The Cyprus field is important because of claims that it was the source of artifacts found in sites around the Mediterranean that were much closer to other potential ore sources. Scaife *et al.* (1996) disputed the normality assumption, but it was only with the publication of a great deal more data from Cyprus, from localized sources, that it became clear that, far from being normally distributed, the field for Cyprus – if it now makes sense to talk of this – was highly multimodal (Stos-Gale *et al.*, 1997). This leads neatly into a discussion of sample-size requirements.

18.2.4 Sample size

Reedy and Reedy (1988) suggested that a desirable minimum sample size for the number of observations from a lead isotope field, for statistical analysis, was 20. Sayre *et al.* (1992a) argued that fewer data still permit useful analysis, but most of the discussants of their paper agreed that 20 was desirable. Pollard and Heron (1996: 328) summarize this consensus by noting that, in characterizing ore fields, 20 geologically well-selected ore samples define 'an agreeable minimum level'. It is, however, obvious from consideration of the history of the Cyprus field noted above that 20 may be seriously inadequate.

That belief that 20 was adequate possibly arose because it is acceptable *if it is assumed that the data are sampled from a normal distribution*, and because sample sizes were frequently not adequate to test this assumption. That 20 will frequently be inadequate can be demonstrated in a variety of ways (Baxter *et al.*, 2000). These include simulating from populations of known structure, or resampling from the data for the largest fields, and testing for normality for each sample generated. For samples of size 20 detecting even quite serious differences from normality can be problematic. It is impossible to set down hard and fast guidelines for what constitutes an adequate sample size, since this will depend on the structure of the data and the purpose to which it is to be put. Samples of size 40 or 50 may be adequate in some cases, but it is straightforward to construct realistic examples where this is inadequate (e.g., Scaife, 1998; Scaife *et al.*, 1999).

As an alternative to using normal-based methods for delineating lead isotope fields Beardah (1999), Scaife (1998), Scaife *et al.* (1999) and Baxter *et al.* (2000) have experimented with the use of two- and three-dimensional kernel density estimation. Though technically feasible, there are a number of difficulties. Three-dimensional KDEs are demanding of data, and the minimum size that is satisfactory for revealing non-normal structure on a regular basis exceeds what is typically available for even the largest data sets. For two-dimensional plots the appearance of a KDE depends critically on the choice of smoothing parameters. These need to reflect the quite strong orientation shown by some of the plots (e.g., Figure 6.1) to be effective, and using a diagonal smoothing matrix $\mathbf{H}$ will fail to capture this. Scaife (1998) experimented with transformation to principal components, use of diagonal $\mathbf{H}$, and retransformation back to the original scale. Beardah (1999) opted to use non-diagonal $\mathbf{H}$, but the choice of parameters is rather *ad hoc* and more guidance is needed on the appropriate choice.

18.3 Conclusion

The above discussion casts doubt on the normality assumption typically made by some analysts, and hence on those procedures that use this assumption. Sample sizes are, however, often too small to test the assumption rigorously. Assertions about the normality of lead isotope data, and the efficacy of procedures that assume it, remain largely a matter of faith.

It can be asked whether the assumption matters much. Baxter *et al.* (2000) construct an example to show that the probabilistic calculations used to determine if an artifact could originate from a field can depend in a non-trivial way on how any non-normality is modeled. However, the assumption will often not be critical. Typically, either a case will be so distant from a field, or so close to it, that any analysis, whether probabilistic or graphical, will reach the same conclusions about the possibility that the field is a potential source for the artifact. Cases that lie outside a field but close to it will lead to uncertain conclusions in either approach, with the uncertainty expressed more precisely but possibly misleadingly in the probabilistic approach. Treatment of outliers is critical here, since the inclusion of statistical outliers in the definition of a field will increase its coverage and modify judgments as to whether a case could belong to a field.

In general, graphical analysis seems preferable, as it is less dependent on assumptions that prejudge the shape of a field and about what constitutes an outlier. This seems now to be the position of one group previously enthusiastic about, and criticized for, the use of multivariate methods of analysis (Gale, 2001: 118). The main advantage of the probabilistic approach is that it facilitates the presentation of results for large numbers of artifacts that need to be compared against a large number of fields (Sayre *et al.*, 2001). In such circumstances such methods could be used as an initial screening device to identify uncertain cases that could then be subjected to more detailed graphical scrutiny.

19

The megalithic yard

It is a sad fact that the megalithic yard hypothesis itself is of negligible inter-est to archaeologists. From what is known of the development and structure of prehistoric societies over the areas and time spans involved in the con-struction of the circles, the hypothesis that a strict mensuration system, based on a common 'brass-edged whalebone yardstick', was in widespread use is not worth entertaining. It belongs to the semi-mystical fringe of archaeology concerned with ley lines, Atlantis and the like. (Fieller, 1993: 283)

19.1 Introduction

Burl (2000: 8) defines megalithic rings as 'approximate circles of spaced standing stones' dating from the Late Neolithic to the Middle Bronze Age, roughly from 3300 to 900 BC. Over 1300 such rings have been recorded in Britain, Ireland and Brittany.

Thom (1955, 1962, 1967, 1971, 1978) proposed a number of controversial theo-ries concerning the geometrical construction, astronomical significance and units of measurements used in the construction of these and other kinds of megalithic sites. Figure 19.1 is similar to Figure 7 of Thom (1955) and is based on the dia-meters of the 45 rings recorded in Tables 2 and 3 of Thom (1955) that are less than 120 feet. Associated with each diameter, y_i, is a standard deviation, s_i. Following Thom, the figure is created in much the same way as a kernel density estimate (Chapter 3) by associating each diameter with a normal distribution centered on y_i and with standard deviation $2s_i$. These are plotted in the figure. Ordinates of overlapping distributions are summed to get the final result.

Thom (1955) noticed that peaks in the figures tended to occur in positions that were integer multiples of 5.44 feet, as illustrated in Figure 19.1. He attempted to confirm this, using statistical methods available at the time, and concluded that a 'universal unit of length was used in setting out [the rings] on the ground' (Thom, 1955: 289). This unit of 5.44 ft is sometimes called the *megalithic fathom*. The *megalithic yard*, the presumptive unit underlying the radii of the rings, is half of this, 2.72 ft. Later work, using statistical theory developed by Broadbent (1955,

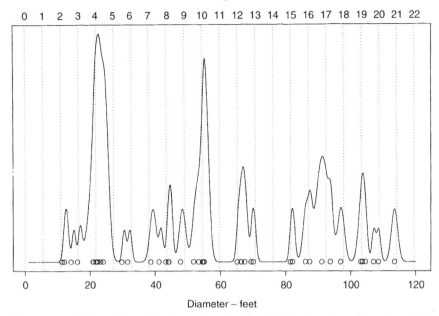

Figure 19.1 A form of kernel density estimate for the 45 circle diameters of less than 120 ft, taken from Table 2 of Thom (1955).

1956) and more accurately surveyed rings, led to the conclusion that 'it is certain that the unit of 2.72 ft was used' (Thom, 1962: 251). This work is summarized in Thom (1967). Other work also claimed that the megalithic yard was used in the construction of structures other than rings, such as stone rows, and outside Britain, in Brittany (e.g., Thom, 1978).

The ubiquity, and precision, claimed for the megalithic yard was at variance with other archaeological evidence. Most archaeologists were reluctant to accept the existence of a common unit of measurement over the temporal and geographical scales involved, and some were actively hostile to the idea (see, for example, the contributions of Patrick and Barber to the discussion of Freeman, 1976). The statistical issues involved in addressing the reality or otherwise of the megalithic yard proved difficult, however, and it was not until the work of Kendall (1974) and Freeman (1976) that the problem was addressed to the satisfaction of at least some archaeologists. This chapter is largely concerned with this work and that of Broadbent (1955, 1956). Thom's theories concerning the astronomical significance of the rings are not dealt with here. Ruggles (1999) provides a critical account based, in part, on resurveys of many of the sites studied by Thom. Heggie (1981), as well as addressing astronomical issues, contains a critical discussion of megalithic geometry and units of length.

Thom also proposed theories concerning the geometry of the rings, and a word or two about this is in order. Burl (2000: 8) describes a ring as an 'approximate circle'. Thom (1967) classified rings into more precise categories such as true circles, ellipses, egg shapes, compound rings and flattened circles, and proposed methods of geometrical construction for the more complex of these. Based on accurate surveys, one or other of these shapes was fitted through the centers of the stones using least squares methods, and diameters were determined from these estimated shapes. It is not always obvious which shape to fit to a data set, and some authors have expressed a concern that there may have been an unconscious bias towards shapes whose dimensions fitted the megalithic yard hypothesis. For this reason some investigators (e.g., Kendall, 1974) have chosen to work with true circles, where there is less possibility of subjective bias in the fitting process. True circles will be referred to simply as 'circles' in what follows.

19.2 Models for the megalithic yard

19.2.1 The basic model

That a common unit of measurement, perturbed by errors, underlies a set of data is an example of a *quantum hypothesis*. A model for such data is

$$y_i = \beta + \delta m_i + \varepsilon_i, \tag{19.1}$$

where the quantum is δ, the m_i are positive integers, and ε_i is an error term with variance σ^2.

In much of the work to be discussed it has been assumed that $\beta = 0$, and this is assumed in what follows. Freeman (1976) noted that the m_i are parameters, so that there are always more parameters than observations. In some treatments this is avoided by estimating m_i by $\hat{m}_i$ to minimize residuals

$$\hat{\varepsilon}_i = |y_i - \tilde{\delta}\hat{m}_i|,$$

where $\tilde{\delta}$ is a proposed value of δ (see below).

19.2.2 Broadbent's method

When Thom undertook his original investigations useful theory to test for the existence of a quantum was not readily available. Broadbent (1955, 1956) tackled this problem, and his results form the basis for Thom's later statistical analyses of the megalithic yard hypothesis. Broadbent (1955, 1956) defined a *lumped variance* as

$$s^2 = \sum_{i=1}^{n} \hat{\varepsilon}_i^2 / n,$$

and proposed tests based on the statistic

$$\frac{s^2}{\tilde{\delta}^2} = \frac{1}{n\tilde{\delta}^2} \sum_{i=1}^{n} \hat{\varepsilon}_i^2. \tag{19.2}$$

Two distinct forms of analysis are possible, termed Type A and Type B analyses by Freeman (1976). In the former case $\tilde{\delta} = \delta$, a known value from which the data are presumed to have arisen. In Type B analyses $\tilde{\delta} = \hat{\delta}$, an estimate determined from the data themselves. In a Type A analysis, if there is a quantum of δ underlying the data, the test statistic (19.2) is expected to be close to zero. If there is no quantum the values of $\hat{\varepsilon}_i$ should be evenly spread over the range 0 to $\delta/2$, and asymptotically the test statistic should have a mean of $1/12$ and variance of $1/180n$. This can be used as a basis for testing whether the presumed quantum exist.

Type B analyses are more difficult since it is not legitimate to determine the value $\hat{\delta}$ that minimizes equation (19.2) and proceed as in a Type A analysis. Allowance must be made for the fact that all possible values have been considered, and resort to simulation is needed to determine the sampling distribution of the test statistic. The then available computer power meant that Broadbent's (1956) simulations were inadequate for the larger data sets subsequently accumulated. Thom's later work has been criticized for switching between Type A and Type B analyses. Later investigators, such as Kendall (1974) and Freeman (1976), developed alternative methodologies that did not assume the prior existence of a quantum.

19.2.3 Kendall's method

In explaining his methodology to a non-statistical audience Kendall (1974) imagined that a set of measurements, diameters, were marked in their correct positions on a long tape that was subsequently wound round a wheel of perimeter θ. If all the ε_i in equation (19.1) are zero, so the data are exactly quantal, and $\theta = \delta$, a little thought shows that the marks corresponding to the diameters will 'pile up' at a single point on the wheel. If the errors are small then the marks will cluster strongly around that point. Clustering will become less marked and eventually disappear as the errors gradually become larger. If the data are quantal but $\theta \neq \delta$ then significant clustering should not occur, and if the data are non-quantal then there will be no significant clustering.

Values of θ that give rise to significant clustering are candidate estimates for a quantum, δ. There are two problems here. The first is to measure the degree of clustering associated with a particular θ, and the second is to determine the significance of any value that suggests strong clustering. For the former problem Kendall (1974) proposed the statistic

$$\phi(\tau) = \sqrt{\frac{2}{n}} \sum_{i=1}^{n} \cos 2\pi \tau y_i = \sqrt{\frac{2}{n}} \sum_{i=1}^{n} \cos \frac{2\pi y_i}{\theta}, \qquad (19.3)$$

where $\tau = 1/\theta$, which should be 'large' if there is clustering and 'small' otherwise. That this is a sensible statistic can be seen by noting that for exactly quantal data (when $y_i = m_i\delta$) and $\theta = \delta$,

$$\sum_{i=1}^{n} \cos \frac{2\pi y_i}{\delta} = \sum_{i=1}^{n} \cos 2\pi m_i = n,$$

so the maximum possible value for $\phi(\tau)$ is $\sqrt{2n}$. When $\theta \neq \delta$ or when there is no underlying quantum, diameters should be randomly spaced around the wheel and positive and negative contributions of $\cos 2\pi \tau y_i$ to the sum will tend to cancel out. The term $\sqrt{2/n}$ is a scaling factor chosen such that, for *fixed* τ, sufficiently large n and not too large θ, $\phi(\tau)$ is approximately a standard normal random variable.

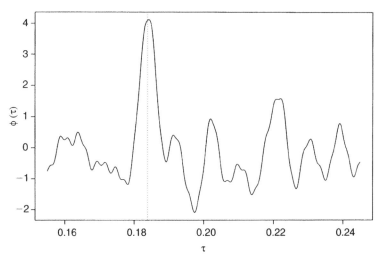

Figure 19.2   A cosine quantogram for the 'good' and 'poor' circle diameters from Thom (1967: Tables 5.1 and 5.2). The dotted line corresponds to a quantum of 5.44 feet.

A plot of $\phi(\tau)$ against τ is a *cosine quantogram*. One such is shown in Figure 19.2, based on the 169 'good' and 'poor' circle diameters for Scotland, England and Wales from Tables 5.1 and 5.2 of Thom (1967), the SEW data set. This is the main data set used by Kendall (1974). A 'good' circle is one whose accuracy of measurement is stated to be ±1 ft or better, while 'poor' circles are those measured with less accuracy. Physical considerations, such as the size of the stones used and minimum circle diameter, place constraints on sensible values of θ to consider, and limits of about 2 ft and 10 or 11 ft translate into the limits for τ of 0.09 to 0.59 used by Kendall (1974). Figure 19.2 is a subsection of the cosine quantogram generated using these figures and shows a clear peak at the quantum of about 5.44 corresponding to the megalithic fathom.

The question of whether or not this is a significant peak now arises. This is a Type B analysis, since 5.44 has been determined by considering all possible values over the range used, rather than being hypothesized in advance of analysis. Kendall (1974) resolved this issue by resorting to Monte Carlo methods (Chapter 12) that involved the generation of random samples that definitely did not involve a quantum effect but were otherwise similar in all statistical respects to the actual data. After some experimentation Kendall concluded that the diameters could be adequately modeled by a shifted half-normal distribution and simulated circle diameters from

$$y_i = 10 + c|z|,$$

where z is a simulated standard normal deviate and c is a constant, 58.8 in the analysis below, designed to equate the average value of the simulated sample with that of the data. Figure 19.3 is similar to that of Kendall's (1974) Figure 11 and is based on 1000 simulations of non-quantal samples of size 169 for each of which the maximum of the cosine quantogram over the range $\tau = 0.09$ to 0.59 has been determined. The distribution shown is a KDE (Chapter 3) of these maxima using a subjectively determined window width of 0.06. The maximum for the real data,

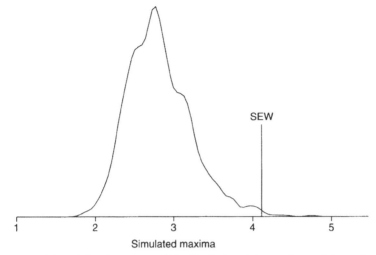

Figure 19.3 A kernel density estimate of the distribution of the maximum of the cosine quantogram for 1000 simulations in which there is no quantum. The vertical marker shows the maximum achieved for the SEW data set.

corresponding to the quantum of 5.44, lies well into the tail of the distribution and suggests that is unlikely to have arisen by chance from a smooth non-quantal distribution. Only four of the simulated maxima exceed that of the SEW data set.

Kendall makes the point that for smaller sample sizes there may be a difficulty in detecting a quantum, even if one exists, because of sample-size effects that make it difficult to disentangle a real effect from general 'noise' (Kendall, 1974: 249). Interpreting the significance of the result for the SEW data is not straightforward, because of the possibility that evidence for a quantum lies in only a subset of the data. A particularly contentious aspect of Thom's claims for the megalithic yard is its ubiquity, so if only a subset of the data is responsible for the significance this has important implications. Kendall (1974) investigates the possibility at some length in the appendix to his paper. This includes analyses of the English and Welsh data only (EW), the Scottish data only, the Scottish data divided into two subsets of similar sample size comparable with the EW data, and mixtures of the Scottish subsamples and EW data. Using similar methods to those described above, and making an allowance for the smaller sample sizes, Kendall concludes that 'my inclination is to suspect that the evidence for the quantum resides in the Scottish data *only*' (Kendall, 1974: 255; emphasis in the original). Freeman's (1976) Bayesian analysis reached broadly similar conclusions.

19.2.4 Freeman's method

Litton and Buck (1995) cited Freeman's (1976) analysis of the megalithic yard as the first published application of Bayesian statistics to archaeological data. Litton and Buck (1995: 10) also noted that the computational methods available at the time imposed a number of constraints on the way the model was formulated.

Starting from model (19.1), and taking $\beta = 0$, Freeman (1976: 25) derived a posterior distribution for δ of the form

$$f(\delta | y_1, \ldots, y_n) \propto \prod_{i=1}^{n} \sum_{m_i=1}^{n} \exp[-(y_i - m_i \delta^2)/2\sigma^2].$$

To obtain this it was assumed that the m_i were uniformly distributed over the interval 1 to N, with the upper limit N assumed to be known. The prior distribution for δ was taken to be constant over the region for which the likelihood was large (i.e., an uninformative prior was used). For computational reasons σ was treated as known, with the posterior being evaluated for several different values of σ.

Freeman (1976: 26–7) presented posterior distributions both for simulated data and subsets of the same data set used by Kendall (1974). A feature of several of these was the flatness and height of the posterior for small values of δ. Freeman reflected that these were *a priori* unreasonable and that a 'real' prior, as opposed to the flat prior used, would downweight this part of the posterior, which was thus ignored in his interpretation.

For the SEW data on good and poor circles used by Kendall, and for $\sigma = 1.5$ and $\sigma = 2$, there was evidence for a quantum at 5.45 (Freeman, 1976: Table 2). For the EW data only there was no such evidence, whereas for the Scottish data there was. This again led to the conclusion that there was no evidence for the ubiquity of a quantum, the evidence for any such quantum residing within the Scottish data. That the results from the two different analyses were broadly similar may be partly explained by a strong mathematical connection between Freeman's posterior distribution and Kendall's cosine quantogram, noted by Silverman (1976) in the discussion of Freeman's paper.

19.3 Discussion

Several of the contributors to the discussion of Freeman's paper clearly regarded his results as conclusively disproving the megalithic yard hypothesis. Freeman (1976: 55), in his reply, viewed this interpretation as too extreme, since the statistical analysis ignores other forms of evidence less readily quantified. It is clear, nevertheless, that most archaeologists regard the non-existence of the megalithic yard as having been settled in the mid-1970s (e.g., Taylor *et al.*, 2000), and many archaeologists never seriously believed in it as a possibility to begin with.

The story of the megalithic yard is interesting as a instance of a specific archaeological problem stimulating the development of statistical methodology that has a value independently of the problem that generated it. That statisticians became interested in the problem is presumably because of the challenge posed, and because the problem is capable of formulation as a statistical model. The megalithic yard or, more generally, the quantum hypothesis, is one of relatively few areas where archaeological problems have stimulated statisticians to develop new methods.

The methodology developed by Kendall (1974) has been used in other archaeological contexts. The Ashanti and related peoples of southern central Ghana used brass weights to weigh out gold dust used in transactions. Hewson (1980) notes written evidence for this practice from around AD 1600. The weights are of two general types, geometric and figurative, with most weighing less than 30 g. Using material from the British Museum and Kendall's (1974) methods, among others, Hewson (1980) found some evidence of a quantum of 1.45 g for the geometric weights, but no evidence of a quantum for the figurative weights. More recently, Pakkanen (2002) has used Kendall's method to investigate ancient foot units using, as a case study, dimensions from the Erechtheion at Athens, Greece. Kernel density estimates were used to display results, as in Figure 19.3. A range of other applications to archaeological quantal problems are described in Fieller (1993). Although the megalithic yard may be dead, the methodology that some regard as having buried it lives on.

20

Comparing assemblage diversity

20.1 Introduction

The statistical comparison of assemblage diversity has been of interest to archaeologists since the early 1980s. The review below is illustrative of how a specific archaeological problem has been approached using a variety of different statistical methods. This chapter is based, in part, on Baxter (2001b).

The following framework is assumed. A data matrix, N, is available in which columns correspond to assemblages, rows correspond to classes (such as artifact types), and entries, n_{ij}, are counts of the occurrence of class i in assemblage j. Each assemblage is conceived of as a sample from a much larger population, and interest centers on comparing these populations in terms of some index of diversity, using the sample data. Though rarely expressed in statistical terminology the implicit null hypothesis is usually that the populations have the same diversity. An example of such a data set is given and discussed in Section 2.2.4.

One possible index of diversity is S_j, the *richness* of the jth population, defined as the number of classes present. The sample richnesses, s_j, cannot be readily used to assess if population richnesses are the same because of differing sample sizes. Specifically, if S_j is bounded and $n_j \ll n_k$, where n_j is the sample size of assemblage j, then even if $S_j = S_k$ it is to be expected that $s_j < s_k$ unless sample sizes are sufficiently large. Much of the archaeological literature on diversity is concerned with dealing with this *sample-size effect*, and different approaches will be illustrated after discussing the concept of diversity in more detail.

20.2 Diversity

It is generally agreed that 'diversity' is a difficult concept to pin down and usually indices, 'borrowed' from the ecological literature (Orton, 2000a: 171), have been employed that measure specific aspects of diversity. A review of some possibilities is given in Bobrowsky and Ball (1989); however, probably the most widely used index, *richness*, as defined above, is also the simplest.

The other aspect of diversity that has attracted attention is *evenness*, where the general idea is that an assemblage in which the sample is evenly distributed among the classes is more diverse that one in which most of the sample is concentrated in a limited number of classes. Evenness has been measured in various ways.

Shannon's index, defined for the *j*th population, is

$$H = -\sum_{i=1}^{S_j} P_{ij} \log P_{ij}$$

where P_{ij} is the proportion of the *i*th class in the *j*th population, and has a maximum of $H_{max} = \log S_j$. In practice it has been common to use sample analogs of H (Conkey, 1980) or $J = H/H_{max}$ (Conkey, 1989; Kintigh, 1989; McCartney and Glass, 1990; Shott, 1997b; Kaufman, 1998) as a measure of evenness, replacing S_j by s_j and P_{ij} by $p_{ij} = n_{ij}/n_j$. These will be denoted by $\hat{H}$ and $\hat{J}$. Conkey (1989) has noted that the index is only valid for infinite populations, and Ringrose (1993b: 281) observes that $\hat{H}$ is a biased estimate of H.

Simpson's index is defined in the population as

$$D = 1 - \sum_{i=1}^{S_j} P_{ij}^2$$

and has an unbiased estimate of the form

$$\hat{D} = 1 - \frac{1}{n_j(n_j - 1)} \sum_{i=1}^{s_j} n_{ij}(n_{ij} - 1)$$

used in studies of diversity by Conkey (1989), Byrd (1997) and Shott (1997b), among others.

Ringrose (1993b: 281) is among those who have observed that both indices are heavily dependent on the most abundant classes and that this, in many cases, is an unwanted characteristic. Both $\hat{H}$ and $\hat{D}$ depend, through s_j, on assemblage richness and are sometimes described as indices of *heterogeneity* which conflate richness and evenness. Such conflation is often considered undesirable. Scaling $\hat{H}$ to get $\hat{J}$ can be viewed as an attempt to remove the dependence on richness. Simpson's index might be similarly scaled (Bobrowsky and Ball, 1989: 7) but this does not appear to be done.

It should be apparent, from this brief description, that indices of evenness have been treated with suspicion by some users. The apparently simpler concept of richness has attracted more attention, though this also poses considerable difficulties. One is that sample richness is not independent of evenness as one would ideally like it to be. Given two populations with the same S_j, one of which has equal probability for all classes, and the other for which most of the probability is concentrated in a small subset of the classes, for small sample sizes the former case will give rise to larger values of s_j than the latter. This is readily demonstrated by simulation (Rhode, 1988). Other difficulties stem from the problems posed by the sample-size effect. Two 'standard' methods have evolved to deal with this based on regression and simulation methodology, and these are the focus of what follows.

20.3   Regression and the sample-size effect

The original regression approach to investigating population richness in the presence of sample-size effects is deceptively simple. The relation between assemblage richness and sample size is typically modeled using a log-linear regression model, such as

$$\log s_j = \log \beta_0 + \beta \log n_j, \tag{20.1}$$

implying

$$s_j = \alpha n_j^{\beta}, \tag{20.2}$$

where $\beta_0 = \log \alpha$ and, for convenience, the error term in the model is omitted (see Section 5.2.2). Commonly, a regression analysis based on model (20.1) that produces a good fit has been deemed to demonstrate the sample-size effect. Cases lying away from the fitted line have been interpreted as assemblages differing in richness from those assemblages that are well fitted.

This kind of usage and interpretation has been subjected to strong criticism (e.g., Rhode, 1988; Cowgill, 1989a; Dunnell, 1989; Ringrose, 1993b). To see why, note that s_j is bounded by S_j, a property not shared by model (20.1) or (20.2) as n_j increases. A better model, which is a generalization of that proposed by Byrd (1997), is

$$s_j = S_j - \alpha \exp(-\beta n_j), \tag{20.3}$$

where S_j is a parameter to be estimated. Assuming that $\beta > 0$, $s_j \to S_j$ as $n_j \to \infty$, which is what is wanted. Byrd (1997) imposed the restriction that $\alpha = S_j$ which ensures that the model behaves properly at low values of n_j; for the purposes of estimating S_j, when behavior at low values of n_j is not a concern, the more general model (20.3) may be preferable.

Figure 20.1 shows some growth models generated from equation (20.3) plotted on a log–log scale, along with six hypothetical data points for which the correlation between $\log s_j$ and $\log n_j$ is 0.99. Several things are obvious from the graph. One is that, whereas a very good linear fit would be obtained from the data, this is perfectly compatible with assemblages that are samples from populations with different values of S_j or with the same value of S_j. In other words, linearity does not demonstrate that all the S_j are the same, as is sometimes suggested. It follows from this that it is not valid to interpret outliers as demonstrating that the associated population has a different richness from other assemblages. A second point is that it is perfectly possible to envisage a set of data points for which the slope of the fitted line is zero or even negative (Ringrose, 1993b). One such example is reported in Thomas (1989).

It is obvious from the explicit inclusion of S_j as a parameter in model (20.3) that there are more parameters to estimate than there are observations, so that the model is not identified unless restrictions are placed on the parameters. It is possible that observed linearity is a consequence of sample-size effects arising when all S_j are equal, but an interpretation in these terms has, in effect, to assume this equality, since the linearity does not demonstrate this.

Byrd (1997) avoided this difficulty in his study of archaeological faunal assemblages from four areas by assuming that, within each area, S_j was a constant for the sites located there. Model (20.3), with $\alpha = S_j$, was used to estimate S_j for each area and these estimates were then compared. Byrd used a 'minimum chi-squared'

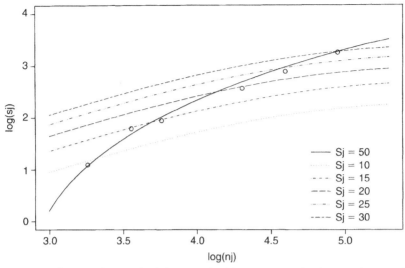

Figure 20.1 Six 'growth curves' relating sample richness to sample size are shown, using model (20.3). The legend shows the richness of the populations used to generate the curves. The five parallel curves are generated using $\beta = -0.015$ and $\alpha = 1$; the solid curve is generated using $\beta = -0.006$ and $\alpha = 1.1$. The hypothetical data points displayed can be interpreted as arising from a single growth curve with $S_j = 50$; several growth curves with constant α and β; or a mixture of the two. In practice it would be impossible to know, from the data alone, what situation applied.

method to estimate S_j, with β determined in an *ad hoc* fashion by relating it to a measure of evenness 'through a laborious process of repeated experimentation with ... experimentally determined data' (Byrd, 1997: 58). Byrd's analysis was also compromised by incorrectly reported results for two of the four sites he investigated. In the analyses to follow non-linear least squares (NLLS) is used, implemented in S-Plus (Venables and Ripley, 1999), to estimate the parameters in model (20.3), with and without the restriction used by Byrd.

The results are shown in Figure 20.2, using the data given in Byrd's paper. Estimates of S_j for four areas, Apalachee Bay, King's Bay, Pensacola Bay, and Cashie, are $(22, 28, 36, 26)$ using model (20.3), and $(22, 26, 33, 28)$ applying the constraint $\alpha = S_j$. This compares with the (corrected) estimates from Byrd's method of $(23, 21, 31, 28)$, and observed maximum richnesses of $(22, 28, 33, 25)$. For Pensacola Bay and, particularly, King's Bay, Byrd's method underestimates the richness, whereas NLLS performs well in the sense that estimates are the same as or greater than the observed maximum richness. Visually, the more general model can be seen to provide a reasonable fit to the data.

To summarize, the regression method for dealing with sample-size effects in the study of assemblage diversity has a long history and continues to be exploited. The analysis given here and in Baxter (2001b) suggests that it is seriously flawed. A non-linear model more in conformity with theoretical constraints on the data is better, and specification of such a model makes clear why the usual regression approach fails. Such a model has more demanding data requirements, in the sense that 'replicate' observations that can be assumed to have the same value for S_j are needed.

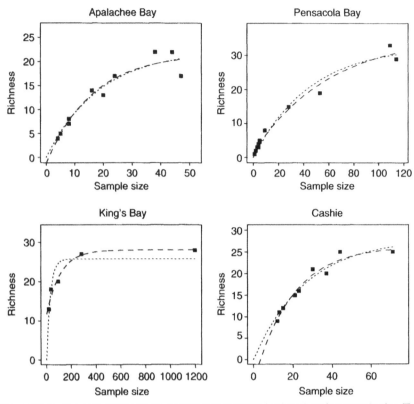

Figure 20.2 For each of four areas, assemblage richness is plotted against sample size. The dashed line shows the estimated model (20.3), and the dotted line is the same model with the constraint $\alpha = S_j$. These are mostly similar, apart from King's Bay where the more complex model leads to a better prediction of population richness but predicts richness in excess of 10 for zero sample size.

20.4 Simulation and the sample-size effect

The simulation methodology of Kintigh (1984, 1989) is the other main method used in archaeology to investigate the relationship between assemblage richness and sample size and, as with the regression approach, continues to be used (e.g., Potter, 1997). In essence the method is simple. Expected richness and associated confidence intervals are generated, for different sample sizes, by repeatedly sampling from a 'background' population whose structure has to be defined. These are compared against the observed values in order to determine whether departures from what is expected under the hypothesis of a common population structure could reasonably be due to chance.

Figure 20.3 illustrates this, using the data of Conkey (1980) discussed in Section 2.2.4. Using 100 samples of size m, for $m = 5$ to 160 in intervals of 5, the solid lines in Figure 2 show the estimated 80% confidence interval for $E(s_m)$, the expected richness for sample size m. Also shown are the actual sample sizes and richness for the five sites used. The 'population' is generated, following

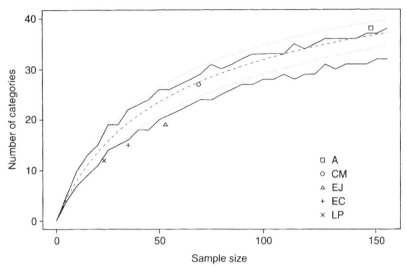

Figure 20.3 The solid lines show the 80% confidence interval obtained using Kintigh's (1984) simulation method, with 100 simulations, applied to Conkey's (1980) data. The dot-dashed line shows the expected richness using equation (20.4) and the dotted lines show the 80% confidence interval associated with this.

Kintigh's (1984, 1989) recommendation, by using the row totals of **N**, and then sampling from these with replacement to get the simulated results. The outcome is essentially the same as that shown in Kintigh (1984) and has been interpreted as showing that Altamira is richer than expected under the 'null' model, whereas Cueto de la Mina does not depart significantly from its expected richness. As with the regression approach, the simulation method has attracted a lot of critical attention, much of which centers on the way in which the background population that determines the simulation results is defined (e.g., Rhode, 1988; McCartney and Glass, 1990; Ringrose, 1993b; Kaufman, 1998; Baxter, 2001b). Define

$$p_i = \frac{\sum_j n_{ij}}{\sum_j n_j}$$

to be the proportion of class i observed across all assemblages and $\mathbf{p} = (p_1 \ p_2 \cdots p_s)$, where s is the total number of classes across the assemblages. In this notation $\mathbf{p}$ is an estimate of some notional population, $\mathbf{P} = (P_1 \ P_2 \cdots P_s)$, that is the background population. Let $\mathbf{P}_j$ define the corresponding population of which the jth assemblage is a sample. The following observations may be made:

1. If $\mathbf{P}_j = \mathbf{P}$ for all j, then $\mathbf{P}$ is readily interpreted as the common population from which the assemblages are sampled. However, as in the regression case, if this view is adopted one is assuming what the method is, in fact, supposed to test.
2. If, for some j, n_j is very much larger than the other sample sizes then $\mathbf{P}$ is dominated by the corresponding $\mathbf{P}_j$ and the simulation approximates to a test of whether or not other assemblages differ from the dominant assemblage. McCartney and Glass (1990) defend this aspect of the method; however, if this situation pertains it seems simpler to equate $\mathbf{P}_j$ with $\mathbf{P}$ and test directly whether other assemblages differ from this.

3. If the P_j do differ by much then P is an artificial construct that cannot be equated with any real population. In particular, it may differ completely from all the P_j, with unpredictable consequences for the simulation (Ringrose, 1993b).
4. Even if P is legitimately defined, the fact that it is estimated by p is ignored in the simulation procedure (Baxter, 2001b). This is taken up below.

Both Rhode (1988) and Ringrose (1993b), citing Smith and Grassle (1977), note that the expectations and confidence intervals generated in Kintigh's (1984) simulation approach can be determined analytically, but do not give full details or examples. Technical details are given in Orton (2000a: 221). Assuming that a random sample of size $m < n$ is drawn from an assemblage containing s classes, *without replacement*, an unbiased estimate of the number of classes expected in an assemblage of that size is

$$E(s_m) = \sum_{j=1}^{s} \left[1 - \frac{(n - n_j)!(n - m)!}{(n - n_j - m)!m!} \right] \qquad (20.4)$$

with variance estimate

$$\text{var}(s_m) = a^{-1} \left[\sum_{j=1}^{s}(a_j(1 - a_j/a) + 2\sum_{j=1}^{s-1}\sum_{i=j+1}^{s}(a_{ij} - a_i a_j)/a \right], \qquad (20.5)$$

where

$$a = n!/(n - m)!m!,$$

$$a_j = (n - n_j)!/(n - n_j - m)!m!$$

and

$$a_{ij} = (n - n_i - n_j)!/(n - n_i - n_j - m)!m!.$$

These expressions are all of the form $n!/(n - x)!m!$ and equal 0 by definition if $x > n$.

Applying these results to Conkey's (1980) data gives the confidence intervals and expectations shown in Figure 20.3. The confidence band is somewhat higher than that generated by Kintigh's (1984) method and, in contrast, leads to the conclusion that Altamira does not differ significantly from expectations under the null model. Computation of equations (20.4) and (20.5) is not always straightforward, and simulation could be used to emulate the results, sampling from p *without* replacement. Though not described in these terms, Kintigh (1984) samples from p with replacement. The former procedure seems preferable, though the issue of whether p, as an estimate of P, defines a legitimate background population remains.

Ringrose (1993b: 282) observed that equation (20.4) is an unbiased estimate of Smith and Grassle's (1977) 'expected species index'. He further argued that this provided an 'intuitive measure of diversity' that had the property that it was dominated by the common classes for small m but became increasingly sensitive to medium-abundance and rare classes as m increased. It was therefore recommended that $E(s_m)$ should be plotted against m, so that if the profile of one assemblage lay consistently above that for another it could be declared the more diverse.

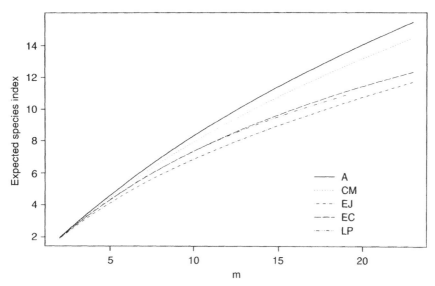

Figure 20.4 Expected richness index for five sites for $m = 2, 3, \ldots, 23$ is shown (see the text for the key to the legend). The consistent separation, with no crossings, suggests an ordering in terms of population richness, with Altamira (A) and Cueto de la Mina (CM) the richest.

This is illustrated in Figure 20.4, where it is seen that Altamira and Cueto de la Mina are the most diverse (richest) sites, and differ somewhat from other sites.

The statistics may also be used to test whether two sites differ significantly in richness. For example, $n_j = 69$ for Cueto de la Mina and $s_j = 27$. If a sample of this size is selected from the Altamira assemblage the expected richness is 28.15, with a standard deviation of 2.05. The observed richness for Cueto de la Mina lies comfortably within one standard deviation of the expected richness, so that the two sites are not significantly different in terms of richness.

Both Ringrose (1993b) and Orton (2000a) observe that use of the expected species index is equivalent to rarefaction analysis. Orton (2000a: 176) considers this to be the best of the approaches suggested to date.

20.5 Resampling approaches

Kaufman (1998), after a critical review of the regression and sampling approaches, proposed the use of jackknifing methodology (Section 12.3). This approach has not been much used, and Baxter (2001b) argued that bootstrap methods were preferable. Lipo *et al.* (1997), in a study primarily concerned with seriation, used bootstrap methods to investigate the relationship between richness and sample size. Their prime concern was to identify assemblages whose size was too small for them to be reasonably included in a seriation. For any given assemblage, 1000 bootstrap samples of size $10, 20, \ldots$ were drawn, and the mean and variance of these bootstrap samples determined for each size category. The 'good' assemblages were deemed to be those for which the mean and variance approached an asymptotic limit before the actual assemblage size.

21

Shorter studies

21.1 Introduction

Had time and space allowed, this could have been a much longer book. The purpose of this chapter is to provide a brief notice of some topics that could have been treated at greater length. There is no implication that I regard these matters as less important than some topics which have been afforded a longer treatment in the book.

21.2 Artifact classification

The importance of applications to artifact classification, for the development of an awareness of statistics in archaeology, is undeniable. Spaulding's (1953) advocacy of attribute clustering, leading ultimately to the idea of using log-linear models (Read, 1974), and early applications of cluster analysis by Hodson *et al.* (1966) and Hodson (1969, 1970) for object clustering introduced important statistical methods to a wider archaeological audience. The success of the enterprise, in developing classifications that archaeologists use (other than, possibly, the originator) is, however, questionable. This was discussed briefly in Section 2.2.3, where it was noted that many commentators (e.g., Dunnell, 1986; Aldenderfer, 1987b; Cowgill, 1990; Adams and Adams, 1991; Ammerman, 1992; Forsyth, 2000) had concluded that archaeological practice was largely unaffected by theoretical (including statistical) endeavors in this area.

My own view is that experienced archaeologists familiar with their material will almost invariably produce more useful classifications than a statistical method, for all but the simplest kinds of artifact. This is, I think, because they are better able to take into account the qualitative as well as quantitative aspects of an artifact; can deal more readily than statistical methods with conditionally present features of an artifact; can better select the important features; and can differentially weight features, in a way that is difficult to emulate in formal analysis. I have, in the past, used methods of cluster analysis for classification using data of mixed type, and including conditionally present features. This did not work well, and I am not aware of many similar and successful published attempts. A general difficulty is that although techniques exist for dealing with data of mixed type, this can result in differential importance being assigned to different types of variable in a manner unintended by the analyst (see the discussion of Gower's coefficient in Section 8.2.1). In principle one may choose to weight variables as one wishes, but it is usually not clear how to do this, and it introduces a subjectivity into the analysis that many think (possibly mistakenly) statistical analysis is designed to avoid.

For a sustained attack on the problem of using statistical methods for developing typologies the work of Read (1974, 1982, 1989) and Read and Russell (1996) can be consulted. The last of these papers makes the important, but neglected, point that measuring too many variables can actually be detrimental to the performance of a clustering algorithm (Read and Russell, 1996: 664). The paper develops a staged approach to typology construction, in which univariate analysis (which can be applied to derived variables, including those defined by multivariate analyses such as principal component analysis) is used to try and identify variables that have a multimodal distribution that defines subgroups in the data. Subgroups so identified are treated in a similar way until the groups that are obtained are judged to be homogeneous. The spirit of this sort of analysis is similar to that of classification trees (Section 9.4), except that the classification is unknown in advance of analysis, and a great deal of exploratory data analysis may be needed to determine the typology.

21.3 Age estimation

Here work on regression and Bayesian approaches to age estimation is surveyed briefly.

Adapting notation for the purposes of this section, write the simple linear regression model as

$$y = \alpha_{yx} + \beta_{yx}x + \varepsilon. \tag{21.1}$$

If this is fitted using training data and the resultant equation used to predict an unknown x from observed y_0, this is variously called classical, conditional, controlled or indirect calibration, or inverse regression (Konigsberg *et al.*, 1998: 68). An alternative approach is to regress x on y,

$$x = \alpha_{xy} + \beta_{xy}y + \varepsilon, \tag{21.2}$$

and use the fitted model to predict x. This approach has been called inverse, uncontrolled, unconditional, direct, natural or random calibration, or simply regression. There has been considerable debate in the literature about which model is appropriate and under what circumstances (Cheng and van Ness, 1999).

In a series of papers Lucy and Pollard (1995), Lucy *et al.* (1995, 1996, 2002) and Aykroyd *et al.* (1996, 1997, 1999) have studied the problem of estimating age at death, x, from skeletal material such as dental age indicators, y. They demonstrate in some of these papers (e.g., Aykroyd *et al.*, 1999) that the common practice of inverse calibration typically leads to results in which younger ages are overestimated and older ages underestimated. They evinced a preference for classical calibration using model (21.1) rather than (21.2), because it was less subject to such bias and correctly recognized that dental condition was dependent on age, rather than vice versa (Lucy and Pollard, 1995; Aykroyd *et al.*, 1997, 1999). Unfortunately, the use of classical calibration also led to greater variability in the age estimates, leading Lucy *et al.* (1996, 2002) and Aykroyd *et al.* (1999) to consider alternative, Bayesian, approaches to age estimation.

In the earliest work (Lucy *et al.*, 1996), Bayes' theorem (14.1) was used in the form

$$P(\theta_i \mid \mathbf{x}) \propto P(\mathbf{x} \mid \theta_i)P(\theta_i),$$

where θ_i is age category i; there are p age indicators, that are categorical variables, with K_j categories for the jth variable; and $\mathbf{x}_j$ is the vector of data for case j. To avoid problems arising from the sparsity of the data, conditional independence of measurements given the age category,

$$P(\mathbf{x} \mid \theta_i) = P(x_1 \mid \theta_i)P(x_2 \mid \theta_i) \cdots P(x_p \mid \theta_i),$$

was assumed where, for example, $P(x_j \mid \theta_i)$ was estimated by the proportion of cases of the jth variable falling into age category i. Even with this assumption, however, sparsity caused problems. An alternative approach (Aykroyd *et al.*, 1996, 1999) is to treat age as a continuous variable and replace $P(\theta_i)$ and $P(\theta_i \mid \mathbf{x})$ by $f(\theta)$ and $f(\theta \mid \mathbf{x})$. As one possibility, $f(\theta)$ may be estimated from the training data using a kernel density estimate (Aykroyd *et al.*, 1996). Similarly, writing

$$P(x_j \mid \theta) = f(\theta \mid x_j)P(x_j)/f(\theta),$$

kernel density estimates of $f(\theta \mid x_j = i)$ for $i = 1, \ldots, K$ may be estimated for each of the K levels of the p variables (Aykroyd *et al.*, 1996).

In Aykroyd *et al.* (1996, 1999), Gustafson's (1950) data were used for illustration. There were $n = 41$ cases scored with respect to $p = 6$ dental indicators for which $K_i = 4$ in each case. In the 1996 paper these were treated as six separate variables. In the 1999 paper, which gives fewer mathematical details, the six indicator variables were summed to provide a single score in the Bayesian analysis, kernel density estimates being obtained for each level of the overall score. Bayesian analysis was contrasted with classical and inverse calibration and found to out-perform them. The average length of 'confidence interval' was about two-thirds of that for either regression method. In the most recent development of the methodology (Lucy *et al.*, 2002) the situation when there are p indicator variables, which may be continuous or discrete, was discussed. Innovations in the methodology compared to the earlier papers include the use of structures more complex than that of conditional independence for the indicator variables given age, and smoothing of discrete as well as continuous variables. The methodology has become quite complex by this stage, and the reader is referred to the paper for details.

A Bayesian approach to age estimation is also used in Gowland and Chamberlain's (2002) study of infanticide in Roman Britain.

21.4 Particle-size analysis

The data in Table 21.1, which are taken from Fieller *et al.* (1992a), show the weight distribution of particles from a sample of sand, trapped on sieves of different mesh size. The sample was one of 226, taken from known beach (110)

Table 21.1 Weights of sand particles trapped on sieves of different mesh size: sand particles within a particular weight category have sizes lying within an interval defined by the corresponding size category and that to its right (Fieller *et al.*, 1992a)

Sieve size (mm)	0.063	0.090	0.125	0.180	0.250	0.355	0.500	0.710
Weight (g)	0.3	0.7	34.4	12.8	3.6	0.4	0.2	–

and dune (39) environments, along with 77 archaeological samples, on the island of Oronsay in the Inner Hebrides, off the coast of Scotland. The archaeological samples were taken from within mesolithic shell middens. It was of interest to see whether the beach and dune samples could be differentiated and, if so, whether the archaeological samples could be classified as one or the other. Fieller *et al.* (1984) provides more detail on the archaeological background, and these data are also discussed in Flenley and Olbricht (1993).

This poses a particular problem of particle-size analysis. One problem is to model the distribution of size for a single sample, such as that in Table 21.1. A second problem is to compare, or classify, the results for different samples. In the data given, the weights rise to and then decline from a single mode. The papers propose using the log-skew-Laplace distribution to model such data, the form of which is given by

$$f(x; \alpha, \beta, \mu) = \begin{cases} (\alpha + \beta)^{-1} \exp[(x - \mu)/\alpha], & x \le \mu, \\ (\alpha + \beta)^{-1} \exp[(\mu - x)/\beta], & x > \mu, \end{cases}$$

a distribution characterized by three parameters, α, β and μ. If the logarithm of relative proportions is plotted against the logarithm of size, this amounts to fitting the data with two straight lines, one of positive slope, determined by α, and another of negative slope, determined by β, that intersect at a mode determined by μ. Estimates of these three parameters characterize a sand sample, and techniques of analysis or display for two- or three-dimensional data can be used to investigate similarities and differences between samples from different sources.

The log-skew-Laplace distribution was proposed as an alternative to competing models, including the log-normal and log-hyperbolic distributions. It is a limiting distribution of the latter, but generally easier to fit.

In Fieller *et al.* (1984) labeled bivariate plots, of the estimates of α and μ, and discriminant function analysis, are used to investigate the distinctions between sands from different locations and environments. Fieller *et al.* (1992b) use three-dimensional plots and Flenley and Olbricht (1993) use principal component analysis and projection pursuit methods to investigate patterns using all three estimated parameters.

This very brief account only scratches the surface of the work that has been done on particle-size analysis, which has general application and not just to archaeological problems. Fieller and Nicholson (1991), for example, adapt the methods to the investigation of grain-size data from archaeological pottery, and Fieller *et al.* (1992a) consider some of the problems posed by multimodal data where there is a mixture of particle size distributions. A package, ShefSize, is available for fitting the log-skew-Laplace and other distributions to particle-size data. Details are given in the appendix.

Appendix

Web resources

A.1 S-Plus **and** R

Most analyses and figures in this book were produced using S-Plus 2000 under Windows (Venables and Ripley, 1999). The web page associated with Venables and Ripley (1999) can be found at

> http://www.stats.ox.ac.uk/pub/MASS3/

During the course of writing the book a substantially updated version of the package, S-Plus 6.0, was released (Venables and Ripley, 2002); see

> http://www.stats.ox.ac.uk/pub/MASS4/

S-Plus is a commercial package, details of which are available from

> http://www.insightful.com/

An open source language, R, that is 'not unlike S', can be investigated at

> http://www.r-project.org/

and is featured in Venables and Ripley (2002). These packages enjoy considerable user support that manifests itself in the production of freely available software libraries. Many of these can be imported painlessly from Professor Brian Ripley's site:

> http://www.stats.ox.ac.uk/~ripley/index.html

The MASS library of Venables and Ripley (1999), bundled with S-Plus 2000, was used extensively for this book.

The sm library of Bowman and Azzalini (1997) was used for kernel density estimation (Chapter 3); see

> http://www.stats.gla.ac.uk/~adrian/sm/

Matt Wand's KernSmooth library was used for non-parametric regression (Section 5.6):

http://www.biostat.harvard.edu/~mwand/software.html

For projection pursuit (Section 7.5) XGobi (Swayne *et al.*, 1998) was used:

http://www.research.att.com/areas/stat/xgobi/

Software updates of the MCLUST library used for model-based clustering (Section 8.4), and associated papers, can be found at

http://www.stat.washington.edu/fraley/

Logistic discrimination (Section 9.3) was undertaken using the nnet library of Venables and Ripley (1999), bundled with S-Plus 2000. Classification tree analysis was undertaken using the rpart library of Therneau and Atkinson (1997), available from Brian Ripley's site, referenced above.

A.2 Other resources

The data used in the case study of Section 8.3, given in Slane *et al.* (1997), are available electronically at

http://web.missouri.edu/~reahn/index.shtml

For those who might wish to emulate the analysis I should note that two specimens, ISR076 and ISR089, labeled as BSP in the database, were treated as ESA in Slane *et al.* (1994), and eight duplicate specimens with the suffix B in the database were not used in the paper. The same source also provided the data used in the first example of Section 9.2.7.

Some readers may find the following software of interest. Other than OxCal and BCal, this software has not been used for this book. A MATLAB package, developed by Christian Beardah, for kernel density estimation, can be downloaded from

http://science.ntu.ac.uk/msor/ccb/densest.html

For CLUSTAN, a commercial package that, in earlier manifestations, was influential in the way cluster analysis was used by archaeologists, see

http://www.clustan.com/

Details of the WinBASP package, written specifically for archaeologists, which includes various options for correspondence analysis and seriation, can be found at

http://www.uni-koeln.de/~al001/basp.html

Another package also aimed at archaeologists, with a variety of facilities for spatial analysis and diversity analysis, among other things, is Keith Kintigh's Tools for Quantitative Archaeology:

http://pages.prodigy.net/keith.kintigh/

Many Bayesian applications are very problem-specific and thus not readily emulated in standard software. The BUGS package (Bayesian inference Using Gibbs Sampling)

http://www.mrc-bsu.cam.ac.uk/bugs/welcome.shtml

provides one resource. The OxCal and BCal packages referenced in Chapter 15 can be found at

http://www.rlaha.ox.ac.uk/orau/

and

http://bcal.sheffield.ac.uk/

Andrew Millard's page

http://www.dur.ac.uk/a.r.millard/BUGS4Arch.html

provides guidance for archaeologists on use of the Windows version, WinBUGS, and includes examples of applications to dating problems.

For the abcml package referenced in Section 17.2.3, see

http://www.anthro.utah.edu/~rogers/abcml/

For the pie-slice package referenced in Section 17.3.3 see

http://www.ucl.ac.uk/archaeology/staff/
profiles/orton.htm

The ShefSize package for particle size analysis, noted in Section 21.4, can be found at

http://www.shef.ac.uk/nickfieller/index.html

A web page for this book is located at

http://science.ntu.ac.uk/msor/mjb/

References

Adams, W.Y. and Adams, E.W. (1991). *Archaeological Typology and Practical Reality*. Cambridge: Cambridge University Press.

Adan-Bayewitz, D., Asaro, F. and Giauque, R.D. (1999). Determining pottery provenance: application of a new high-precision X-ray fluorescence method and comparison with instrumental neutron activation analysis. *Archaeometry* **41**, 1–24.

Aitchison, J. (1976). Discussion of Freeman, P.R. (1976). A Bayesian analysis of the megalithic yard. *Journal of the Royal Statistical Society A* **139**, 36–37.

Aitchison, J. (1986). *The Statistical Analysis of Compositional Data*. London: Chapman & Hall.

Aitchison, J. (2001). Simplicial inference. In Viana, M. and Richards, D. (eds), *Algebraic Methods in Statistics*, Contemporary Mathematics **287**. Providence, RI: American Mathematical Society, 1–22.

Aitchison, J., Barceló-Vidal, C. and Pawlowsky-Glahn, V. (2002). Some comments on compositional data analysis in archaeometry, in particular the fallacies in Tangri and Wright's dismissal of logratio analysis. *Archaeometry* **44**, 295–304.

Aitchison, T., Ottaway, B. and Al-Ruzaiza, A.S. (1991). Summarizing a group of ^{14}C dates on the historical time scale: with a worked example from the Late Neolithic of Bavaria. *Antiquity* **65**, 108–16.

Aitken, M.J. (1990). *Science-based Dating in Archaeology*. London: Longman.

Aldenderfer, M.S. (1982). Methods of cluster validation for archaeology. *World Archaeology* **14**, 61–72.

Aldenderfer, M.S. (ed.) (1987a). *Quantitative Research in Archaeology*. Newbury Park, CA: Sage.

Aldenderfer, M.S. (1987b). Assessing the impact of quantitative thinking on archaeological research. In Aldenderfer, M.S. (ed.), *Quantitative Research in Archaeology*. Newbury Park, CA: Sage, 9–29.

Aldenderfer, M. (1991). The analytical engine: computer simulation and archaeological research. In Schiffer, M.B. (ed.), *Studies in Archaeological Method and Theory 3*. Tucson: University of Arizona Press, 195–247.

Aldenderfer, M.S. (1998). Quantitative methods in archaeology: a review of recent trends and developments. *Journal of Archaeological Research* **6**, 91–120.

Aldenderfer, M.S. and Maschner, H.D.G. (eds) (1996). *Anthropology, Space and Geographic Information Systems*. Oxford: Oxford University Press.

Allen, K.M.S., Green, S. and Zubrow, E.B.W. (eds) (1990). *Interpreting Space: GIS and Archaeology*. London: Taylor and Francis.

Allum, G.T., Aykroyd, R.G. and Haigh, J.G.B. (1995). A new statistical approach to reconstruction from area magnetometry data. *Archaeological Prospection* **2**, 197–205.

Allum, G.T., Aykroyd, R.G. and Haigh, J.G.B. (1996). Restoration of magnetometry data using inverse-data methods. *Analecta Praehistorica Leidensia* **28**, 111–19.

Allum, G.T., Aykroyd, R.G. and Haigh, J.G.B. (1997). Seeing beneath the surface: imaging in archaeological geophysics. In Mardia, K.V., Gill, C.A. and Aykroyd, R.G. (eds), *Proceedings in the Art and Science of Bayesian Image Analysis*. Leeds: Leeds University Press, 213–14.

Allum, G.T., Aykroyd, R.G. and Haigh, J.G.B. (1999). Empirical Bayes estimation for archaeological stratigraphy. *Applied Statistics* **48**, 1–14.

Ammerman, A.J. (1992). Taking stock of quantitative archaeology. *Annual Review of Anthropology* **21**, 231–55.

Ammerman, A.J., Kintigh, K.W. and Simek, J.F. (1987). Recent developments in the application of the *k*-means approach to spatial analysis. In Sieveking, G. De G. and Newcomer, M. (eds), *The Human Uses of Flint and Chert*. Cambridge: Cambridge University Press, 210–16.

Andresen, J., Madsen, T. and Scollar, I. (eds) (1993). *Computing the Past: CAA92*. Aarhus: Aarhus University Press.

Arbia, G. and Espa, G. (1996). Forecasting statistical models of archaeological site location. *Archeologia e Calcolatori* **7**, 365–72.

Arnold, D.E., Neff, H. and Glascock, M.D. (2000). Testing assumptions of neutron activation analysis: communities, workshops and paste preparation in Yucatán, Mexico. *Archaeometry* **42**, 302–16.

Atkinson, A.C. (1994). Fast very robust methods for the detection of multiple outliers. *Journal of the American Statistical Association* **89**, 1328–39.

Atkinson, A.C. and Mulira, H.-M. (1993). The stalactite plot for the detection of multivariate outliers. *Statistics and Computing* **3**, 27–35.

Attanasio, D.A., Armiento, G., Brilli, M., Emanuele, M.C., Platania, R. and Turi, B. (2000). Multi-method marble provenance determinations: the Carrara marbles as a case study for the combined use of isotopic, electron spin resonance and petrographic data. *Archaeometry* **42**, 257–72.

Aykroyd, R.G. and Al-Gezeri, S.M. (2002). Extensions to the single-layer model for archaeological survey data. Paper presented at the 21st Leeds Annual Statistics Workshop (http://www.amsta.leeds.ac.uk/~robert/).

Aykroyd, R.G., Lucy, D. and Pollard, A.M. (1996). Statistical methods for the estimation of human age at death. Report No. STAT 96/08, Department of Statistics, University of Leeds, UK.

Aykroyd, R.G., Lucy, D., Pollard, A.M. and Solheim, T. (1997). Regression analysis in adult age estimation. *American Journal of Physical Anthropology* **104**, 259–65.

Aykroyd, R.G., Lucy, D., Pollard, M. and Roberts, C.A. (1999). Nasty, brutish, but not necessarily short: a reconsideration of the statistical methods used to calculate age at death from adult human skeletal and dental age indicators. *American Antiquity* **64**, 55–70.

Aykroyd, R.G., Haigh, J.G.B. and Allum, G.T. (2001). Bayesian methods applied to data from archaeological magnetometry. *Journal of the American Statistical Association* **453**, 64–76.

Bailey, T.C. and Gatrell, A.C. (1995). *Interactive Spatial Data Analysis*. Harlow: Prentice Hall.

Banfield, J.D. and Raftery, A.E. (1993). Model-based Gaussian and non-Gaussian clustering. *Biometrics* **49**, 803–21.

Banning, E.B. (2000). *The Archaeologist's Laboratory: The Analysis of Archaeological Data*. New York: Kluwer Academic/Plenum.

Baquedano, E. and Orton, C. (1990). Similarities between sculptures using Jaccard's coefficient in the study of Aztec Tlaltecuhtli. *Papers from the Institute of Archaeology* **1**, 16–23.

Barceló, J.A. (1991). Some theoretical consequences of the use of advanced statistics in archaeology. In Waldren, W.H., Ensenyat, J.J. and Kennard, R.C. (eds), *2nd Deya International Conference on Prehistory: Volume II Archaeological Technology and Theory*, BAR International Series 574. Oxford: Tempus Reparatum, 267–99.

Barceló, J., Briz, I. and Vila, A. (eds) (1999). *New Techniques for Old Times: CAA98*, BAR International Series 757. Oxford: Archaeopress.

Baringhaus, L. and Henze, N. (1988). A consistent test for multivariate normality based on the empirical characteristic function. *Metrika* **35**, 339–48.

Barnett, S.M. (2000). Luminescence dating of pottery from later prehistoric Britain. *Archaeometry* **42**, 431–57.

Baxter, M.J. (1988). The morphology and evolution of post-medieval wine bottles revisited. *Science and Archaeology* **30**, 10–14.

Baxter, M.J. (1989). Multivariate analysis of data on glass compositions: a methodological note. *Archaeometry* **31**, 45–53.

Baxter, M.J. (1991). An empirical study of principal component and correspondence analysis of glass compositions. *Archaeometry* **33**, 29–41.

Baxter, M.J. (1992). Archaeological uses of the biplot – a neglected technique? In Lock, G. and Moffett, J. (eds), *Computer Applications and Quantitative Methods in Archaeology 1991*, BAR International Series 577. Oxford: Tempus Reparatum, 141–8.

Baxter, M.J. (1993). Comment on D. Tangri and R.V.S. Wright, 'Multivariate analysis of compositional data . . .'. *Archaeometry* **35**, 112–15.

Baxter, M.J. (1994a). *Exploratory Multivariate Analysis in Archaeology*. Edinburgh: Edinburgh University Press.

Baxter, M.J. (1994b). Principal component analysis in archaeometry. *Archeologia e Calcolatori* **5**, 23–38.

Baxter, M.J. (1994c). Stepwise discriminant analysis in archaeometry: a critique. *Journal of Archaeological Science* **21**, 659–66.

Baxter, M.J. (1995). Standardization and transformation in principal component analysis, with applications to archaeometry. *Applied Statistics* **44**, 513–27.

Baxter, M.J. (1999a). On the multivariate normality of data arising from lead isotope fields. *Journal of Archaeological Science* **26**, 117–24.

Baxter, M.J. (1999b). Testing multivariate normality with applications to lead isotope data analysis in archaeology. In Dingwall, L., Exon, S., Gaffney, V., Laflin, S. and van Leusen, M. (eds), *Archaeology in the Age of the Internet: CAA97*, BAR International Series 750. Oxford: Archaeopress, 107.

Baxter, M.J. (1999c). Detecting multivariate outliers in artefact compositional data. *Archaeometry* **41**, 321–38.

Baxter, M.J. (2001a). Statistical modelling of artefact compositional data. *Archaeometry* **43**, 131–47.

Baxter, M.J. (2001b). Methodological issues in the study of assemblage diversity. *American Antiquity* **66**, 715–25.

Baxter, M.J. (2003). *K*-means spatial clustering revisited. Submitted for publication.

Baxter, M.J. and Beardah, C.C. (1995). Graphical presentation of results from principal components analysis. In Huggett, J. and Ryan, N. (eds), *Computer Applications and Quantitative Methods in Archaeology 1994*, BAR International Series 565. Oxford: Tempus Reparatum, 63–7.

Baxter, M.J. and Beardah, C.C. (1996). Beyond the histogram: improved approaches to simple data display in archaeology using kernel density estimates. *Archeologia e Calcolatori* **7**, 397–408.

Baxter, M.J. and Buck, C.E. (2000). Data handling and statistical analysis. In Ciliberto, E. and Spoto, G. (eds), *Modern Analytical Methods in Art and Archaeology*. New York: Wiley, 681–746.

Baxter, M.J. and Cool, H.E.M. (1995). Notes on some statistical aspects of pottery quantification. *Medieval Ceramics* **19**, 89–98.

Baxter, M.J. and Gale, N.H. (1998). Testing for multivariate normality via univariate tests: a case study using lead-isotope ratio data. *Journal of Applied Statistics* **25**, 671–83.

Baxter, M.J. and Jackson, C.M. (2001). Variable selection in artefact compositional studies. *Archaeometry* **43**, 253–68.

Baxter, M.J., Cool, H.E.M. and Heyworth, M.P. (1990). Principal component and correspondence analysis of compositional data: some similarities. *Journal of Applied Statistics* **17**, 229–35.

Baxter, M.J., Cool, H.E.M., Heyworth, M.P. and Jackson, C.M. (1995). Compositional variability in colourless Romano-British glass. *Archaeometry* **37**, 129–41.

Baxter, M.J., Beardah, C.C. and Wright, R.V.S. (1997). Some archaeological applications of kernel density estimates. *Journal of Archaeological Science* **24**, 347–54.

Baxter, M.J., Beardah, C.C. and Westwood, S. (2000). Sample size and related issues in the analysis of lead isotope data. *Journal of Archaeological Science* **27**, 973–80.

Bayes, T.R. (1763). An essay towards solving a problem in the doctrine of chances. *Philosophical Transactions of the Royal Society of London* **53**, 370–418. Reprinted (1958) in *Biometrika* **45**, 296–315.

Bayliss, A., Bronk Ramsey, C. and McCormac, G. (1997). Dating Stonehenge. *Proceedings of the British Academy* **92**, 39–59.

Beale, E.M.L. and Little, R.J.A. (1975). Missing values in multivariate analysis. *Journal of the Royal Statistical Society B* **37**, 129–45.

Beardah, C.C. (1999). Uses of multivariate kernel density estimates in archaeology. In Dingwall, L., Exon, S., Gaffney, V., Laflin, S. and van Leusen, M. (eds), *Archaeology in the Age of the Internet: CAA97*, BAR International Series 750. Oxford: Archaeopress, 107.

Beardah, C.C. and Baxter, M.J. (1996a). The archaeological use of kernel density estimates. *Internet Archaeology* 1 (http://intarch.ac.uk/journal/issue1/beardah_index.html).

Beardah, C.C. and Baxter, M.J. (1996b). MATLAB routines for kernel density estimation and the graphical presentation of archaeological data. *Analecta Praehistorica Leidensia* 28, 179–84.

Beardah, C.C. and Baxter, M.J. (1999). Three-dimensional data display using kernel density estimates. In Barceló, J., Briz, I. and Vila, A. (eds), *New Techniques for Old Times: CAA98*. Oxford: Archaeopress, 163–9.

Beardah, C.C., Baxter, M.J., Papageorgiou, I. and Cau, M.A. (2002). 'Mixed-mode' approaches to the grouping of ceramic artefacts using S-Plus. Paper presented at Computer Applications and Quantitative Methods in Archaeology 2002, Heraklion, Crete. Submitted for publication.

Beier, T. and Mommsen, H. (1994). Modified Mahalanobis filters for grouping pottery by chemical composition. *Archaeometry* 36, 287–306.

Bell, M., Fletcher, M. and Lock, G. (1990). The application of Monte Carlo methods in archaeology. In Voorrips, A. and Ottaway, B.S. (eds), *New Tools from Mathematical Archaeology*. Warsaw: Scientific Information Centre of the Polish Academy of Sciences, 37–44.

Bellhouse, D.R. (1980). Sampling studies in archaeology. *Archaeometry* 22, 123–32.

Benfer, R.A. and Benfer, A.N. (1981). Automatic classification of inspectional categories: multivariate theories of archaeological data. *American Antiquity* 46, 381–96.

Bernardo, J.M. and Smith, A.F.M. (1994). *Bayesian Theory*. Chichester: Wiley.

Besag, J.P. (1974). Spatial interaction and the statistical analysis of lattice systems. *Journal of the Royal Statistical Society B* 36, 192–235.

Bieber, A.M., Brooks, D.W., Harbottle, G. and Sayre, E.V. (1976). Application of multivariate techniques to analytical data on Aegean ceramics. *Archaeometry* 18, 59–74.

Binford, L.R.B. (1964). A consideration of archaeological research design. *American Antiquity* 29, 429–51.

Binford, L.R. (1978). Dimensional analysis of behavior and site structure: learning from an Eskimo hunting stand. *American Antiquity* 34, 330–61.

Binford, L.R.B. and Binford, S.R. (1966). A preliminary analysis of functional variability in the Mousterian of Levallois facies. *American Anthropologist* 68, 239–95.

Bishop, Y.M.M., Fienberg, S.E. and Holland, P.W. (1975). *Discrete Multivariate Analysis*. Cambridge, MA: MIT Press.

Blankholm, H.P. (1991). *Intrasite Spatial Analysis in Theory and Practice*. Aarhus: Aarhus University Press.

Bobrowsky, P.T. and Ball, B.F. (1989). The theory and mechanics of ecological diversity in archaeology. In Leonard, R.D. and Jones, G.T. (eds), *Quantifying Diversity in Archaeology*. Cambridge: Cambridge University Press, 4–12.

Bollong, C.A., Jacobson, L., Peisach, M., Pineda, C.A. and Sampson, C.G. (1997). Ordination versus clustering of elemental data from PIXE analysis of herder-hunter pottery: a comparison. *Journal of Archaeological Science* 24, 319–27.

Bølviken, E.E., Helskog, K., Holm-Olsen, I., Solheim, L. and Bertelsen, R. (1982). Correspondence analysis: an alternative to principal components. *World Archaeology* **14**, 41–60.

Bookstein, F.L. (1991). *Morphometric Tools for Landmark Data*. Cambridge: Cambridge University Press.

Bove, F. (1981). Trend surface analysis and the Lowland Classic Maya collapse. *American Antiquity* **46**, 93–112.

Bowman, A.W. and Azzalini, A. (1997). *Applied Smoothing Techniques for Data Analysis*. Oxford: Clarendon Press.

Bowman, A.W. and Foster, P.J. (1993a). Density based exploration of bivariate data. *Statistics and Computing* **3**, 171–7.

Bowman, A.W. and Foster, P.J. (1993b). Adaptive smoothing and density-based tests of multivariate normality. *Journal of the American Statistical Association* **88**, 529–37.

Bowman, A.W., Jones, M.C. and Gijbels, I. (1998). Testing monotonicity of regression. *Journal of Computational and Graphical Statistics* **7**, 489–500.

Bowman, S. (1990). *Radiocarbon Dating*. London: British Museum Publications.

Bradley, R. and Small, C. (1985). Looking for circular structures in post hole distributions: quantitative analysis of two settlement plans from Bronze Age England. *Journal of Archaeological Science* **12**, 285–97.

Brainerd, G.W. (1951). The place of chronological ordering in archaeological analysis. *American Antiquity* **16**, 301–13.

Braun, D.P. (1981). A critique of some recent North American mortuary studies. *American Antiquity* **46**, 398–416.

Breiman, L., Friedman, J.H., Olshen, R.A. and Stone, C.J. (1984). *Classification and Regression Trees*. Monterey, CA: Wadsworth and Brooks/Cole.

Broadbent, S.S. (1955). Quantum hypotheses. *Biometrika* **42**, 45–57.

Broadbent, S.S. (1956). Examination of a quantum hypothesis based on a single set of data. *Biometrika* **43**, 32–44.

Bronk Ramsey, C. (1995). Radiocarbon calibration and analysis of stratigraphy: the OxCal program. *Radiocarbon* **37**, 425–30.

Bronk Ramsey, C. (1999). An introduction to the use of Bayesian statistics in the interpretation of radiocarbon dates. In *Proceedings of the International Workshop on Frontiers in Accelerator Mass Spectometry*. Sakura, Japan: National Museum of Japanese History, 151–60.

Bronk Ramsey, C. (2000). Comment on 'The use of Bayesian statistics for ^{14}C dates of chronologically ordered samples: a critical analysis'. *Radiocarbon* **42**, 199–202.

Bronk Ramsey, C. (2001). Development of the radiocarbon calibration program. *Radiocarbon* **43**, 355–63.

Bronk Ramsey, C. and Bayliss, A. (2000). Dating Stonehenge. In Lockyear, K., Sly, T.J.T. and Mihăilescu-Bîrliba, V. (eds), *CAA(96). Computer Applications and Quantitative Methods in Archaeology*, BAR International Series 845. Oxford: Archaeopress, 29–39.

Bronk Ramsey, C., Van Der Plicht, J. and Weninger, B. (2001). 'Wiggle matching' radiocarbon dates. *Radiocarbon* **43**, 381–9.

Brooks, S.P. (1998). Markov chain Monte Carlo method and its application. *The Statistician* **47**, 69–100.

Brothwell, D.R. and Krzanowski, W.J. (1974). Evidence of biological differences between early British populations from Neolithic to medieval times, as revealed by eleven commonly available cranial vault measurements. *Journal of Archaeological Science* **1**, 249–60.

Brown, C.T. (2001). The fractal dimensions of lithic reduction. *Journal of Archaeological Science* **28**, 619–31.

Brown, J.A. (1987). Quantitative burial analysis as interassemblage comparison. In Aldenderfer, M.S. (ed.), *Quantitative Research in Archaeology*. Newbury Park, CA: Sage, 294–308.

Buck, C.E. (1993). The provenancing of archaeological ceramics: a Bayesian approach. In Andresen, J., Madsen, T. and Scollar, I. (eds), *Computing the Past: CAA92*. Aarhus: Aarhus University Press, 293–301.

Buck, C.E. and Christen, J.A. (1998). A novel approach to selecting samples for radiocarbon dating. *Journal of Archaeological Science* **25**, 303–10.

Buck, C.E. and Litton, C.D. (1991). A computational Bayes approach to some common archaeological problems. In Lockyear, K. and Rahtz, S.P.Q. (eds), *Computer Applications and Quantitative Methods in Archaeology 1990*, BAR International Series 565. Oxford: Tempus Reparatum, 93–9.

Buck, C.E. and Litton, C.D. (1996). Mixtures, Bayes and archaeology. In Bernardo, J.M., Berger, J.O., Dawid, A.P. and Smith, A.F.M. (eds), *Bayesian Statistics 5*. Oxford: Oxford University Press, 499–506.

Buck, C.E. and Sahu, S.K. (2000). Bayesian models for relative archaeological chronology building. *Applied Statistics* **49**, 423–40.

Buck, C.E., Kenworthy, J.A., Litton, C.D. and Smith, A.F.M. (1991). Combining archaeological and radiocarbon information: a Bayesian approach to calibration. *Antiquity* **65**, 808–21.

Buck, C.E., Litton, C.D. and Smith, A.F.M. (1992). Calibration of radiocarbon results pertaining to related archaeological events. *Journal of Archaeological Science* **19**, 497–512.

Buck, C.E., Christen, J.A., Kenworthy, J.A. and Litton, C.D. (1994a). Estimating the duration of archaeological activity using ^{14}C determinations. *Oxford Journal of Archaeology* **13**, 229–40.

Buck, C.E., Litton, C.D. and Shennan, S.J. (1994b). A case study in combining radiocarbon and archaeological information: the early Bronze-Age settlement of St. Veit-Klinglberg, Land Salzburg, Austria. *Germania* **2**, 427–47.

Buck, C.E., Cavanagh, W.G. and Litton, C.D. (1996). *Bayesian Approach to Interpreting Archaeological Data*. Chichester: Wiley.

Buck, C.E., Christen, J.A. and James, G.N. (1999). BCal: an on-line Bayesian radiocarbon calibration tool. *Internet Archaeology* **7** (http://intarch.ac.uk/journal/issue7/buck_index.html).

Buck, C., Cummings, V., Henley, C., Mills, S. and Trick, S. (eds) (2000). *U.K. Chapter of Computer Applications and Quantitative Methods in Archaeology*, BAR International Series 844. Oxford: Archaeopress.

Budd, P., Gale, D., Pollard, A.M., Thomas, R.G. and Williams, P.A. (1993). Evaluating lead isotope data: further observations. *Archaeometry* **35**, 262–3.

Budd, P., Pollard, A.M., Scaife, B. and Thomas, R.G. (1995). Oxhide ingots, recycling and the Mediterranean metals trade. *Journal of Mediterranean Archaeology* **8**, 1–32.

Burenhult, G. (ed.) (2002). *Archaeological Informatics: Pushing the Envelope, CAA 2001*, BAR International Series 1016. Oxford: Archaeopress.

Burl, A. (2000). *The Stone Circles of Britain, Ireland and Brittany.* New Haven, CT and London: Yale University Press.

Buxeda i Garrigós, J. (1999). Alteration and contamination of archaeological ceramics: the perturbation problem. *Journal of Archaeological Science* **26**, 295–313.

Byrd, J.E. (1997). The analysis of diversity in archaeological faunal assemblages: Complexity and subsistence strategies in the southeast during the Middle Woodland period. *Journal of Anthropological Archaeology* **16**, 49–72.

Byrd, J.E. and Owens, D.D. (1997). A method for measuring relative abundance of fragmented archaeological ceramics. *Journal of Field Archaeology* **24**, 315–20.

Campbell, N.A. (1985). Updating formula for allocation of individuals. *Applied Statistics* **34**, 235–6.

Caroni, C. and Prescott, P. (1992). Sequential application of Wilk's multivariate outlier test. *Applied Statistics* **41**, 355–64.

Carr, C. (1984). The nature of organization of intrasite archaeological records and spatial analytical approaches to their investigation. In Schiffer, M.B. (ed.), *Advances in Archaeological Method and Theory 7*. New York: Academic Press, 103–222.

Carr, C. (ed.) (1985a). *For Concordance in Archaeological Analysis.* Kansas City, MO: Westport Publishers.

Carr, C. (1985b). Alternative models, alternative techniques: variable approaches to intrasite spatial analysis. In Carr, C. (ed.), *For Concordance in Archaeological Analysis.* Kansas City, MO: Westport Publishers, 297–459.

Carr, C. (1987). Removing discordance from quantitative analysis. In Aldenderfer, M.S. (ed.), *Quantitative Research in Archaeology.* Newbury Park, CA: Sage, 185–243.

Cau, M.A. (1999). Importaciones de cerámicas tardorromanas de cocina en les Illes Balears: el caso de Can Sora (Eivissa). In *Monografías de Arte y Arqueología.* Granada: Universidad de Granada, 198–219.

Cau, M.A., Day, P.M., Baxter, M.J., Papageorgiou, I., Iliopoulos, I. and Montana, G. (2003). Exploring automatic grouping procedures in ceramic petrology. Submitted for publication.

Cavanagh, W.G. and Laxton, R.R. (1981). The structural mechanics of the Mycenaean Tholos tomb. *Annual of the British School at Athens* **76**, 109–40.

Cavanagh, W.G. and Laxton, R.R. (1982). Corbelled vaulting in the late Minoan tholos tombs of Crete. *Annual of the British School at Athens* **77**, 65–77.

Chapdelaine, C., Millaire, J.-F. and Kennedy, G. (2001). Compositional analysis and provenance study of spindle whorls from the Moche Site, North Coast of Peru. *Journal of Archaeological Science* **28**, 795–806.

Chen, L. and Shapiro, S.S. (1995). An alternative test for normality based on normalized spacings. *Journal of Statistical Computation and Simulation* **53**, 269–87.

Chen, T., Rapp Jr, G., Jing, Z. and He, N. (1999). Provenance studies of the earliest Chinese protoporcelain using neutron activation analysis. *Journal of Archaeological Science* **26**, 1003–15.

Cheng, C.-L. and van Ness, J.W. (1999). *Statistical Regression with Measurement Error*. London: Arnold.

Cherry, J.F., Gamble, C. and Shennan, S. (eds) (1978). *Sampling in Contemporary British Archaeology*, BAR British Series 50. Oxford: BAR.

Christen, J.A. (1994a). Bayesian interpretation of ^{14}C results. Unpublished PhD thesis, University of Nottingham, UK.

Christen, J.A. (1994b). Summarizing a set of radiocarbon determinations: a robust approach. *Applied Statistics* **43**, 489–503.

Christen, J.A. and Buck, C.E. (1998). Sample selection in radiocarbon dating. *Applied Statistics* **47**, 543–57.

Christen, J.A. and Litton, C.D. (1995). A Bayesian approach to wiggle-matching. *Journal of Archaeological Science* **22**, 719–25.

Christenson, A.L. and Read, D.W. (1977). Numerical taxonomy, R-mode factor analysis and archaeological classification. *American Antiquity* **42**, 163–79.

Ciliberto, E. and Spoto, G. (eds) (2000). *Modern Analytical Methods in Art and Archaeology*. New York: Wiley.

Clark, G.A. (1974). On the analysis of multidimensional contingency tables using log-linear models. In Wilcock, J. and Laflin, S. (eds), *Computer Applications in Archaeology*. Birmingham: University of Birmingham Computer Centre, 47–58.

Clark, G.A. (1976). More on contingency table analysis, decision-making criteria and the use of log-linear models. *American Antiquity* **41**, 259–73.

Clark, G.A. (1982). Quantifying archaeological research. In Schiffer, M.B. (ed.), *Advances in Archaeological Method and Theory 5*. New York: Academic Press, 217–73.

Clark, G.A. and Stafford, C.R. (1982). Quantification in American archaeology: a historical perspective. *World Archaeology* **14**, 98–119.

Clark, L.A. and Pregibon, D. (1992). Tree-based models. In Chambers, J.M. and Hastie, T.J. (eds), *Statistical Models in S*. New York: Chapman & Hall.

Clark, P.J. and Evans, F.C. (1954). Distance to nearest neighbour as a measure of spatial relationships in populations. *Ecology* **35**, 445–53.

Clark, R.M. (1979). Calibration, cross-validation and carbon-14. *I. Journal of the Royal Statistical Society A* **142**, 47–62.

Clark, R.M. (1980). Calibration, cross-validation and carbon-14. *II. Journal of the Royal Statistical Society A* **143**, 177–94.

Clarke, D.L. (1962). Matrix analysis and archaeology with particular reference to British Beaker pottery. *Proceedings of the Prehistoric Society* **28**, 371–82.

Clarke, D. (1968). *Analytical Archaeology*. London: Methuen.

Cleveland, W.S. (1979). Robust locally weighted regression and smoothing scatterplots. *Journal of the American Statistical Association* **74**, 829–36.

Cliff, A.D. and Ord, J.K. (1973). *Spatial Autocorrelation*. London: Pion.

Cliff, A.D. and Ord, J.K. (1975). The comparison of means when samples consist of spatially autocorrelated observations. *Environment and Planning A* **7**, 725–34.

Cliff, A.D. and Ord, J.K. (1981). *Spatial Processes*. London: Pion.

Clouse, R.A. (1999). Interpreting archaeological data through correspondence analysis. *Historical Archaeology* **33**, 90–107.

Cochran, W.G. (1977). *Sampling Techniques* (3rd edition). New York: Wiley.

Conkey, M.W. (1980). The identification of prehistoric hunter-gatherer aggregation sites: the case of Altamira. *Current Anthropology* **21**, 609–30.

Conkey, M.W. (1989). The use of diversity in stylistic analysis. In Leonard, R.D. and Jones, G.T. (eds), *Quantifying Diversity in Archaeology*. Cambridge: Cambridge University Press, 118–29.

Cook, D., Buja, A. and Cabrera, J. (1993). Projection pursuit indexes based on orthonormal function expansions. *Journal of Computational and Graphical Statistics* **2**, 225–50.

Cool, H.E.M. (1994). The quantification of Roman vessel glass assemblages. Unpublished report: English Heritage.

Cool, H.E.M. and Baxter, M.J. (1995). Finds from the fortress: artefacts, buildings and correspondence analysis. In Wilcock, J. and Lockyear, K. (eds), *Computer Applications and Quantitative Methods in Archaeology 1993*, BAR International Series 598. Oxford: Tempus Reparatum, 177–82.

Cool, H.E.M. and Baxter, M.J. (1996). Quantifying glass assemblages. *Annales du 13e Congrès de l'Association Internationale pour l'Histoire du Verre*, 93–101.

Cool, H.E.M. and Baxter, M.J. (1999). Peeling the onion: an approach to comparing vessel glass assemblages. *Journal of Roman Archaeology* **12**, 72–100.

Cowgill, G.L. (1968). Archaeological applications of factor, cluster and proximity analysis. *American Antiquity* **33**, 367–75.

Cowgill, G. (1972). Models, methods and techniques for seriation. In Clarke, D. (ed.), *Models in Archaeology*. London: Methuen, 381–424.

Cowgill, G.L. (1975). A selection of samplers: comments on archaeo-statistics. In Mueller, J.W. (ed.), *Sampling in Archaeology*. Tucson: University of Arizona Press, 258–74.

Cowgill, G.L. (1977a). The trouble with significance tests and what we can do about it. *American Antiquity* **42**, 350–68.

Cowgill, G.L. (1977b). Review of *Mathematics and Computers in Archaeology* by J.E. Doran and F.R. Hodson. *American Antiquity* **42**, 126–9.

Cowgill, G.L. (1989a). The concept of diversity in archaeological theory. In Leonard, R.D. and Jones, G.T. (eds), *Quantifying Diversity in Archaeology*. Cambridge: Cambridge University Press, 131–41.

Cowgill, G.L. (1989b). Formal approaches in archaeology. In Lamberg-Karlovsky, C.C. (ed.), *Archaeological Thought in America*. Cambridge: Cambridge University Press, 74–88.

Cowgill, G.L. (1990). Artifact classification and archaeological purposes. In Voorrips, A. (ed.), *Mathematics and Information Science in Archaeology: A Flexible Framework*, Studies in Modern Archaeology 3. Bonn: Holos, 61–78.

Cowgill, G.L. (1993). Distinguished lecture in archaeology: beyond criticising New Archaeology. *American Anthropologist* **95**, 551–73.

Cowgill, G.L. (2001). Past, present and future of quantitative methods in United States archaeology. In Stančič, Z. and Veljanovski, T. (eds), *Computer Archaeology for Understanding the Past: CAA 2000*. Oxford: Archaeopress, 35–40.

Cox, D.R. and Snell, E.J. (1989). *Analysis of Binary Data* (2nd edition). London: Chapman & Hall.

Cox, T.F. and Cox, M.A.A. (2001). *Multidimensional Scaling* (2nd edition). Boca Raton, FL: Chapman & Hall/CRC.

Cressie, N.A.C. (1993). *Statistics for Spatial Data* (revised edition). New York: Wiley.

Curet, L.A. (1998). New formulae for estimating prehistoric populations for lowland South America and the Caribbean. *Antiquity* **72**, 358–75.

Dacey, M.F. (1973). Statistical tests of spatial association in the locations of tool types. *American Antiquity* **38**, 320–8.

D'Agostino, R.B. (1986). Tests for the normal distribution. In Stephens, M.A. and D'Agostino, R.B. (eds), *Goodness-of-Fit Techniques*. New York: Dekker, 367–419.

Davison, A.C. and Hinkley, D.V. (1997). *Bootstrap Methods and Their Application*. Cambridge: Cambridge University Press.

Delicado, P. (1999). Statistics in archaeology: new directions. In Barceló, J., Briz, I. and Vila, A. (eds), *New Techniques for Old Times: CAA98*. Oxford: Archaeopress, 29–37.

Dellaportas, P. (1998). Bayesian classification of Neolithic tools. *Applied Statistics* **47**, 279–97.

Dempster, A.P., Laird, N.M. and Rubin, D.B. (1977). Maximum likelihood from incomplete data via the EM algorithm. *Journal of the Royal Statistical Society B* **39**, 1–38.

Dibble, H.L. and Pelchin, A. (1995). The effect of hammer mass and velocity on flake mass. *Journal of Archaeological Science* **22**, 429–39.

Digby, P.G.N. and Kempton, R.A. (1987). *Multivariate Analysis of Ecological Communities*. London: Chapman & Hall.

Diggle, P.J. (1983). *Statistical Analysis of Spatial Point Patterns*. London: Academic Press.

Dingwall, L., Exon, S., Gaffney, V., Laflin, S. and van Leusen, M. (eds) (1999). *Archaeology in the Age of the Internet: CAA97*, BAR International Series 750. Oxford: Archaeopress.

Djindjian, F. (1985). Seriation and toposeriation by correspondence analysis. *PACT* **11**, 119–35.

Djindjian, F. (1989). Fifteen years of contributions of the French school of data analysis. In Rahtz, S. and Richards, J. (eds), *Computer Applications and Quantitative Methods in Archaeology 1989*, BAR International Series 548. Oxford: BAR, 193–204.

Djindjian, F. (1990). Ordering and structuring in archaeology. In Voorrips, A. (ed.), *Mathematics and Information Science in Archaeology: A Flexible Framework*, Studies in Modern Archaeology 3. Bonn: Holos, 79–92.

Djindjian, F. (1991). *Méthodes pour l'Archéologie*. Paris: Armand Colin.

Djindjian, F. and de Croisset, E. (1976a). Un essai de reconnaissance de formes sur une série de deux cents bifaces Mousteriéns de Tabaterie (Dordogne) par l'analyse des donnés. *Cahiers du Centre de Recherches Préhistoriques, Université de Paris I* **5**, 39–60.

Djindjian, F. and de Croisset, E. (1976b). Etude typométrique d'une série de deux cents bifaces Mousteriéns de Tabaterie (Dordogne) par l'analyse des donnés. In *Union Internationale des Sciences Préhistoriques et Protohistoriques, 9e Congrés: Thèmes Spécialisés*. Nice: Osseux, 38–50.

Djingova, R. and Kuleff, I. (1992). An archaeometric study of medieval glass from the first Bulgarian capital, Pliska (ninth to tenth century AD). *Archaeometry* **34**, 53–61.

Dobson, A.J. (1990). *An Introduction to Generalized Linear Models*. London: Chapman & Hall.

Dolukhanov, P., Sokoloff, D. and Shukurov, A. (2001). Radiocarbon chronology of Upper Palaeolithic sites in eastern Europe at improved resolution. *Journal of Archaeological Science* **28**, 699–712.

Donnelly, K. (1978). Simulations to determine the variance and edge-effect of total nearest neighbour distance. In Hodder, I. (ed.), *Simulation Methods in Archaeology*. Cambridge: Cambridge University Press, 91–5.

Doran, J.E. and Hodson, F.R. (1966). A digital computer analysis of Palaeolithic flint assemblages. *Nature* **210**, 688–9.

Doran, J.E. and Hodson, F.R. (1975). *Mathematics and Computers in Archaeology*. Edinburgh: Edinburgh University Press.

Draper, N.R. and Smith, H. (1998). *Applied Regression Analysis* (3rd edition). New York: Wiley.

Drennan, R.D. (1996). *Statistics for Archaeologists*. New York: Plenum Press.

Dryden, I.L. (2001). Statistical shape analysis in archaeology. Technical Report, University of Nottingham, UK.

Dryden, I.L. and Mardia, K.V. (1998). *Statistical Shape Analysis*. Chichester: Wiley.

Duff, A.I. (1996). Ceramic micro-seriation: types or attributes? *American Antiquity* **61**, 89–101.

Dunnell, R.C. (1970). Seriation method and its evaluation. *American Antiquity* **35**, 305–19.

Dunnell, R.C. (1986). Methodological issues in Americanist artifact classification. In Schiffer, M.B. (ed.), *Advances in Archaeological Method and Theory 9*. New York: Academic Press, 149–207.

Dunnell, R.C. (1989). Diversity in archaeology: a group of measures in search of application? In Leonard, R.D. and Jones, G.T. (eds), *Quantifying Diversity in Archaeology*. Cambridge: Cambridge University Press, 142–9.

Dunnell, R. (2000). Seriation. In Ellis, L. (ed.), *Archaeological Method and Theory*. New York: Garland Publishing, 548–50.

Earle, T.K. (1976). A nearest neighbor analysis of two formative settlement systems. In Flannery, K.V. (ed.), *The Early Mesoamerican Village*. New York: Academic Press, 196–223.

Ebert, D. (2002). The potential of geostatistics in the analysis of fieldwalking data. In Wheatley, D., Earl, G. and Poppy, S. (eds), *Contemporary Themes in Archaeological Computing*. Oxford: Oxbow Books, 82–9.

Efron, B. (1979). Bootstrap methods: another look at the jackknife. *Annals of Statistics* **7**, 1–26.

Efron, B. and Tibshirani, R.J. (1993). *An Introduction to the Bootstrap*. London: Chapman & Hall.

Englestad, E. (1988). Pit-houses in Arctic Norway – an investigation of their typology using multiple correspondence analysis. In Madsen, T. (ed.), *Multivariate Archaeology*. Aarhus: Aarhus University Press, 71–84.

Everitt, B.S. (1992). *The Analysis of Contingency Tables*. London: Chapman & Hall.

Everitt, B.S. (1998). *The Cambridge Dictionary of Statistics*. Cambridge: Cambridge University Press.

Everitt, B.S. and Dunn, G. (2001). *Applied Multivariate Data Analysis* (2nd edition). London: Arnold.

Everitt, B.S., Landau, S. and Leese, M. (2001). *Cluster Analysis* (4th edition). London: Arnold.

Fan, Y. and Brooks, S.P. (2000). Bayesian modelling of prehistoric corbelled domes. *The Statistician* **49**, 339–54.

Fieller, N.R.J. (1993). Archaeostatistics – old statistics in ancient contexts. *The Statistician* **42**, 279–95.

Fieller, N.R.J. and Nicholson, P.T. (1991). Grain size analysis in archaeological pottery: the use of statistical models. In Middleton, A. (ed.), *Recent Developments in Ceramic Petrology*, Occasional Paper 81. London: British Museum, 161–71.

Fieller, N.R.J. and Turner, A. (1982). Number estimation in vertebrate samples. *Journal of Archaeological Science* **9**, 49–62.

Fieller, N.R.J., Gilbertson, D.D. and Olbricht, W. (1984). A new method for environmental analysis of particle size distribution data from shoreline sediments. *Nature* **311**, 648–51.

Fieller, N.R.J., Flenley, E.C. and Olbricht, W. (1992a). Statistics of particle-size data. *Applied Statistics* **41**, 127–46.

Fieller, N.R.J., Gilbertson, D.D., Griffin, C.M., Briggs, D.J. and Jenkinson, R.D.S. (1992b). The statistical modelling of the grain-size distributions of cave sediments using log skew Laplace distributions – Creswell Crags, near Sheffield, England. *Journal of Archaeological Science* **19**, 129–50.

Fisher, P., Farrelly, C., Maddocks, A. and Ruggles, C. (1997). Spatial analysis of visible areas from the Bronze Age cairns of Mull. *Journal of Archaeological Science* **24**, 581–92.

Fisher, R.A. (1936). The use of multiple measurements in taxonomic problems. *Annals of Eugenics* **7**, 179–88.

Flenley, E.C. and Olbricht, W. (1993). Classification of archaeological sands by particle size analysis. In Opitz, O., Lausen, B. and Klar, R. (eds), *Information and Classification. Concepts, Methods and Applications. Proceedings of the 16th Annual Conference of the Gesellschaft für Klassifikation e.V.* Berlin: Springer, 478–89.

Fletcher, M.F. and Lock, G.R. (1981). Computerised pattern perception within post hole distributions. *Science and Archaeology* **23**, 15–20.

Fletcher, M.F. and Lock, G.R. (1984a). Post built structures at Danebury hillfort: an analytical search method with statistical discussion. *Oxford Journal of Archaeology* **3**, 175–96.

Fletcher, M.F. and Lock, G.R. (1984b). A mathematical model to predict patterning within post hole distributions. *Science and Archaeology* **26**, 5–8.

Fletcher, M. and Lock, G.R. (1990). How random is random: searching for patterns in point distributions. *Journal of Quantitative Anthropology* **2**, 1–16.

Fletcher, M. and Lock, G.R. (1991). *Digging Numbers*. Oxford: Oxford University Committee for Archaeology.

Flury, B. and Riedwyl, H. (1988). *Multivariate Statistics: A Practical Approach*. London: Chapman & Hall.

Ford, J.A. (1962). *A Quantitative Method for Deriving Cultural Chronology*, Technical Manual 1. Washington, DC: Pan America Union.

Forsyth, H. (2000). Mathematics and computers: the classifier's ruse. In Lock, G. and Brown, K. (eds), *On the Theory and Practice of Archaeological Computing*. Oxford: Oxford University Committee for Archaeology, 31–9.

Fraley, C. and Raftery, A.E. (2000). Model-based clustering, discriminant analysis and density estimation. Technical Report no. 380, Department of Statistics, University of Washington, Seattle, WA.

Freeman, P.R. (1976). A Bayesian analysis of the megalithic yard (with discussion). *Journal of the Royal Statistical Society A* **139**, 20–55.

Freeman, P.R. (1987). How to simulate if you must. In Ruggles, C.L.N. and Rahtz, S.P.Q. (eds), *Computer and Quantitative Methods in Archaeology 1987*, BAR International Series 393. Oxford: BAR, 139–46.

Friedman, J.H. and Rubin, J. (1967). On some invariant criteria for grouping data. *Journal of the American Statistical Association* **82**, 249–66.

Gaffney, V. and Stančič, Z. (1991). *GIS Approaches to Regional Analysis: A Case Study of the Island of Hvar*. Ljubljana: Znanstveni Inštitut Filozofska Fakulteta.

Galanidou, N. (1993). Quantitative methods for spatial analysis at rockshelters: the case of Klithi. In Andresen, J., Madsen, T. and Scollar, I. (eds), *Computing the Past: CAA92*. Aarhus: Aarhus University Press, 357–66.

Galanidou, N. (1998a). The spatial organization of Klithi. In Bailey, G. (ed.), *Klithi: Palaeolithic Settlement and Quaternary Landscapes in Northern Greece, Volume 1*. Cambridge: McDonald Institute Monographs, 275–304.

Galanidou, N. (1998b). Lithic refitting and site structure at Kastritsa. In Bailey, G. (ed.), *Klithi: Palaeolithic Settlement and Quaternary Landscapes in Northern Greece, Volume 2*. Cambridge: McDonald Institute Monographs, 497–520.

Gale, N.H. (2001). Archaeology, science-based archaeology and the Mediterranean Bronze Age metals trade: a contribution to the debate. *European Journal of Archaeology* **4**, 113–30.

Gale, N.H. and Stos-Gale, Z. (1993). Evaluating lead isotope data: further observations. Comments on Budd *et al.* (1993) – Comments II. *Archaeometry* **35**, 252–9.

Gale, N.H. and Stos-Gale, Z. (2000). Lead isotope analysis applied to provenance studies. In Ciliberto, E. and Spoto, G. (eds), *Modern Analytical Methods in Art and Archaeology*. New York: Wiley, 503–84.

García-Heras, M., Blackman, M.J., Fernández-Ruiz, R. and Bishop, R.L. (2001). Assessing ceramic compositional data: a comparison of total reflection X-ray fluorescence and instrumental neutron activation analysis on Late Iron Age Spanish Celtiberian ceramics. *Archaeometry* **43**, 325–47.

Gardin, J.C. (ed.) (1970). *Archéologie et Calculateurs: Problèmes Sémiologiques et Mathématiques*. Paris: Editions du CNRS.

Gauch, H.G. (1982). *Multivariate Analysis in Community Ecology*. Cambridge: Cambridge University Press.

Gelfand, A.E. (1971a). Rapid seriation methods with archaeological applications. In Hodson, F.R., Kendall, D.G. and Tautu, P. (eds), *Mathematics in the Archaeological and Historical Sciences*. Edinburgh: Edinburgh University Press, 185–201.

Gelfand, A.E. (1971b). Seriation methods for archaeological materials. *American Antiquity* **36**, 263–74.

Gilks, W.R., Richardson, S. and Spiegelhalter, D.J. (eds) (1996). *Markov Chain Monte Carlo in Practice*. London: Chapman & Hall.

Glascock, M.D. (1992). Characterization of archaeological ceramics at MURR by neutron activation analysis and multivariate statistics. In Neff, H. (ed.), *Chemical Characterization of Ceramic Pastes in Archaeology*. Madison, WI: Prehistory Press, 11–26.

Glover, D.M. and Hopke, P.K. (1992). Exploration of multivariate chemical data by projection pursuit. *Chemometrics and Intelligent Laboratory Systems* **16**, 45–59.

Glover, D.M. and Hopke, P.K. (1994). Exploration of multivariate atmospheric particulate compositional data by projection pursuit. *Atmospheric Environment* **28**, 1411–24.

Gómez Portugal Aguilar, D., Litton, C.D. and O'Hagan, A. (2002). Novel statistical model for a piece-wise linear radiocarbon calibration curve. *Radiocarbon* **44**, 195–212.

Gordon, A.D. (1999). *Classification* (2nd edition). London: Chapman & Hall/CRC.

Gould, R.A. and Yellen, J. (1987). Man the hunted: determinants of household spacing in desert and tropical foraging societies. *Journal of Anthropological Archaeology* **6**, 77–103.

Gower, J.C. (1971). A general coefficient of similarity and some of its properties. *Biometrics* **27**, 857–74.

Gower, J.C. and Hand, D.J. (1996). *Biplots*. London: Chapman & Hall.

Gowland, R.L. and Chamberlain, A.T. (2002). A Bayesian approach to aging perinatal skeletal material from archaeological sites: implications for the evidence for infanticide in Roman-Britain. *Journal of Archaeological Science* **29**, 677–85.

Graham, I. (1970). Discrimination of British lower and middle Palaeolithic handaxe groups using canonical variates. *World Archaeology* **1**, 321–37.

Graham, I. (1980). Spectral analysis and distance methods in the study of archaeological distributions. *Journal of Archaeological Science* **7**, 105–29.

Graham, I., Galloway, P. and Scollar, I. (1976). Model studies in computer seriation. *Journal of Archaeological Science* **3**, 1–30.

Graybill, F.A. (1983). *Matrices with Applications in Statistics* (2nd edition). Belmont, CA: Wadsworth.

Grayson, D.K. (1981). The effects of sample size on some derived measures in vertebrate faunal analysis. *Journal of Archaeological Science* **8**, 77–88.

Grayson, D.K. (1984). *Quantitative Zooarchaeology*. Orlando, FL: Academic Press.

Greenacre, M.J. (1984). *Theory and Applications of Correspondence Analysis*. London: Academic Press.

Gregg, S.A., Kintigh, K.W. and Whallon, R. (1991). Linking ethno-archaeological interpretation and archaeological data: the sensitivity of spatial analytical methods to post-depositional disturbance. In Kroll, E. and Price, T.D. (eds), *The Interpretation of Archaeological Spatial Patterning*. New York: Plenum Press, 149–96.

Grier, C. and Savelle, J.M. (1994). Intrasite spatial patterning and Thule Eskimo social organization. *Arctic Anthropology* **31**, 95–107.

Gustafson, G. (1950). Age determination on teeth. *Journal of the American Dental Association* **41**, 45–54.

Halekoh, U. and Vach, W. (1999). Bayesian seriation as a tool in archaeology. In Dingwall, L., Exon, S., Gaffney, V., Laflin, S. and van Leusen, M. (eds), *Archaeology in the Age of the Internet: CAA97*, BAR International Series 750. Oxford: Archaeopress, 107.

Hall, M.E. (2001). Pottery styles during the Early Jomon period: geochemical perspectives on the Moroiso and Ukishima pottery styles. *Archaeometry* **43**, 59–75.

Hall, M.E. and Minyaev, S. (2002). Chemical analyses of Xiong-nu pottery: a preliminary study of exchange on the Inner Asian steppes. *Journal of Archaeological Science* **29**, 135–44.

Hall, M.E., Brimmer, S.P., Li, F.-H. and Yablonsky, L. (1998). ICP-MS and ICP-OES studies of gold from a late Sarmatian burial. *Journal of Archaeological Science* **25**, 545–52.

Hall, M., Amraatuvshin, Ch. and Erdenbat, E. (1999). X-ray fluorescence analysis of pottery from Northern Mongolia. *Journal of Radioanalytical and Nuclear Chemistry* **240**, 763–73.

Hand, D.J. (1997). *Construction and Assessment of Classification Rules.* Chichester: Wiley.

Harbottle, G. (1976). Activation analysis in archaeology. *Radiochemistry* **3**, 33–72.

Harrington, J.C. (1954). Dating stem fragments of seventeenth and eighteenth century tobacco pipes. *Quarterly Bulletin of the Archaeological Society of Virginia* **9**, 6–8.

Hedges, R.E.M. (2000). Radiocarbon dating. In Ciliberto, E. and Spoto, G. (eds), *Modern Analytical Methods in Art and Archaeology.* New York: Wiley, 465–502.

Heidke, J.M. and Miksa, E.J. (2000). Correspondence and discriminant analysis of sand and sand temper compositions, Tonto Basin, Arizona. *Archaeometry* **42**, 273–99.

Heizer, R.F. and Cook, S.F. (1956). Some aspects of the quantitative approach in archaeology. *Southwestern Journal of Anthropology* **12**, 229–48.

Heizer, R.F. and Cook, S.F. (eds) (1960). *The Application of Quantitative Methods in Archaeology.* New York: Wenner-Gren Foundation.

Heggie, D.C. (1981). *Megalithic Science.* London: Thames and Hudson.

Hempel, C.G. (1962). Deductive-nomological vs. statistical explanation. In Feigl, H. and Maxwell, G. (eds), *Scientific Explanation, Space and Time.* Minneapolis: University of Minnesota Press, 98–169.

Hempel, C.G. (1965). *Aspects of Scientific Explanation.* New York: Free Press.

Henze, N. and Zirkler, B. (1990). A class of invariant consistent tests for multivariate normality. *Communications in Statistics: Theory and Methods* **19**, 3595–617.

Hewson, A.D. (1980). The Ashanti weights – a statistical evaluation. *Journal of Archaeological Science* **7**, 363–70.

Hietala, H. (ed.) (1984). *Intrasite Spatial Analysis in Archaeology.* Cambridge: Cambridge University Press.

Higham, C.F.W., Kijngam, A. and Manly, B.F.J. (1982). Site location and site hierarchy in Thailand. *Proceedings of the Prehistoric Society* **48**, 1–27.

Hill, J.B. (2000). Decision making at the margins: settlement trends, temporal scale, and ecology in the Wadi al Hasa, west-central Jordan. *Journal of Anthropological Archaeology* **19**, 221–41.

Hill, M.O. (1974). Correspondence analysis: a neglected multivariate technique. *Applied Statistics* **23**, 340–54.

Hill, M.O. and Gauch, H.G. (1980). Detrended correspondence analysis: an improved ordination technique. *Vegetatio* **42**, 47–58.

Hodder, I. (1971). The use of nearest neighbour analysis. *Cornish Archaeology* **10**, 35–6.

Hodder, I. (ed.) (1978). *Simulation Studies in Archaeology.* Cambridge: Cambridge University Press.

Hodder, I. (ed.) (2001). *Archaeological Theory Today.* Cambridge: Polity Press.

Hodder, I. and Okell, E. (1978). An index for assessing the association between distributions of points in archaeology. In Hodder, I. (ed.), *Simulation Studies in Archaeology.* Cambridge: Cambridge University Press, 97–107.

Hodder, I. and Orton, C. (1976). *Spatial Analysis in Archaeology.* Cambridge: Cambridge University Press.

Hodson, F.R. (1969). Searching for structure within multivariate archaeological data. *World Archaeology* **1**, 90–105.

Hodson, F.R. (1970). Cluster analysis and archaeology: some new developments and applications. *World Archaeology* **1**, 299–320.

Hodson, F.R. (1971). Numerical typology and prehistoric archaeology. In Hodson, F.R., Kendall, D.G. and Tautu, P. (eds), *Mathematics in the Archaeological and Historical Sciences.* Edinburgh: Edinburgh University Press, 30–45.

Hodson, F.R., Sneath, P.H.A. and Doran, J.E. (1966). Some experiments in the numerical analysis of archaeological data. *Biometrika* **53**, 311–24.

Hodson, F.R., Kendall, D.G. and Tautu, P. (1971). *Mathematics in the Archaeological and Historical Sciences.* Edinburgh: Edinburgh University Press.

Hole, B. (1980). Sampling in archaeology: a critique. *Annual Review of Anthropology* **9**, 217–34.

Holm-Olsen, I.M. (1985). Farm mounds and land registers in Helgøy, north Norway: an investigation of trends in site location by correspondence analysis. *American Archaeology* **5**, 27–34.

Hopkins, J.A. (1999). Multivariate resampling techniques for assessing sample sizes for biplots. In Barceló, J., Briz, I. and Vila, A. (eds), *New Techniques for Old Times: CAA98.* Oxford: Archaeopress, 185–8.

Horton, D.R. (1984). Minimum numbers: a consideration. *Journal of Archaeological Science* **11**, 255–71.

Howell, T.L. and Kintigh, K.W. (1996). Archaeological identification of kin groups using mortuary and biological data: an example from the American Southwest. *American Antiquity* **61**, 537–56.

Huggett, J. and Ryan, N. (eds) (1995). *Computer Applications and Quantitative Methods in Archaeology 1994,* BAR International Series 565. Oxford: Tempus Reparatum.

Ihm, P. (1978). *Statistik in der Archäologie.* Cologne and Bonn: Rheinland Verlag Rudolf Habelt.

Ihm, P. (1981). The Gaussian model in chronological seriation. In Cowgill, G.L., Whallon, R. and Ottaway, B.S. (eds), *Coloquio Manejo de Datos y Métodos Matemáticos de Arqueología, X Congreso UISPP*, Mexico City (UISPP Congress X, Commission IV), 108–24.

Ihm, P. (1990). Stochastic models and data analysis in archaeology. In Voorrips, A. (ed.), *Mathematics and Information Science in Archaeology: A Flexible Framework*, Studies in Modern Archaeology 3. Bonn: Holos, 115–34.

Impey, O.R. and Pollard, A.M. (1985). A multivariate metrical study of ceramics made by three potters. *Oxford Journal of Archaeology* 4, 157–64.

Jackson, J.E. (1991). *A User's Guide to Principal Components*. New York: Wiley.

Jacobi, R.M., Laxton, R.R. and Switsur, V.R. (1980). Seriation and dating of Mesolithic sites in southern England. *Revue d'Archéometrie* 4, 165–73.

Jensen, C.K. and Nielsen, K.H. (eds) (1997). *Burial and Society: The Chronological and Social Analysis of Archaeological Burial Data*. Aarhus: Aarhus University Press.

Johnson, I. (1976). Contribution méthodologique à l'étude de la répartition des vestiges dans des niveaux archéologiques. DES Thesis, University of Bordeaux I.

Johnson, I. (1984). Cell frequency recording and analysis of artifact distribution. In Hietala, H. (ed.), *Intrasite Spatial Analysis in Archaeology*. Cambridge: Cambridge University Press, 75–96.

Johnson, I. and North, M. (1997). *Archaeological Applications of GIS: Proceedings of Colloquium II, UISPP XIIIth Congress, Forli, Italy, September, 1996*. Sydney: Sydney University [CD-ROM].

Johnson, M. (1999). *Archaeological Theory*. Oxford: Blackwell Publishers.

Johnson, N.L. and Kotz, S. (1970). *Distributions in Statistics – Continuous Univariate Distributions 2*. New York: Wiley.

Jolliffe, I.T. (1989). Rotation of ill-defined principal components. *Applied Statistics* 38, 139–47.

Jolliffe, I.T. (1995). Rotation of principal components: choice of normalization constraints. *Journal of Applied Statistics* 22, 29–35.

Jolliffe, I.T. (2002). *Principal Component Analysis* (2nd edition). New York: Springer.

Jones, M.C. (1993). Simple boundary correction for kernel density estimation. *Statistics and Computing* 3, 135–46.

Jones, M.C. and Sibson, R. (1987). What is projection pursuit? (with discussion). *Journal of the Royal Statistical Society A* 150, 1–36.

Jones, M.C. and Williams-Thorpe, O. (2001). An illustration of the use of an atypicality index in provenancing British stone axes. *Archaeometry* 43, 1–18.

Jones, M.C., Marron, J.S. and Sheather, S.J. (1996). Progress in data-based bandwidth selection for kernel density estimation. *Computational Statistics* 11, 337–81.

Jørgensen, L. (ed.) (1992). *Chronological Studies of Anglo-Saxon England, Lombard Italy and Vendel Period Switzerland*. Copenhagen: Institute of Prehistoric and Classical Archaeology, University of Copenhagen, 94–112.

Judge, W.J. and Sebastian, L. (eds) (1988). *Quantifying the Present and Predicting the Past: Theory, Method and Application of Archaeological Predictive Modeling*. Washington, DC: US Government Printing Office.

Kaiser, H.F. (1958). The varimax criterion for analytic rotation in factor analysis. *Psychometrika* **23**, 187–200.

Kamermans, H. and Fennema, K. (eds) (1996). *Interfacing the Past: CAA95. Analecta Praehistorica Leidensia* **28**.

Kaufman, D. (1998). Measuring archaeological diversity: an application of the jackknife technique. *American Antiquity* **63**, 73–85.

Kendall, D.G. (1963). A statistical approach to Flinders Petrie's sequence dating. *Bulletin of the International Statistical Institute* **40**, 657–80.

Kendall, D.G. (1969a). Some problems and methods in statistical archaeology. *World Archaeology* **1**, 68–76.

Kendall, D.G. (1969b). Incidence matrices, interval graphs and seriation in archaeology. *Pacific Journal of Mathematics* **28**, 565–70.

Kendall, D.G. (1970). A mathematical approach to seriation. *Philosophical Transactions of the Royal Society of London A* **269**, 125–34.

Kendall, D.G. (1971a). Seriation from abundance matrices. In Hodson, F.R., Kendall, D.G. and Tautu, P. (eds), *Mathematics in the Archaeological and Historical Sciences*. Edinburgh: Edinburgh University Press, 215–52.

Kendall, D.G. (1971b). Abundance matrices and seriation in archaeology. *Zeitschrift für Wahrscheinlichkeitstheorie und verwandte Gebiete* **17**, 104–12.

Kendall, D.G. (1974). Hunting quanta. *Philosophical Transactions of the Royal Society of London A* **276**, 231–66.

Kendall, D.G., Barden, D., Carne, T.K. and Le, H. (1999). *Shape and Shape Theory*. Chichester: Wiley.

Kimball, L.R. (1987). A consideration of the role of quantitative archaeology in theory construction. In Aldenderfer, M.S. (ed.), *Quantitative Research in Archaeology*. Newbury Park, CA: Sage, 114–25.

Kintigh, K.W. (1984). Measuring archaeological diversity by comparison with simulated assemblages. *American Antiquity* **49**, 44–54.

Kintigh, K.W. (1987). Quantitative methods designed for archaeological problems. In Aldenderfer, M.S. (ed.), *Quantitative Research in Archaeology*. Newbury Park, CA: Sage, 126–34.

Kintigh, K.W. (1988). The effectiveness of subsurface testing: a simulation approach. *American Antiquity* **53**, 686–707.

Kintigh, K.W. (1989). Sample size, significance and measures of diversity. In Leonard, R.D. and Jones, G.T. (eds), *Quantifying Diversity in Archaeology*. Cambridge: Cambridge University Press, 25–36.

Kintigh, K.W. (1990). Intrasite spatial analysis: a commentary on major methods. In Voorrips, A. (ed.), *Mathematics and Information Science in Archaeology: A Flexible Framework*, Studies in Modern Archaeology 3. Bonn: Holos, 165–200.

Kintigh, K.W. and Ammerman, A.J. (1982). Heuristic approaches to spatial analyis in archaeology. *American Antiquity* **47**, 31–63.

Klecka, W.R. (1980). *Discriminant Analysis*. Thousand Oaks, CA: Sage.

Knapp, A.B. (2000). Archaeology, science-based archaeology and the Mediterranean Bronze Age metals trade. *European Journal of Archaeology* **3**, 31–56.

Knutsson, K., Dahlquist, B. and Knutsson, H. (1988). Patterns of tool use: the microwear analysis of the quartz and flint assemblages from the Bjurselet site, Västerbotten, northern Sweden. In Beyries, S. (ed.), *Industries Lithiques: Tracéologie et Technologie*, BAR International Series 411(1). Oxford: BAR, 253–94.

Koetje, T.A. (1987). *Spatial Patterns in Magdelenian Open Air Site from Isle Valley, Southwestern France*, BAR International Series 346. Oxford: BAR.

Kohler, T.A. and Parker, S.C. (1986). Predictive models for archaeological resource location. In Schiffer, M.B. (ed.), *Advances in Archaeological Method and Theory, Volume 9*. New York: Academic Press, 397–452.

Konigsberg, L.W., Hens, S.M., Jantz, L.M. and Jungers, W.L. (1998). Stature estimation and calibration: Bayesian and maximum likelihood perspectives in physical anthropology. *Yearbook of Physical Anthropology* 41, 65–92.

Koziol, J.A. (1986). Assessing multivariate normality – a compendium. *Communications in Statistics – Theory and Methods* 15, 2763–83.

Koziol, J.A. (1987). An alternative formulation of Neyman's smooth goodness of fit tests under composite hypotheses. *Metrika* 34, 17–24.

Krakker, J.J., Shott, M.J. and Welch, P.D. (1983). Design and evaluation of shovel-test sampling in regional archaeological survey. *Journal of Field Archaeology* 10, 469–80.

Kroll, E.M. and Price, T.D. (eds) (1991). *The Interpretation of Archaeological Spatial Patterning*. New York: Plenum Press.

Kruskal, J.B. (1971). Multi-dimensional scaling in archaeology: time is not the only dimension. In Hodson, F.R., Kendall, D.G. and Tautu, P. (eds), *Mathematics in the Archaeological and Historical Sciences*. Edinburgh: Edinburgh University Press, 119–32.

Krzanowski, W.J. (1988). *Principles of Multivariate Analysis*. Oxford: Clarendon Press.

Krzanowski, W.J. and Marriott, F.H.C. (1994). *Multivariate Analysis: Distributions, Ordination and Inference*. London: Edward Arnold.

Krzanowski, W.J. and Marriott, F.H.C. (1995). *Multivariate Analysis: Classification, Covariance Structures and Repeated Measurements*. London: Edward Arnold.

Kuttruff, J.T. (1993). Mississippian status differentiation through textile analysis: a Caddoan example. *American Antiquity* 58, 125–45.

Kvamme, K.L. (1983). Computer processing techniques for regional modeling of archaeological site location. *Advances in Computer Archaeology* 1, 26–32.

Kvamme, K.L. (1985). Determining empirical relationships between the natural environment and prehistoric site locations: a hunter-gatherer example. In Carr, C. (ed.), *For Concordance in Archaeological Analysis*. Kansas City, MO: Westport Publishers, 208–38.

Kvamme, K.L. (1990a). Spatial autocorrelation and the Classic Maya collapse revisited: refined techniques and new conclusions. *Journal of Archaeological Science* 17, 197–207.

Kvamme, K.L. (1990b). The fundamental principles and practice of predictive archaeological modeling. In Voorrips, A. (ed.), *Mathematics and Information Science in Archaeology: A Flexible Framework*, Studies in Modern Archaeology 3. Bonn: Holos, 257–95.

Kvamme, K.L. (1992). A predictive site location model on the High-Plains: an example with an independent test. *Plains Anthropologist* **37**, 19–40.

Kvamme, K.L. (1993). Spatial statistics and GIS: an integrated approach. In Andresen, J., Madsen, T. and Scollar, I. (eds), *Computing the Past: CAA92*. Aarhus: Aarhus University Press, 91–103.

Kvamme, K.L. (1996). Investigating chipping debris scatters: GIS as an analytical machine. In Maschner, H.D.G. (ed.), *New Methods, Old Problems: Geographic Information Systems in Modern Archaeological Research*. Carbondale: Southern Illinois University at Carbondale, 38–71.

Kvamme, K.L. (1997). A wider view of the relationship between settlement size and population in the Peruvian Andes. *American Antiquity* **62**, 719–22.

Kvamme, K.L. (1999). Recent directions and developments in geographical information systems. *Journal of Archaeological Research* **7**, 153–201.

Kvamme, K.L., Stark, M.T. and Longacre, W.A. (1996). Alternative procedures for assessing standardization in ceramic assemblages. *American Antiquity* **61**, 116–26.

Lachenbruch, P.A. (1975). *Discriminant Analysis*. New York: Hafner Press.

Lance, G.N. and Williams, W.T. (1967). A general theory of classificatory sorting strategies: 1. Hierarchical systems. *Computer Journal* **9**, 373–80.

Laxton, R.R. (1976). A measure of pre-Q-ness with applications to archaeology. *Journal of Archaeological Science* **3**, 43–54.

Laxton, R.R. (1987). Some results on mathematical seriation with applications. In Ruggles, C.L.N. and Rahtz, S.P.Q. (eds), *Computer and Quantitative Methods in Archaeology 1987*, BAR International Series 393. Oxford: BAR, 39–44.

Laxton, R.R. (1990). Methods of chronological ordering. In Voorrips, A. and Ottaway, B.S. (eds), *New Tools from Mathematical Archaeology*. Warsaw: Scientific Information Centre of the Polish Academy of Sciences, 37–44.

Laxton, R.R. and Restorick, J. (1989). Seriation by similarity and consistency. In Rahtz, S. and Richards, J. (eds), *Computer Applications and Quantitative Methods in Archaeology 1989*, BAR International Series 548. Oxford: BAR, 215–25.

Le, H. and Small, C.G. (1999). Multidimensional scaling of simplex shapes. *Pattern Recognition* **32**, 1601–13.

LeBlanc, S.A. (1975). Micro-seriation: a method for fine chronologic differentiation. *American Antiquity* **40**, 22–38.

Leese, M.N. (1988). Statistical treatment of stable isotope data. In Herz, N. and Waelkens, M. (eds), *Classical Marbles: Geochemistry, Technology, Trade*. Dordrecht: Kluwer Academic Publishers.

Leese, M.N. (1992). Evaluating lead isotope data. Comments on Sayre *et al.* (1992a) – Comments II. *Archaeometry* **34**, 318–22.

Leese, M.N. and Main, P.L. (1994). The efficient computation of unbiased Mahalanobis distances and their interpretation in archaeometry. *Archaeometry* **36**, 307–16.

Leese, M.N. and Needham, S.P. (1986). Frequency table analysis: examples from Early Bronze Age axe decoration. *Journal of Archaeological Science* **13**, 1–12.

Leese, M.N., Hughes, M.J. and Stopford, J. (1989). The chemical composition of tiles from Bordesley. In Rahtz, S. and Richards, J. (eds), *Computer*

Applications and Quantitative Methods in Archaeology 1989, BAR International Series 548. Oxford: BAR, 241–9.

Lele, S.R. and Richtsmeier, J.T. (2001). *An Invariant Approach to the Statistical Analysis of Shape*. London: Chapman & Hall/CRC.

Lentfer, C., Therin, M. and Torrence, R. (2002). Starch grains and environmental reconstruction: a modern test case from West New Britain, Papua New Guinea. *Journal of Archaeological Science* **29**, 687–98.

Leonard, R.D. and Jones, G.T. (eds) (1989). *Quantifying Diversity in Archaeology*. Cambridge: Cambridge University Press.

Lewis, B. (1986). The analysis of contingency tables in archaeology. In Schiffer, M.B. (ed.), *Advances in Archaeological Method and Theory 9*. New York: Academic Press, 277–310.

Lightfoot, K.G. (1986). Regional surveys in the Eastern United States: the strengths and weaknesses of implementing sub-surface testing programs. *American Antiquity* **51**, 484–504.

Lightfoot, K.G. (1989). A defense of shovel-test sampling: a reply to Shott. *American Antiquity* **54**, 413–16.

Lindley, D.V. (2000). The philosophy of statistics (with comments). *The Statistician* **49**, 293–337.

Lipo, C.P., Madsen, M.E., Dunnell, R.C. and Hunt, T. (1997). Population structure, cultural transmission and frequency seriation. *Journal of Anthropological Archaeology* **16**, 301–33.

Litton, C.D. and Buck, C.E. (1995). The Bayesian approach to the interpretation of archaeological data. *Archaeometry* **37**, 1–24.

Litton, C.D. and Buck, C.E. (1996). An archaeological example: radiocarbon dating. In Gilks, W., Richardson, S. and Spiegelhalter, D. (eds), *Markov Chain Monte Carlo in Practice*. London: Chapman & Hall, 465–80.

Litton, C.D. and Leese, M.N. (1991). Some statistical problems arising in radiocarbon calibration. In Lockyear, K. and Rahtz, S.P.Q. (eds), *Computer Applications and Quantitative Methods in Archaeology 1990*, BAR International Series 565. Oxford: Tempus Reparatum, 101–9.

Litton, C.D. and Restorick, J. (1983). Computer analysis of post hole distributions. In Laflin, S. (ed.), *Computer Applications in Archaeology 1983*. Birmingham: University of Birmingham Computer Centre, 85–92.

Litton, C.D. and Zainodin, H.J. (1991). Statistical models of dendrochronology. *Journal of Archaeological Science* **18**, 429–40.

Liversidge, H.M. and Molleson, T.I. (1999). Developing permanent tooth length as an estimate of age. *Journal of Forensic Sciences* **44**, 917–20.

Lock, G. (ed.) (2000). *Beyond the Map: Archaeology and Spatial Technologies*. Amsterdam: IOS Press.

Lock, G. and Moffett, J. (eds) (1992). *Computer Applications and Quantitative Methods in Archaeology 1991*, BAR International Series 577. Oxford: Tempus Reparatum.

Lock, G.R. and Stančič, Z. (eds) (1995). *Archaeology and Geographical Information Systems: A European Perspective*. London: Taylor and Francis.

Lockyear, K. (1996). Dmax based cluster analysis and the supply of coinage to Iron Age Dacia. *Analecta Praehistorica Leidensia* **28**, 165–78.

Lockyear, K. (1999). Coins, copies and kernels – a note on the potential of kernel density estimates. In Dingwall, L., Exon, S., Gaffney, V., Laflin, S. and van Leusen, M. (eds), *Archaeology in the Age of the Internet: CAA97*, BAR International Series 750. Oxford: Archaeopress, 85–90.

Lockyear, K. (2000a). Experiments with detrended correspondence analysis. In Lockyear, K., Sly, T.J.T. and Mihăilescu-Bîrliba, V. (eds), *CAA(96). Computer Applications and Quantitative Methods in Archaeology*, BAR International Series 845. Oxford: Archaeopress, 9–17.

Lockyear, K. (2000b). Site finds in Roman Britain: a comparison of techniques. *Oxford Journal of Archaeology* **19**, 397–423.

Lockyear, K. and Rahtz, S.P.Q. (eds) (1991). *Computer Applications and Quantitative Methods in Archaeology 1990*, BAR International Series 565. Oxford: Tempus Reparatum.

Lockyear, K., Sly, T.J.T. and Mihăilescu-Bîrliba, V. (eds) (2000). *CAA(96). Computer Applications and Quantitative Methods in Archaeology*, BAR International Series 845. Oxford: Archaeopress.

Lotwick, H.W. and Silverman, B.W. (1982). Methods for analysing spatial processes of several types of points. *Journal of the Royal Statistical Society B* **44**, 406–13.

Lucy, D. and Pollard, A.M. (1995). Further comments on the estimation of error associated with the Gustafson dental age estimation method. *Journal of Forensic Sciences* **40**, 222–7.

Lucy, D., Pollard, A.M. and Roberts, C.A. (1995). A comparison of three dental techniques for estimating age at death in humans. *Journal of Archaeological Science* **22**, 417–28.

Lucy, D., Aykroyd, R.G., Pollard, A.M. and Solheim, T. (1996). A Bayesian approach to adult human age estimation from dental observations by Johanson's age changes. *Journal of Forensic Sciences* **41**, 189–94.

Lucy, D., Aykroyd, R.G. and Pollard, A.M. (2002). Nonparametric calibration for age estimation. *Applied Statistics* **51**, 183–96.

Lucy, S. (2000). *The Anglo-Saxon Way of Death*. Stroud: Sutton Publishing.

Lukesh, S.S. and Howe, S. (1978). Proto-appennine vs. Sub-appennine: mathematical distinction between two ceramics phases. *Journal of Field Archaeology* **5**, 339–47.

Lyman, R.L. (1994). Quantitative units and terminology in zooarchaeology. *American Antiquity* **59**, 36–71.

Madsen, T. (ed.) (1988a). *Multivariate Archaeology*. Aarhus: Aarhus University Press.

Madsen, T. (1988b). Multivariate distances and archaeology. In Madsen, T. (ed.), *Multivariate Archaeology*. Aarhus: Aarhus University Press, 7–27.

Malkovich, J.F. and Afifi, A.A. (1973). On tests for multivariate normality. *Journal of the American Statistical Association* **68**, 176–9.

Mallory-Greenhough, L.M., Greenhough, J.D. and Owen, J.V. (1998). New data for old pots: trace-element characterization of ancient Egyptian pottery using ICP-MS. *Journal of Archaeological Science* **25**, 85–97.

Mameli, L., Barceló, J.A. and Estevez, J. (2002). The statistics of archaeological deformation processes: an archaeozoological experiment. In Burenhult, G. (ed.), *Archaeological Informatics: Pushing the Envelope, CAA 2001*, BAR International Series 1016. Oxford: Archaeopress, 221–30.

Manly, B.J.F. (1994). *Multivariate Statistical Methods – A Primer* (2nd edition). London: Chapman & Hall.

Manly, B.F.J. (1996). The statistical analysis of artefacts in graves: presence and absence data. *Journal of Archaeological Science* **23**, 473–84.

Manly, B.F.J. (1997). *Randomization, Bootstrap and Monte Carlo Methods in Biology* (2nd edition). London: Chapman & Hall.

Mardia, K.V. and Kent, J.T. (1991). Rao score tests for goodness of fit and independence. *Biometrika* **78**, 355–63.

Mardia, K.V., Kent, J.T. and Bibby, J.M. (1979). *Multivariate Analysis*. London: Academic Press.

Marquardt, W.M. (1978). Advances in archaeological seriation. In Schiffer, M. (ed.), *Advances in Archaeological Method and Theory 1*. New York: Academic Press, 257–314.

Maschner, H.D.G. (ed.) (1996). *New Methods, Old Problems: Geographic Information Systems in Modern Archaeological Research*. Carbondale: Southern Illinois University at Carbondale.

Maschner, H.D.G. and Stein, J.W. (1995). Multivariate approaches to site location on the Northwest coast of North America. *Antiquity* **69**, 61–73.

McCartney, P.H. and Glass, M.F. (1990). Simulation models and the interpretation of archaeological diversity. *American Antiquity* **55**, 521–36.

McClellan, T.L. (1979). Chronology of the 'Philistine' burials at Tell el-Farah (South). *Journal of Field Archaeology* **6**, 57–73.

McCorriston, J. and Weisberg, S. (2002). Spatial and temporal variation in Mesopotamian agricultural practices in the Khabur Basin, Syrian Jazira. *Journal of Archaeological Science* **29**, 485–98.

McCullagh, P. and Nelder, J.A. (1989). *Generalized Linear Models* (2nd edition). London: Chapman & Hall.

McHugh, F. (1999). *Theoretical and Quantitative Approaches to the Study of Mortuary Practice*, BAR International Series 785. Oxford: Archaeopress.

McKern, T.W. and Munro, E.H. (1959). A statistical technique for classifying human remains. *American Antiquity* **24**, 375–82.

McLachlan, G.J. (1992). *Discriminant Analysis and Pattern Recognition*. New York: Wiley.

Millard, A. (2002). A Bayesian approach to sapwood estimates and felling dates in dendrochronology. *Archaeometry* **44**, 137–43.

Mommsen, H., Beier, Th. and Hein, A. (2002). A complete chemical grouping of the Berkeley neutron activation analysis data on Mycenaean pottery. *Journal of Archaeological Science* **29**, 613–37.

Moran, P.A.P. (1950). Notes on continuous stochastic phenomena. *Biometrika* **37**, 17–23.

Moreno-García, M., Orton, C. and Rackham, J. (1996). A new statistical tool for comparing animal bone assemblages. *Journal of Archaeological Science* **23**, 437–53.

Morris, E.L. (1994). Production and distribution of pottery and salt in Iron Age Britain. *Proceedings of the Prehistoric Society* **60**, 371–93.

Mueller, J.W. (ed.) (1975). *Sampling in Archaeology*. Tucson: University of Arizona Press.

Müller, J. and Zimmerman, A. (eds) (1997). *Archäologie und Korrespondenzanalyse*. Espelkamp: Verlag Marie Leidorf.

Müller, P., Erkanli, A. and West, M. (1996). Bayesian curve fitting using multivariate normal mixtures. *Biometrika* **83**, 67–79.

Myers, O.M. (1950). *Some Applications of Statistics to Archaeology*. Cairo: Government Press.

Nance, J.D. (1990). Statistical sampling in archaeology. In Voorrips, A. (ed.), *Mathematics and Information Science in Archaeology: A Flexible Framework*, Studies in Modern Archaeology 3. Bonn: Holos, 135–63.

Nance, J.D. (1994). Statistical sampling, estimation, and analytic procedures in archaeology. *Journal of Quantitative Anthropology* **4**, 221–48.

Nance, J.D. and Ball, B.F. (1986). No surprises? The reliability and validity of test-pit sampling. *American Antiquity* **51**, 457–83.

Nance, J.D. and Ball, B.F. (1989). A shot in the dark: Shott's comments on Nance and Ball. *American Antiquity* **54**, 405–12.

Nason, G.P. (1995). Three-dimensional projection pursuit. *Applied Statistics* **44**, 411–30.

Naylor, J.C. and Smith, A.F.M. (1988). An archaeological inference problem. *Journal of the American Statistical Association* **83**, 588–95.

Needham, S. (1991). *Runnymede Bridge: Excavation Report*. London: British Museum Publications.

Neff, H. (ed.) (1992). *Chemical Characterization of Ceramic Pastes in Archaeology*. Madison, WI: Prehistory Press.

Neff, H. (1994). RQ-mode principal components analysis of ceramic compositional data. *Archaeometry* **36**, 115–30.

Neiman, F.D. (1995). Stylistic variation in evolutionary perspective: inferences from decorative diversity and interassemblage distance in Illinois Woodland ceramic assemblages. *American Antiquity* **60**, 7–36.

Neiman, F.D. (1997). Conspicuous consumption as wasteful advertising: a Darwinian perspective on spatial patterns in Classic Maya terminal monument dates. In Barton, M.C. and Clark, G.A. (eds), *Rediscovering Darwin: Evolutionary Theory in Archeological Explanation*. Arlington, VA: American Anthropological Association, 267–90.

Nicholls, G.K. and Jones, M. (2001). Radiocarbon dating with temporal order constraints. *Applied Statistics* **50**, 503–21.

Nicholson, M. and Barry, J. (1995). Inferences from spatial surveys about the presence of an unobserved species. *Oikos* **72**, 74–8.

O'Brien, M.J. and Lyman, R.L. (2000). *Seriation, Stratigraphy and Index Fossils*. New York: Kluwer Academic/Plenum Publishers.

O'Connor, T. (2000). *The Archaeology of Animal Bones*. Stroud: Sutton Publishing.

O'Connor, T.P. (2001). Animal bone quantification. In Brothwell, D.R. and Pollard, A.M. (eds), *Handbook of Archaeological Sciences*. New York: Wiley, 703–10.

O'Hagan, A. (1994). *Bayesian Inference*. London: Arnold.

O'Hare, G.B. (1990). A preliminary study of polished stone artefacts in prehistoric southern Italy. *Proceedings of the Prehistoric Society* **56**, 123–52.

Ortman, S.G. (2000). Conceptual metaphor in the archaeological record: methods and an example from the American southwest. *American Antiquity* **65**, 613–45.

Orton, C. (1975). Quantitative pottery studies: some progress, problems and prospects. *Science and Archaeology* **16**, 30–5.

Orton, C. (1980). *Mathematics in Archaeology*. London: Collins.

Orton, C. (1982). Computer simulation experiments to assess the performance of measures of quantity of pottery. *World Archaeology* **14**, 1–20.

Orton, C. (1992). Quantitative methods in the 1990s. In Lock, G. and Moffett, J. (eds), *Computer Applications and Quantitative Methods in Archaeology 1991*, BAR International Series 577. Oxford: Tempus Reparatum, 137–40.

Orton, C. (1993a). How many pots make 5 – an historical review of pottery quantification. *Archaeometry* **35**, 169–84.

Orton, C. (1993b). What lies behind the quantification debate? In Andresen, J., Madsen, T. and Scollar, I. (eds), *Computing the Past: CAA92*. Aarhus: Aarhus University Press, 273–8.

Orton, C. (1996). Dem dry bones. In Bird, J., Hassall, M. and Sheldon, H. (eds), *Interpreting Roman London*. Oxford: Oxbow Books, 199–208.

Orton, C. (1997). Testing significance or testing credulity? *Oxford Journal of Archaeology* **16**, 219–25.

Orton, C. (1999). Plus ça change? 25 years of statistics in archaeology. In Dingwall, L., Exon, S., Gaffney, V., Laflin, S. and van Leusen, M. (eds), *Archaeology in the Age of the Internet: CAA97*. Oxford: Archaeopress, 25–34.

Orton, C. (2000a). *Sampling in Archaeology*. Cambridge: Cambridge University Press.

Orton, C. (2000b). A Bayesian approach to a problem of archaeological site evaluation. In Lockyear, K., Sly, T.J.T. and Mihăilescu-Bîrliba, V. (eds), *CAA(96). Computer Applications and Quantitative Methods in Archaeology*, BAR International Series 845. Oxford: Archaeopress, 1–7.

Orton, C. (2000c). Spatial analysis. In Ellis, L. (ed.), *Archaeological Method and Theory*. New York: Garland Publishing, 584–8.

Orton, C. and Tyers, P.A. (1990). Statistical analysis of ceramic assemblages. *Archeologia e Calcolatori* **1**, 81–110.

Orton, C. and Tyers, P.A. (1991). A technique for reducing the size of sparse contingency tables. In Lockyear, K. and Rahtz, S.P.Q. (eds), *Computer Applications and Quantitative Methods in Archaeology 1990*, BAR International Series 565. Oxford: Tempus Reparatum, 121–6.

Orton, C. and Tyers, P.A. (1992). Counting broken objects: the statistics of ceramic assemblages. *Proceedings of the British Academy* **77**, 163–84.

Orton, C. and Tyers, P.A. (1993). *A User's Guide to Pie-slice*. University of London: Institute of Archaeology.

Orton, C., Tyers, P. and Vince, A. (1993). *Pottery in Archaeology*. Cambridge: Cambridge University Press.

Orwell, G. (1945). *Animal Farm*. London: Secker & Warburg.

O'Shea, J.M. (1984). *Mortuary Variability: An Archaeological Investigation*. New York: Academic Press.

O'Shea, J.M. (1985). Cluster analysis and mortuary patterning: an experimental assessment. *PACT* **11**, 91–110.

O'Shea, J.M. and Zvelebil, M. (1984). Oleneostroviski Mogilnik: reconstructing the social and economic organisation of prehistoric foragers in Northern Russia. *Journal of Anthropological Archaeology* **3**, 1–40.

Ottaway, B. (1981). Mixed data classification in archaeology. *Revue d'Archéométrie* **5**, 139–44.

Pakkanen, J. (1998). *The Temple of Athena Alea at Tegea*. Helsinki: Department of Art History, University of Helsinki.

Pakkanen, J. (2002). Deriving ancient foot units from building dimensions: a statistical approach employing cosine quantogram analysis. In Burenhult, G. (ed.), *Archaeological Informatics: Pushing the Envelope, CAA 2001*, BAR International Series 1016. Oxford: Archaeopress, 501–6.

Palumbo, G. (1987). 'Egalitarian' or 'Stratified' society. Some notes on mortuary practices and social structure at Jericho in EB IV. *Bulletin of the American Schools of Oriental Research* **267**, 43–59.

Papageorgiou, I., Baxter, M.J. and Cau, M.A. (2001). Model-based cluster analysis of artefact compositional data. *Archaeometry* **43**, 571–88.

Parker, S. (1985). Predictive modeling of site settlement systems using multivariate logistics. In Carr, C. (ed.), *For Concordance in Archaeological Analysis.* Kansas City, MO: Westport Publishers, 173–207.

Parker-Pearson, M. (1999). *The Archaeology of Death and Burial.* Gloucester: Sutton.

Patterson, L.W. (1990). Characteristics of bifacial-reduction flake-size distribution. *American Antiquity* **55**, 550–8.

Pearson, R., Lee, J.-W. and Koh, W. (1989). Social ranking in the Kingdom of Old Silla, Korea: analysis of burials. *Journal of Anthropological Archaeology* **8**, 1–50.

Peebles, C.S. (1972). Monothetic-divisive analysis of the Moundville burials: an initial report. *Newsletter of Computer Archaeology* **8**, 1–13.

Pelchin, A. (1996). Controlled experiments in the production of flake attributes. Unpublished PhD thesis, University of Pennsylvania, USA.

Penny, K.I. (1996). Appropriate critical values when testing for a single multivariate outlier by using the Mahalanobis distance. *Applied Statistics* **45**, 73–81.

Pernicka, E. (1993). Evaluating lead isotope data: further observations. Comments on Budd *et al.* (1993) – Comments II. *Archaeometry* **35**, 259–62.

Perry, D.W., Buckland, P.C. and Snæsdóttir, M. (1985). The application of numerical techniques to insect assemblages from the site of Storaborg, Iceland. *Journal of Archaeological Science* **12**, 335–45.

Petrie, W.M.F. (1899). Sequences in prehistoric remains. *Journal of the Anthropological Institute* **29**, 295–301.

Phillip, G. and Ottaway, B.S. (1983). Mixed data cluster analysis: an illustration using Cypriot hooked-tang weapons. *Archaeometry* **25**, 119–33.

Plog, S. (1976). Relative efficiencies of sampling techniques for archaeological surveys. In Flannery, K.V. (ed.), *The Early Mesoamerican Village.* New York: Academic Press, 136–58.

Pollard, A.M. (1986). Multivariate methods of data analysis. In Jones, R.E. (ed.), *Greek and Cypriot Pottery: A Review of Scientific Studies*, Fitch Laboratory Occasional Paper 1. Athens: British School at Athens, 56–83.

Pollard, A.M. and Hatcher, H. (1986). The chemical analysis of oriental body compositions: part 2 – Greenwares. *Journal of Archaeological Science* **13**, 261 87.

Pollard, A.M. and Heron, C. (1996). *Archaeological Chemistry.* Cambridge: Royal Society of Chemistry.

Poplin, F. (1976). Remarques théoriques et pratiques sur les unités utilisées dans les études d'ostéologie quantitative, particulièrement en archaéologie préhistorique. In *Union Internationale des Sciences Préhistoriques et Protohistoriques, 9e Congrès: Thèmes Specialisées.* Nice: Osseux, 124–41.

Potter, J.M. (1997). Communal ritual and faunal remains: an example from the Dolores Anasazi. *Journal of Field Archaeology* **24**, 353–64.

Pyszczyk, H.W. (1989). Consumption and ethnicity: an example from the fur trade in Western Canada. *Journal of Anthropological Archaeology* **8**, 213–49.

Rackham, D.J. (1986). Assessing the relative frequencies of species by the application of a stochastic model to a zooarchaeological database. In van Wijngaarden-Bakker, L.H. (ed.), *Database Management and Zooarchaeology*, Journal of the European Group of Physical, Chemical, Biological and Mathematical Techniques Applied to Archaeology, Research Volume 40. Amsterdam: PACT, 185–93.

Rahtz, S.P.Q. (ed.) (1988). *Computer Applications and Quantitative Methods in Archaeology 1988*, BAR International Series 446. Oxford: BAR.

Rahtz, S. and Richards, J. (eds) (1989). *Computer Applications and Quantitative Methods in Archaeology 1989*, BAR International Series 548. Oxford: BAR.

Rao, C.R. (1948). The utilization of multiple measurements in problems of biological classification. *Journal of the Royal Statistical Society B* **10**, 159–203.

Read, D.W. (1974). Some comments on typology in archaeology and an outline of a methodology. *American Antiquity* **36**, 216–42.

Read, D.W. (1986). Sampling procedures for regional surveys: a problem of representativeness and effectiveness. *Journal of Field Archaeology* **13**, 477–91.

Read, D.W. (1989). Statistical methods and reasoning in archaeological research: a review of praxis and promise. *Journal of Quantitative Anthropology* **1**, 5–78.

Read, D.W. and Russell, G. (1996). A method for taxonomic typology construction and an example: utilized flakes. *American Antiquity* **61**, 663–84.

Reece, R. (1995). Site-finds in Roman Britain. *Britannia* **26**, 179–206.

Reedy, C.L. and Reedy, C.R. (1988). Lead isotope analysis for provenance studies in the Aegean region: a re-evaluation. In Sayre, E.V., Vandiver, P., Druzik, J. and Stevenson, C. (eds), *Materials Issues in Art and Archaeology*. Pittsburgh: Materials Research Society, 65–70.

Reedy, C.L. and Reedy, C.R. (1992). Evaluating lead isotope data. Comments on Sayre *et al.* (1992a) – Comments IV. *Archaeometry* **34**, 327–9.

Renfrew, C. and Bahn, P. (2000). *Archaeology: Theory, Methods and Practice* (3rd edition). London: Thames and Hudson.

Renfrew, C. and Cooke, K.L. (1974). *Transformations: Mathematical Approaches to Culture Change*. New York: Academic Press.

Renfrew, C. and Sterud, G. (1969). Close-proximity analysis: a rapid method for the ordering of archaeological materials. *American Antiquity* **34**, 265–77.

Rhode, D. (1988). Measurement of archaeological diversity and the sample-size effect. *American Antiquity* **53**, 708–16.

Rice, P.M. and Saffer, M.E. (1982). Cluster analysis of mixed-level data: pottery provenience as an example. *Journal of Archaeological Science* **9**, 395–409.

Ridings, R. and Sampson, G.G. (1990). There's no percentage in it: intersite spatial analysis of Bushman (San) pottery decorations. *American Antiquity* **55**, 766–80.

Ringrose, T.J. (1992). Bootstrapping and correspondence analysis in archaeology. *Journal of Archaeological Science* **19**, 615–29.

Ringrose, T.J. (1993a). Bone counts and statistics – a critique. *Journal of Archaeological Science* **20**, 121–57.

Ringrose, T.J. (1993b). Diversity indices and archaeology. In Andresen, J., Madsen, T. and Scollar, I. (eds), *Computing the Past: CAA92*. Aarhus: Aarhus University Press, 279–85.

Ripley, B.D. (1976). The second-order analysis of stationary point processes. *Journal of Applied Probability* **13**, 255–66.

Ripley, B.D. (1977). Modelling spatial patterns. *Journal of the Royal Statistical Society B* **39**, 172–212.

Ripley, B.D. (1979). Tests of 'randomness' for spatial point patterns. *Journal of the Royal Statistical Society B* **41**, 368–74.

Ripley, B.D. (1981). *Spatial Statistics*. New York: Wiley.

Ripley, B.D. (1988). *Statistical Inference for Spatial Processes*. Cambridge: Cambridge University Press.

Ripley, B.D. (1996). *Pattern Recognition and Neural Networks*. Cambridge: Cambridge University Press.

Robertson, I.G. (1999). Spatial and multivariate analysis, random sampling error, and analytical noise: empirical Bayesian methods at Teotihuacan, Mexico. *American Antiquity* **64**, 137–52.

Robinson, W.R. (1951). A method for chronologically ordering archaeological deposits. *American Antiquity* **16**, 293–301.

Rogers, A.R. (2000a). Analysis of bone counts by maximum likelihood. *Journal of Archaeological Science* **27**, 111–25.

Rogers, A.R. (2000b). On equifinality in faunal assemblages. *American Antiquity* **65**, 709–23.

Rogers, A.R. (2000c). On the value of soft bones in faunal analysis. *Journal of Archaeological Science* **27**, 635–9.

Rogers, A.R. and Broughton, J.M. (2001). Selective transport of animal parts by ancient hunters: a new statistical method and an application to the Emeryville shellmound fauna. *Journal of Archaeological Science* **28**, 763–73.

Rothschild, N.A. (1979). Mortuary behavior and social organization at Indian Knoll and Dickson Mounds. *American Antiquity* **44**, 658–75.

Rousseeuw, P.J. and van Zomeren, B.C. (1990). Unmasking multivariate outliers and leverage points. *Journal of the American Statistical Association* **85**, 633–9.

Royston, J.P. (1982). An extension to Shapiro and Wilk's *W* test for normality for large samples. *Applied Statistics* **31**, 115–24.

Royston, J.P. (1983). Some techniques for assessing multivariate normality based on the Shapiro–Wilk *W. Applied Statistics* **32**, 121–33.

Ruggles, C. (1986). You can't have one without the other? I.T. and Bayesian statistics and their possible impact within archaeology. *Science and Archaeology* **28**, 9–15.

Ruggles, C. (1999). *Astronomy in Prehistoric Britain and Ireland*. New Haven, CT and London: Yale University Press.

Ruggles, C.L.N. and Rahtz, S.P.Q. (eds) (1987). *Computer and Quantitative Methods in Archaeology 1987*, BAR International Series 393. Oxford: BAR.

Ruppert, D., Sheather, S.J. and Wand, M.P. (1995). An effective bandwidth selector for local least squares regression. *Journal of the American Statistical Association* **90**, 1257–70.

Sain, S.R., Baggerly, K.A. and Scott, D.W. (1994). Cross-validation of multivariate densities. *Journal of the American Statistical Association* **89**, 807–17.

Sanford, R.F., Pierson, C.T. and Crovelli, R.A. (1993). An objective replacement method for censored geochemical data. *Mathematical Geology* **25**, 59–80.

Savage, S.H. (1997a). Assessing departures from log-normality in the rank-size rule. *Journal of Archaeological Science* **24**, 233–44.

Savage, S.H. (1997b). Descent group competition and economic strategies in Predynastic Egypt. *Journal of Anthropological Archaeology* **16**, 226–68.

Sayre, E.V. (1975). Brookhaven procedures for statistical analyses of multivariate archaeometric data. Unpublished manuscript.

Sayre, E.V., Yener, K.A., Joel, E.C. and Barnes, I.L. (1992a). Statistical evaluation of the presently accumulated lead isotope data from Anatolia and surrounding regions. *Archaeometry* **34**, 73–105.

Sayre, E.V., Yener, K.A. and Joel, E.C. (1992b). Evaluating lead isotope data. Comments on Sayre *et al.* (1992a) – Reply. *Archaeometry* **34**, 330–6.

Sayre, E.V., Yener, K.A. and Joel, E.C. (1993). Evaluating lead isotope data: further observations. Comments on Budd *et al.* (1993) – Comments I. *Archaeometry* **35**, 247–52.

Sayre, E.V., Joel, E.C., Blackman, M.J., Yener, K.A. and Özbal, H. (2001). Stable lead isotope studies of Black Sea Anatolian ore sources and related Bronze Age and Phrygian artifacts from nearby archaeological sites. Appendix: new Central Taurus ore data. *Archaeometry* **43**, 77–115.

Scaife, B. (1998). Lead isotope analysis in archaeology. Unpublished PhD thesis, University of Bradford, UK.

Scaife, B., Budd, P., McDonnell, J.G., Pollard, A.M. and Thomas, R.G. (1996). A reappraisal of statistical techniques used in lead isotope analysis. In Demirci, S., Özer, A.M. and Summers, G.D. (eds), *Archaeometry 94: Proceedings of the 29th International Symposium on Archaeometry, Ankara, 9–14 May 1994*. Ankara: Tubitak, 301–7.

Scaife, B., Budd, P., McDonnell, J.G. and Pollard, A.M. (1999). Lead isotope analysis, oxhide ingots and the presentation of scientific data in archaeology. In Young, S.M.M., Pollard, A.M., Budd, P. and Ixer, R.A. (eds), *Metals in Antiquity*, BAR International Series 792. Oxford: Archaeopress, 122–33.

Schafer, J.L. (1997). *Analysis of Incomplete Multivariate Data*. London: Chapman & Hall.

Scheps, S. (1982). Statistical blight. *American Antiquity* **47**, 836–51.

Schreiber, K.J. and Kintigh, K.W. (1996). A test of the relationship between site size and population. *American Antiquity* **61**, 573–9.

Schutowski, H., Herrmann, B., Wiedemann, F., Bocherens, H. and Grupe, G. (1999). Diet, status and decomposition at Weingarten: trace element and isotope analysis on early medieval skeletal material. *Journal of Archaeological Science* **26**, 675–85.

Scollar, I. (1999). 25 years of computer applications in archaeology. In Dingwall, L., Exon, S., Gaffney, V., Laflin, S. and van Leusen, M. (eds), *Archaeology in the Age of the Internet: CAA97*. Oxford: Archaeopress, 5–10.

Scollar, I., Herzog, I. and Greenacre, M.J. (1993). Colour and graphic display aids for correspondence analysis. In Andresen, J., Madsen, T. and Scollar, I. (eds), *Computing the Past: CAA92*. Aarhus: Aarhus University Press, 325–7.

Scott, A. (1993). A parametric approach to seriation. In Andresen, J., Madsen, T. and Scollar, I. (eds), *Computing the Past: CAA92*. Aarhus: Aarhus University Press, 317–24.

Scott, A. and Hillson, S. (1988). An application of the EM algorithm to archaeological data analysis. In Rahtz, S.P.Q. (ed.), *Computer Applications and Quantitative Methods in Archaeology 1988*, BAR International Series 446(i). Oxford: BAR, 43–52.

Scott, D.W. (1992). *Multivariate Density Estimation*. New York: Wiley.

Seber, G.A.F. (1982). *The Estimation of Animal Abundance* (2nd edition). London: Charles Griffin.

Seber, G.A.F. (1984). *Multivariate Observations*. New York: Wiley.

Shanks, M. and Tilley, C. (1992). *Re-constructing Archaeology: Theory and Practice* (2nd edition). London: Routledge.

Sheather, S.J. and Jones, M.C. (1991). A reliable data-based bandwidth selection method for kernel density estimation. *Journal of the Royal Statistical Society B* **53**, 683–90.

Shennan, S. (1988). *Quantifying Archaeology*. Edinburgh: Edinburgh University Press.

Shennan, S. (1997). *Quantifying Archaeology* (2nd edition). Edinburgh: Edinburgh University Press.

Shott, M.J. (1985). Shovel-test sampling as a site discovery technique: a case study from Michigan. *Journal of Field Archaeology* **12**, 457–68.

Shott, M.J. (1987). Feature discovery and the sampling requirements of archaeological evaluations. *Journal of Field Archaeology* **14**, 359–71.

Shott, M.J. (1989). Shovel-test sampling in archaeological survey: a comment on Nance and Ball, and Lightfoot. *American Antiquity* **54**, 396–404.

Shott, M.J. (1994). Size and form in the analysis of flake debris: review and recent approaches. *Journal of Archaeological Method and Theory* **1**, 69–110.

Shott, M.J. (1997a). Stones and shafts redux: the metric discrimination of chipped-stone dart and arrow points. *American Antiquity* **62**, 86–101.

Shott, M.J. (1997b). Activity and formation as sources of variation in Great Lakes Paleoindian assemblages. *Midcontinental Journal of Archaeology* **22**, 197–236.

Shott, M.J. (2000). The quantification problem in stone-tool assemblages. *American Antiquity* **65**, 725–38.

Shott, M.J. (2001). Quantification of broken objects. In Brothwell, D.R. and Pollard, A.M. (eds), *Handbook of Archaeological Sciences*. Chichester: Wiley, 711–21.

Shott, M.J., Bradbury, A.P., Carr, P.J. and Odell, G.H. (2000). Flake size from platform attributes: predictive and empirical approaches. *Journal of Archaeological Science* **27**, 877–94.

Silverman, B.W. (1976). Discussion of Freeman (1976). *Journal of the Royal Statistical Society A* **139**, 44–5.

Silverman, B.W. (1986). *Density Estimation for Statistics and Data Analysis*. London: Chapman & Hall.

Simek, J.F. (1984). *A K-means Approach to the Analysis of Spatial Structure in Upper Palaeolithic Habitation Sites: Le Flageolet I and Pincevent Section 36*, BAR International Series 205. Oxford: BAR.

282   *References*

Simek, J.F., Ammerman, A.J. and Kintigh, K.W. (1985). Explorations in heuristic spatial analysis: analyzing the structure of material accumulation over space. *PACT* **11**, 229–47.

Simonoff, J.S. (1996). *Smoothing Methods in Statistics*. New York: Springer.

Slane, K.W., Elam, J.M., Glascock, M.D. and Neff, H. (1994). Compositional analysis of Eastern Sigillata A and related wares from Tel-Anafa (Israel). *Journal of Archaeological Science* **21**, 51–64.

Slane, K.W., Elam, J.M., Glascock, M.D. and Neff, H. (1997). Appendix 1: Results of neutron activation analysis at MURR. *Journal of Roman Archaeology Supplement* **10-II(i)**, 394–401.

Small, C.G. (1996). *The Statistical Theory of Shape*. New York: Springer.

Smith, A.F.M. and Roberts, G.O. (1993). Bayesian computation via the Gibbs sampler and related Markov chain Monte Carlo methods. *Journal of the Royal Statistical Society B* **55**, 3–24.

Smith, W. and Grassle, J.F. (1977). Sampling properties of a family of diversity measures. *Biometrics* **33**, 288–92.

Soggnes, K. (1987). Rock art and settlement pattern in the Bronze Age. Example from Stordal, Trondeleg, Norway. *Norweigan Archaeological Review* **20**, 110–19.

Sokal, R.R. and Sneath, P.H.A. (1963). *Principles of Numerical Taxonomy*. San Francisco: Freeman.

Sokal, R.R., Lengyel, I.A., Derish, P.A., Wooton, M.C. and Oden, N. (1987). Spatial autocorrelation of ABO serotypes in medieval cemeteries as an indicator of ethnic and familial structure. *Journal of Archaeological Science* **14**, 615–33.

Spaulding, A.C. (1953). Statistical techniques for the discovery of artefact types. *American Antiquity* **18**, 305–13.

Spaulding, A.C. (1960). Statistical description and comparison of artifact assemblages. In Heizer, R.F. and Cook, S.F. (eds), *The Application of Quantitative Methods in Archaeology*. New York: Wenner-Gren Foundation, 60–92.

Spaulding, A.C. (1976). Multifactor analysis of association: an application to Owasco ceramics. In Cleland, C.E. (ed.), *Cultural Continuity and Change*. New York: Academic Press, 59–68.

Spaulding, A.C. (1977). On growth and form in archaeology: multivariate analysis. *Journal of Anthropological Research* **33**, 1–15.

Spaulding, A.C. (1982). Structure in archaeological data: nominal variables. In Whallon, R.W. and Brown, J.A. (eds), *Essays on Archaeological Typology*. Evanston, IL: Center for American Archaeology Press, 1–20.

Stančič, Z. and Veljanovski, T. (eds) (2001). *Computer Archaeology for Understanding the Past: CAA 2000*, BAR International Series 931. Oxford: Archaeopress.

Stark, B.L. and Young, D.L. (1981). Linear nearest neighbor analysis. *American Antiquity* **46**, 284–300.

Steele, T.E. and Weaver, T.D. (2002). The modified triangular graph: a refined method for comparing mortality profiles in archaeological samples. *Journal of Archaeological Science* **29**, 317–22.

Steier, P. and Rom, W. (2000). The use of Bayesian statistics for ^{14}C dates of chronologically ordered samples: a critical analysis. *Radiocarbon* **42**, 183–98.

Steier, P., Rom, W. and Puchegger, S. (2001). New methods and critical aspects in Bayesian mathematics for ^{14}C calibration. *Radiocarbon* **43**, 373–80.

Stos-Gale, Z.A., Gale, N.H. and Annets, N. (1996). Lead isotope data from the Isotrace Laboratory, Oxford: *Archaeometry* data base 3, ores from the Aegean, part 1. *Archaeometry* **38**, 381–90.

Stos-Gale, Z.A., Maliotis, G., Gale, N.H. and Annets, N. (1997). Lead isotope characteristics of the Cyprus copper ore deposits applied to provenance studies of copper oxhide ingots. *Archaeometry* **39**, 83–123.

Struyf, A., Hubert, M. and Rousseeuw, P.J. (1996). Clustering in an object-oriented environment. *Journal of Statistical Software* **1**, 1–30.

Stuiver, M., Reimer, P.J., Bard, E., Beck, J.W., Burr, G.S., Hughen, K.A., Kromer, B., McCormac, F.G., v. d. Plicht, J. and Spurk, M. (1998). INTCAL98 Radiocarbon age calibration 24,000–0 cal BP. *Radiocarbon* **40**, 1041–83.

Sundstrom, L. (1993). A simple mathematical procedure for estimating the adequacy of site survey strategies. *Journal of Field Archaeology* **20**, 91–6.

Swayne, D.F., Cook, D. and Buja, A. (1998). XGobi: interactive dynamic data visualization in the X window system. *Journal of Computational and Graphical Statistics* **7**, 113–30.

Tainter, J.A. (1975). Social inference and mortuary practices: an experiment in numerical classification. *World Archaeology* **7**, 1–16.

Tangri, D. and Wright, R.V.S. (1993). Multivariate analysis of compositional data: applied comparisons favour standard principal components analysis over Aitchison's loglinear contrast method. *Archaeometry* **35**, 103–12.

Taylor, T., Bond, J., Heron, C., Pollard, A.M. and Dockrill, S. (2000). Letter to *The Guardian*, 16 November.

Tennessen, D., Blanchette, R.A. and Windes, T.C. (2002). Differentiating aspen and cottonwood in prehistoric wood from Chacoan great house ruins. *Journal of Archaeological Science* **29**, 521–7.

Therneau, T.M. and Atkinson, E.J. (1997). An introduction to recursive partitioning using the RPART routines. Technical Report, Mayo Foundation.

Thom, A. (1955). A statistical examination of the megalithic sites in Britain. *Journal of the Royal Statistical Society A* **118**, 275–91.

Thom, A. (1962). The megalithic unit of length. *Journal of the Royal Statistical Society A* **125**, 243–51.

Thom, A. (1967). *Megalithic Sites in Britain*. Oxford: Clarendon Press.

Thom, A. (1971). *Megalithic Lunar Observatories*. Oxford: Clarendon Press.

Thom, A. (1978). The distances between stones in stone rows. *Journal of the Royal Statistical Society A* **141**, 253–7.

Thomas, D.H. (1976). *Figuring Anthropology*. New York: Holt, Rinehart and Winston.

Thomas, D.H. (1978). The awful truth about statistics in archaeology. *American Antiquity* **43**, 231–44.

Thomas, D.H. (1986). *Refiguring Anthropology*. Prospect Heights, IL: Waveland Press.

Thomas, D.H. (1989). Diversity in hunter-gatherer cultural geography. In Leonard, R.D. and Jones, G.T. (eds), *Quantifying Diversity in Archaeology*. Cambridge: Cambridge University Press, 85–91.

Thompson, R.E., Voit, E.O. and Scott, G.I. (2000). Statistical modeling of sediment and oyster PAH contamination data collected at South Carolina Estuary (complete and left-censored samples). *Environmetrics* **11**, 99–119.

Thompson, S.K. and Seber, G.A.F. (1996). *Adaptive Sampling*. New York: Wiley.

Trigger, B.G. (1989). *A History of Archaeological Thought*. Cambridge: Cambridge University Press.

Trompetter, W.J. and Coote, G.E. (1993). Aging and seasonal dating of Tuatua shells using strontium markings. In Fankhauser, B.L. and Bird, J.R. (eds), *Archaeometry: Current Australasian Research*. Canberra: Department of Prehistory, Research School of Pacific Studies, Australian National University, 91–7.

Truncer, J., Glascock, M.D. and Neff, H. (1998). Steatite source characterization in eastern north America: new results using instrumental neutron activation analysis. *Archaeometry* **40**, 23–44.

Tubb, A., Parker, A.J. and Nickless, G. (1980). The analysis of Romano-British pottery by atomic absorption spectrophotometry. *Archaeometry* **22**, 153–71.

Tukey, J.W. (1977). *Exploratory Data Analysis*. Reading, MA: Addison-Wesley.

Tyers, P.A. (1996). *Roman Pottery in Britain*. London: Routledge.

Underhill, L.G. and Peisach, M. (1985). Correspondence analysis and its application in multi-elemental trace analysis. *Journal of Trace and Microprobe Techniques* **3**, 41–65.

Upton, G.J.G. and Fingleton, B. (1985). *Spatial Data Analysis by Example. Volume 1: Point Pattern and Quantitative Data*. Chichester: Wiley.

Valdés, L., Arenal, I. and Pujana, I. (eds) (1995). *Aplicaciones Informáticas en Arqueología: Teorías y Sistemas, 1: Saint-Germain-en-Laye 1991; 2: Bilbao 1993*. Bilbao: Denboraren Argia.

Vaquero, M. (1999). Intrasite spatial organization of lithic production in the Middle Palaeolithic: the evidence of the Abric Romaní (Capellades, Spain). *Antiquity* **73**, 493–504.

Venables, W.N. and Ripley, B.D. (1999). *Modern Applied Statistics with S-PLUS* (3rd edition). New York: Springer.

Venables, W.N. and Ripley, B.D. (2002). *Modern Applied Statistics with S* (4th edition). New York: Springer.

Vescelius, G.S. (1960). Archaeological sampling: a problem in statistical inference. In Dole, G.E. and Caneiro, R.L. (eds), *Essays in the Science of Culture in Honor of Leslie A. White*. New York: Cromwell, 457–70.

Vierra, R.K. and Carlson, D.L. (1981). Factor analysis, random data, and patterned results. *American Antiquity* **46**, 272–83.

Voorrips, A. (ed.) (1990). *Mathematics and Information Science in Archaeology: A Flexible Framework*, Studies in Modern Archaeology 3. Bonn: Holos.

Voorrips, A. and Loving, S.H. (eds) (1985). *To Pattern the Past*. PACT 11.

Voorrips, A. and Ottaway, B.S. (eds) (1990). *New Tools from Mathematical Archaeology*. Warsaw: Scientific Information Centre of the Polish Academy of Sciences.

Wand, M.P. and Jones, M.C. (1995). *Kernel Smoothing*. London: Chapman & Hall.

Wandsnider, L. (1996). Describing and comparing archaeological spatial structures. *Journal of Archaeological Method and Theory* **3**, 319–84.

Ward, G.K. and Wilson, S.R. (1978). Procedures for comparing and combining radiocarbon age determinations; a critique. *Archaeometry* **20**, 19–31.

Warren, R.W. (1990a). Predictive modelling in archaeology: a primer. In Allen, K., Green, S. and Zubrow, E. (eds), *Interpreting Space: GIS and Archaeology*. London: Taylor and Francis, 90–111.

Warren, R.W. (1990b). Predictive modelling of archaeological site location: a case study in the midwest. In Allen, K., Green, S. and Zubrow, E. (eds), *Interpreting Space: GIS and Archaeology*. London: Taylor and Francis, 201–15.

Warren, R.W. and Asch, D.L. (2000). A predictive model of archaeological site location in the Eastern Prarie peninsular. In Westcott, K.L. and Brandon, R.J. (eds), *Practical Applications of GIS for Archaeologists*. London: Taylor and Francis, 5–32.

Westcott, K.L. and Brandon, R.J. (eds) (2000). *Practical Applications of GIS for Archaeologists: A Predictive Modelling Kit*. London: Taylor and Francis.

Westwood, S. and Baxter, M.J. (2000). Exploring archaeometric data using projection pursuit methodology. In Buck, C., Cummings, V., Henley, C., Mills, S. and Trick, S. (eds), *U.K. Chapter of Computer Applications and Quantitative Methods in Archaeology*, BAR International Series 844. Oxford: Archaeopress, 81–90.

Whallon, R.W. (1972). A new approach to pottery typology. *American Antiquity* **37**, 13–33.

Whallon, R.W. (1973). Spatial analysis of occupation floors I: Applications of dimensional analysis of variance. *American Antiquity* **38**, 320–8.

Whallon, R.W. (1974). Spatial analysis of occupation floors II: The application of nearest neighbor analysis. *American Antiquity* **39**, 16–34.

Whallon, R.W. (1984). Unconstrained clustering for the analysis of spatial distributions in archaeology. In Hietala, H. (ed.), *Intrasite Spatial Analysis in Archaeology*. Cambridge: Cambridge University Press, 242–77.

Whallon, R. (1987). Simple statistics. In Aldenderfer, M.S. (ed.), *Quantitative Research in Archaeology*. Newbury Park, CA: Sage, 135–50.

Whallon, R.W. (1990). Defining structure in clustering dendrograms with multilevel clustering. In Voorrips, A. and Ottaway, B.S. (eds), *New Tools from Mathematical Archaeology*. Warsaw: Scientific Information Centre of the Polish Academy of Sciences, 1–13.

Whallon, R.W. and Brown, J.A. (1982). *Essays on Archaeological Typology*. Evanston, IL: Center for American Archaeology Press.

Wheatley, D. and Gillings, M. (2002). *Spatial Technology and Archaeology*. London: Taylor and Francis.

Wheatley, D., Earl, G. and Poppy, S. (eds) (2002). *Contemporary Themes in Archaeological Computing*. Oxford: Oxbow Books.

Whitley, D.S. and Clark, W.A.V. (1985). Spatial autocorrelation tests and the Classic Maya collapse: methods and inferences. *Journal of Archaeological Science* **12**, 377–95.

Wilcock, J.D. (1999). Getting the best fit? 25 year of statistical techniques in archaeology. In Dingwall, L., Exon, S., Gaffney, V., Laflin, S. and van Leusen, M. (eds), *Archaeology in the Age of the Internet: CAA97*. Oxford: Archaeopress, 35–51.

Wilcock, J.D. and Lockyear, K. (eds) (1995). *Computer Applications and Quantitative Methods in Archaeology 1993*, BAR International Series 598. Oxford: Tempus Reparatum.

Wilcock, J.D. and Sanie, S. (2000). Retrospect on 1970 looking back on the developments of computing archaeology in Romania since the Mamaia conference. In Lockyear, K., Sly, T.J.T. and Mihăilescu-Bîrliba, V. (eds), *CAA(96). Computer Applications and Quantitative Methods in Archaeology*, BAR International Series 845. Oxford: Archaeopress, 157–67.

Wild, C.J. and Nichol, R.K. (1983). Estimation of the original number of individuals from paired counts using estimators of the Krantz type. *Journal of Field Archaeology* **10**, 337–44.

Wilkinson, E.M. (1971). Archaeological seriation and the traveling salesman problem. In Hodson, F.R., Kendall, D.G. and Tautu, P. (eds), *Mathematics in the Archaeological and Historical Sciences*. Edinburgh: Edinburgh University Press, 276–83.

Wilkinson, E.M. (1974). Techniques of data analysis: seriation theory. *Archaeophysika* **5**, 1–142.

Williams, J.T. (1993). Spatial autocorrelation and the Classic Maya collapse: one technique, one conclusion. *Journal of Archaeological Science* **20**, 705–9.

Williams-Thorpe, O., Jones, M.C., Tindle, A.G. and Thorpe, R.S. (1996). Magnetic susceptibility variations at Mons Claudianus and in Roman columns: a method of provenancing to within a single quarry. *Archaeometry* **38**, 15–41.

Williams-Thorpe, O., Jones, M.C., Webb, P.C. and Rigby, I.J. (2000). Magnetic susceptibility thickness corrections for small artifacts and comments on the effects of 'background' materials. *Archaeometry* **42**, 101–8.

Wilson, M. (1998). Pacific rock-art and cultural genesis: a multivariate approach. In Chippindale, C. and Taçon, P.S.C. (eds), *The Archaeology of Rock-Art*. Cambridge: Cambridge University Press, 163–84.

Winder, N. (1991). How many bones make five? The art and science of guesstimation in archaeozoology. *International Journal of Osteoarchaeology* **1**, 111–26.

Winder, N. (1993). Using modern bone assemblages to estimate ancient populations. *Circea* **10**, 63–8.

Winder, N. (1997). Release–recapture statistics in osteoarchaeology. In Anderson, S. (ed.), *Computing and Statistics in Osteoarchaeology*. Oxford: Oxbow, 3–11.

Wishart, D. (1987). *CLUSTAN User Manual*. St Andrews: University of St Andrews.

Woodman, P.E. and Woodward, M. (2002). The use and abuse of statistical methods in archaeological site location modelling. In Wheatley, D., Earl, G. and Poppy, S. (eds), *Contemporary Themes in Archaeological Computing*. Oxford: Oxbow Books, 22–7.

Wright, R.V.S. (1985). Detecting patterns in tabled archaeological data by principal components and correspondence analysis: programs in BASIC for portable microcomputers. *Science and Archaeology* **27**, 35–8.

Wright, R.V.S. (1989). *Doing Multivariate Archaeology and Prehistory: Handling Large Data Sets with MV-Arch*. Sydney: Department of Anthropology, University of Sydney.

Yellen, J.E. (1977). *Archaeological Approaches to the Present: Models for Reconstructing the Past*. London: Academic Press.

Zeidler, J.A., Buck, C.E. and Litton, C.D. (1998). The integration of archaeological phase information and radiocarbon results from the Jama River Valley, Ecuador: a Bayesian approach. *Latin American Antiquity* **9**, 160–79.

Zipf, G.K. (1949). *Human Behavior and the Principle of Least Effort*. New York: Hafner.

Index

age estimation 245–6
analysis of variance (ANOVA) 58–9, 110, 131–2
animal bone 46, 139, 211–15, 219–20
applications
  associations between grave artifacts 155–7
  blood type patterns 172–3
  Bronze Age Mull viewshed analysis 155
  Classic Maya collapse 57, 169–72
  clustering medieval tin-glazed wares 184–5
  coins from Romano-British sites 140–1
  corbelled dome shapes 54, 62–3
  correspondence analysis in North America 142–3
  dating phases at Runnymede Bridge 194
  environmental degradation in Jordan 166
  flake size analysis 54–6
  Hallstatt grave distributions 168
  identifying building at Danebury 154–5
  Inuit site location 165–6
  Iron Age brooch shapes 175
  Israeli pottery provenance 98–9
  magnetic susceptibility 36, 63
  Neolithic stone tool classification 182–4
  Paleolithic stone tool assemblages 76
  Pincevent artifact distributions 168–9
  population size prediction 52–4, 56
  pottery compositional outliers 124–6
  post-hole patterns 154–5, 173–5
  Roman republican coin hoards 139
  sand-temper compositions 114
  scatterplot smoothing 64
  sampling for 'nothing' 181–2
  stone axe provenance 72
  stone circle diameters 232–3
  Teotihuacan artifact types 179–81
  see also data sets/analyses
artifact typology/classification 8–9, 11, 92, 128, 244–5
assemblage comparison 23–27
assemblage diversity 236–43
  evenness 237
  richness 236

Bayesian methods 2, 6, 11, 13, 176–186
  age estimation 246
  Bayes theorem 176–7
  clustering 182–5
  estimating proportions 179–82
  Megalithic yard 233–4
  missing values 12
  radiocarbon calibration 191–9
  seriation 207
Bayes information criterion 101–3
Bayes rule 106
binary (0–1) data 57–8, 60, 93–4, 110
biplot see principal component analysis
bone 23, 27–8, 35, 190–1
  see also animal bone, human bone
bootstrapping 144, 148–54, 215, 243
Bronze Age 10, 33, 54, 110, 133, 155, 174, 194, 222, 228
burial analysis 96, 155–7, 168, 172–3, 203–6

calibration 245
capture-recapture methods 212–3

cemetery analysis *see* burial analysis
censored data 119, 121–2
ceramics *see* pottery
chi-squared 9
  and tabular data 128–31
  distance 143–4, 205
  distribution 70, 98, 124, 164, 190
  goodness of fit 215, 238
chronological ordering *see* seriation
Clark/Evans statistic 163–6
cluster analysis 10–11, 27, 90–104, 125,
    244
  average-link 90–2, 125
  centroid-link 93, 125
  complete-link 92, 125
  Bayesian 182–5
  hierarchical agglomerative 92–5
  hierarchical divisive 95
  k-means 91, 97
  k-means spatial clustering 102–4, 160–1
  model-based 93, 99–104, 160–1
  single-link 92
  Ward's method 91, 93
  Whallon's unconstrained 161–2
  with Mahalanobis distance 70, 97–9
classification maximum likelihood 99–100,
    184
classification trees 116–18, 245
compositional data 75–7
computer intensive methods 147–58
  see also bootstrapping, cross-validation,
    EM algorithm, jackknife, Markov
    chain Monte
Carlo, simulation
coins 139, 140–1
computer packages
  abcml 214–15, 250
  BCal 186, 196, 198, 249–50
  BUGS 250
  CLUSTAN 93, 249
  MATLAB 33, 249
  MINITAB 16, 111
  OxCal 13, 186, 191, 194, 197, 249–50
  pie-slice 219, 250
  R 2, 248
  ShefSize 247, 250
  S-Plus 2, 18, 33, 65, 87, 100, 116, 184,
    239, 248–9
  SPSS 16, 111, 135
  Tools for Quantitative
    Archaeology 249
  WinBASP 202, 249

concordance 2, 7–8
confidence ellipsoids 71–2, 223
confidence intervals 148–51, 177
correspondence analysis (CA) 12–13, 77,
    136–46, 208, 219–20
  and bootstrapping 151–2
  detrended 204
  horseshoe effect 137, 204–5
  in spatial analysis 168
  inertia 137
  mathematics of 143–6
  multiple 87
countries/regions (discussion or analyses of
    material from)
  Austria 168
  Cyprus 225–6
  Denmark 203
  Ecuador 195
  Egypt 36
  England 20–1, 37, 50–2, 56, 133–5,
    138–9, 154, 173–5, 190, 194, 201,
    218–19
  France 54, 168–9
  Ghana 234–5
  Greece 19–20, 54, 148–51, 182–4, 223,
    225–6, 235
  Hungary 172–3
  Israel 75, 98–9
  Italy 22–3, 33–5, 54, 80–3, 110, 139, 184
  Jordan 166
  Mesoamerica 57, 169–72, 179–81
  Mesopotamia 62
  New Zealand 198
  North America 27–8, 62, 103–4, 112–14,
    135–6, 142–3, 165–6, 239–40
  Norway 140
  Pacific 140
  Palestine 24–7
  Peru 53–4
  Romania 139
  Scotland 155, 232–3, 247
  South America 56
  Spain 23–4, 86–8, 184
  Switzerland 175
  Tanzania 214–15
  Thailand 61–2, 156
cross-validation 109–10, 112–13, 117, 171
culture history 3–4

data sets/analyses
  assemblage diversity 239–40
  burial assemblage 24–7, 137–8

data sets/analyses *contd.*
　　column drum heights 148–51
　　engraved bone artefacts 23–4
　　Bronze Age axe decoration 133–5
　　Bronze Age cup measurements 33–5
　　dating phases from the Jama River Valley
　　　195–6
　　lead isotope ratios 19–20, 71–2
　　Mask Site 27–8, 35–6, 103–4, 161, 165,
　　　168
　　Mesolithic flint tool assemblages 201,
　　　206–7
　　outlier detection 124–5
　　Paleolithic assemblage composition 75–6
　　pottery distribution 50–2, 60
　　pottery petrography 86–8, 140–1
　　pottery typology 130–1
　　Romano-British glass assemblages
　　　138–9
　　Romano-British pottery compositions
　　　20–2, 78–9, 90–1, 125–6
　　site location in Thailand 61–2
　　site location in Alaska 135–6
　　site population and size 53–4
　　steatite quarry sources 108, 112–14,
　　　116–18
　　stone axe measurements 22–3, 81–3
　　Stonehenge antler-pick dates 190–1
　　see also applications
data sparsity 89, 98, 125, 129, 219
data transformation 74–5
　　centering 74
　　in correspondence analysis 140–2
　　logarithmic 51–3, 55, 74, 121, 223
　　log-ratio 74–7, 114, 121, 161
　　ratio 74, 124
　　square-root 141
dendrochronology 185–6
deviance statistic 61, 132–3
dimensional ANOVA 12
discriminant analysis (DA) 69–70, 105–16
　　and missing data 120, 123
　　and regression 57, 110–1
　　classification success 107–10
　　linear 106–7, 162
　　logistic 114–16
　　quadratic 106, 111–12
　　stepwise 111, 118
distance and similarity 68–72
　　chi-squared distance 143–4, 205
　　Euclidean distance 69, 92, 156
　　Gower's coefficient 94, 245

information statistic 95
Jaccard coefficient 94, 96
Mahalanobis distance 20, 69–72, 106–8,
　　120–2, 222
Manhattan distance 68–9, 92, 204
simple matching coefficient 94
distance decay 50, 54
distributions
　　binomial 60–1
　　chi-square 70
　　F 70
　　log-skew-Laplace 247
　　log-normal 121, 247
　　multinomial 131, 213, 219
　　multivariate normal 106, 214
　　Pearson Type III 164
　　Poisson 131
　　power law 55
　　uniform 192–3
diversity 23, 54, 236–43

EM algorithm 101–3, 120, 122–3, 203
English beer-drinking 220
estimated vessel equivalent (EVE) 216–19
expected species index 242
exploratory data analysis (EDA) 7

factor analysis 11, 12, 27, 83–5, 168
faunal data 211–15, 238–40
Fisher's exact test 129

generalized linear models 59–62
　　logistic regression 60–62
　　log-linear model 131–6
geographical information systems (GIS) 6,
　　155, 160, 162
geophysical prospection 185
Gibbs sampling *see* Markov chain Monte
　　Carlo
Gini index 117
glass 77, 111, 115, 121–2, 138–9, 218,
　　220–1
goodness-of-fit 51, 56–7, 207, 215
Gower's coefficient *see* distance and
　　similarity

Hotelling's T 70, 222–3
human bone 56, 120, 172, 245

image analysis 185
incidence matrix 27, 200
inertia *see* correspondence analysis
Iron Age 27, 50, 154, 175, 203

jackknife 109–10, 153–4, 243

kernel density estimates 29–37, 153, 214,
    232–3, 235
  adaptive 32
  and age estimation 246
  and lead isotope analysis 225–7
  multivariate 32–3, 225, 227
  regression 64–5
K-function 166–8
k-means (spatial) clustering *see* cluster
    analysis

lead isotope analysis 19–20, 37, 70–2,
    222–7
least squares 50, 190, 230, 239
  see also regression
leave-one-out estimates 71, 109–10
Lincoln/Peterson index 212–13
local density analysis 167–8
logistic regression *see* regression
log-linear model 9, 59–60, 129, 131–6,
    219, 244
log-ratio analysis 76–7, 114, 122
lowess smoother 65, 171–2

Mahalanobis distance *see* distance and
    similarity
Mantel test 156, 172
Markov chain Monte Carlo (MCMC) 3,
    157–8, 199
  Gibbs sampling 183, 193, 195, 198
Material/artifact categories
  see bone, coins, glass, pottery/ceramics,
      metal, stone, wood
matrix
  correlation 67
  covariance 66–7
  data 66–7
  P- 201
  Q- 201
  similarity 203–4
maximum likelihood 59–61, 85
medieval 75, 172, 184
megalithic yard 2–3, 9–10, 178, 228–35
Mesolithic 201, 247
metal 175
  Bronze Age axes 133–5
  see also lead isotope analysis
minimal animal units (MAU) 212
minimum number of elements (MNE) 212
minimum number of individuals (MNI)
    212, 214, 221

missing data 119–21, 182
mixed-mode data 94–5, 118, 182, 244–5
mixture model 100–1, 182
model-based methods 7, 93
model selection 135
Moran's I statistic 170
multidimensional scaling 85–8, 140, 146,
    168, 175
  and seriation 202–6, 208

nearest neighbor analysis 12, 163–5, 168,
    174
Neolithic 10, 54, 182, 229
neural networks 105
New Archaeology 3–4, 6, 38
Neyman-Pearson approach 6, 11
normality, tests of 224–6
number of identified specimens (NISP) 211

outliers
  in cluster analysis 123–7
  in correspondence analysis 139, 141
  in lead isotope analysis 224, 227
  in radiocarbon data 198
  in regression 50–1, 238
  in seriation 204

Paleolithic 55, 76, 105, 162
particle-size analysis 246–7
Poisson process 28, 165, 174, 197
pottery/ceramics 50–2, 86–8, 114, 180–1
  composition 20–2, 75, 98–9, 102, 124–6,
      184
  dimensions 33–5, 152–4
  in burials 24–7, 137–8
  quantification 215–19
pottery information equivalent (PIE)
    218–19
positivism 4, 6
post-medieval 114
post-processual archaeology 3–5
predictive modelling 57, 162–3
principal component analysis 73–83, 120,
    215, 247
  biplot 73, 78–9
  correspondence analysis as 144
  for outlier detection 125–7
  in shape analysis 175
  in testing for normality 223–5
  rotation 11, 81–3
principal coordinate analysis 86
processual archaeology 3–6
projection pursuit 88–9, 225, 247

provenance 19
  of British stone axes 72
  of marble 112
  of pottery 78–9, 114
  of steatite 112–14

quantification 210–21
  of bone 46, 211–15, 219–20
  of glass 220–1
  of pottery 215–19
  of stone tools 221
quantogram 232
quantum hypothesis 230

radiocarbon dating 13, 63–5, 187–99
  combining dates 189–90
  prior assumptions 196–8
randomization tests 155–6
rank-size law 54
rarefication analysis 243
regression 50–65, 216, 238–40
  and discriminant analysis 110–1
  inverse 245
  logistic 60–62, 115, 162–3
  multiple 55–6, 162
  non-linear 62–3, 238–40
  non-parametric 63–5
  polynomial 56
  see also least squares
residuals 215, 230
  chi-squared 129, 134
  deviance 132, 134, 136
  in regression 51, 57, 65, 171–2
Roman 20, 75, 86, 111, 121, 138–40,
      219–20, 246

sampling methods 10, 38–49
  adaptive 46–7
  cluster 41–4
  non-random 43–5
  simple random 40–1
  stratified random 41
  systematic 43
  test-pit 48–9
sample selection 198–9
sample size effects 40–1, 143, 164, 180, 233
  and bootstrapping 149

and cluster analysis 99
and diversity 236
and KDEs 33
and lead isotope analysis 222–3, 226–7
and projection pursuit 89
and seriation 209
scatterplot smoother 64
scree plot 80–1
seriation 8, 27, 137–8, 200–9, 216, 243
sherd family 216
Shannon's index 237
shape analysis 173–5
Shapiro-Wilk test 89, 225
similarity coefficients *see* distance and
      similarity
Simpson's index 237
simulation 12, 231, 237, 240–3
  Monte Carlo 164–5, 173, 232
singular value decomposition (SVD) 68
spatial analysis 159–75
  intersite 163
  intrasite 12, 27–8, 163
  point pattern analysis 163–9
spatial autocorrelation 169–73
spectral analysis 168
standardization of variables 67
stone 72, 75–6, 105, 112, 221
  axe typology 22–3, 81–3
  columns 36–7, 148–51
  flakes 54–6
  flint tools 201, 206–7
  steatite 108, 112–14, 116
  tool typology 182–4

tabular data 128–46
taphonomy 211
ternary diagrams 75–6, 153
trace operator 68
trend surface analysis 56–7, 170

variable selection 111, 245
variogram 172

wood 62, 185–6

zooarchaeology 212
  see also animal bone

Printed and bound by CPI Group (UK) Ltd, Croydon, CR0 4YY

16/04/2025

14658544-0005